COMPUTERS and TELECOMMUNICATIONS:
Issues in Public Policy

Prentice-Hall
Series in Automatic Computation
George Forsythe, editor

ARBIB, *Theories of Abstract Automata*
BATES AND DOUGLAS, *Programming Language/One*
BAUMANN, FELICIANO, BAUER, AND SAMELSON, *Introduction to ALGOL*
BLUMENTHAL, *Management Information Systems*
BOBROW AND SCHWARTZ, editors, *Computers and the Policy-Making Community: Applications to International Relations*
BOWLES, editor, *Computers in Humanistic Research*
CESCHINO AND KUNTZMANN, *Numerical Solution of Initial Value Problems*
CRESS, DIRKSEN, AND GRAHAM, *FORTRAN IV with WATFOR*
DESMONDE, *Computers and Their Uses*
DESMONDE, *A Conversational Graphic Data Processing System: The IBM 1130/2250*
DESMONDE, *Real-Time Data Processing Systems: Introductory Concepts*
EVANS, WALLACE, AND SUTHERLAND, *Simulation Using Digital Computers*
FIKE, *Computer Evaluation of Mathematical Functions*
FIKE, *PL/1 for Scientific Programmers*
FORSYTHE AND MOLER, *Computer Solution of Linear Algebraic Systems*
GAUTHIER AND PONTO, *Designing Systems Programs*
GOLDEN, *FORTRAN IV: Programming and Computing*
GOLDEN AND LEICHUS, *IBM 360: Programming and Computing*
GORDON, *System Simulation*
GREENSPAN, *Lectures on the Numerical Solution of Linear, Singular and Nonlinear Differential Equations*
GRISWOLD, POAGE, AND POLONSKY, *The SNOBOL4 Programming Language*
GRUENBERGER, editor, *Computers and Communications—Toward a Computer Utility*
GRUENBERGER, editor, *Critical Factors in Data Management*
HARTMANIS AND STEARNS, *Algebraic Structure Theory of Sequential Machines*
HULL, *Introduction to Computing*
HUSSON, *Microprogramming: Principles and Practices*
JOHNSON, *System Structure in Data, Programs, and Computers*
KIVIAT, VILLANUEVA, AND MARKOWITZ, *The SIMSCRIPT II Programming Language*
LOUDEN, *Programming the IBM 1130 and 1800*
MARTIN, *Design of Real-Time Computer Systems*
MARTIN, *Programming Real-Time Computer Systems*
MARTIN, *Telecommunications and the Computer*
MARTIN, *Teleprocessing Network Organization*
MARTIN, *The Computerized Society*
MATHISON AND WALKER, *Computers and Telecommunications: Issues in Public Policy*
MC KEEMAN, ET AL., *A Compiler Generator*
MINSKY, *Computation: Finite and Infinite Machines*
MOORE, *Interval Analysis*
PYLYSHYN, *Perspectives on the Computer Revolution*
PRITSKER AND KIVIAT, *Simulation with GASP II*
SAMMET, *Programming Languages: History and Fundamentals*
SNYDER, *Chebyshev Methods in Numerical Approximation*
STERLING AND POLLACK, *Introduction to Statistical Data Processing*
STROUD AND SECREST, *Gaussian Quadrature Formulas*
TAVISS, *The Computer Impact*
TRAUB, *Iterative Methods for the Solution of Equations*
VARGA, *Matrix Iterative Analysis*
VAZSONYI, *Problem Solving by Digital Computers with PL/1 Programming*
WILKINSON, *Rounding Errors in Algebraic Processes*
ZIEGLER, *Time-Sharing Data Processing Systems*

Stuart L. Mathison
Arthur D. Little, Inc.

Philip M. Walker
Office of the Secretary of Defense
U. S. Department of Defense

COMPUTERS and TELECOMMUNICATIONS: Issues in Public Policy

Prentice-Hall, Inc.
Englewood Cliffs, New Jersey

To Joyce and June

© 1970 by
Prentice-Hall, Inc.
Englewood Cliffs, N. J.

All rights reserved. No part of this book may be reproduced in any form or by any means without permission in writing from the publisher.

13-165910-3

Library of Congress Catalog Card No.: 79-109108

Printed in the United States of America

Current printing (last digit):

10 9 8 7 6 5 4 3 2 1

PRENTICE-HALL INTERNATIONAL, INC., *London*
PRENTICE-HALL OF AUSTRALIA, PTY. LTD., *Sydney*
PRENTICE-HALL OF CANADA, LTD., *Toronto*
PRENTICE-HALL OF INDIA PRIVATE LTD., *New Delhi*
PRENTICE-HALL OF JAPAN, INC., *Tokyo*

FOREWORD

The Computer Utility Industry Can Be Viewed as the Most Basic Tool of the Last Third of the Twentieth Century. So commented J.-J. Servan-Schreiber in his recent book, *The American Challenge.* Certainly the components of information networks such as data banks, computer centers, terminal devices, and communication lines exist or are in the process of coming into existence. Whether computers are employed for time-sharing, remote inquiry, or computational services, increasingly the question is not whether remote "computer utility" services will proliferate but, rather, how quickly they will do so.

Attributing these new services to continued changes in technology is a relatively easy task. A more difficult problem is assessing the implications which that technology holds for the computer industry, the communications industry, and the general public. The announcement in 1966 by the Federal Communications Commission of an Inquiry into the interdependence of computers and communications, which at the time of this writing was still in progress, marked a first step in this endeavor. The following year a Presidential task force on communications policy began a broad study also encompassing this subject. The issues at stake in these investigations remain today largely unresolved, and, due to continual advances in technology and the growth of data communications, can be expected to remain with us for the foreseeable future. Ultimate resolution of these important issues will shape the structure of the remote-access computing industry as it matures. The issues focused upon by the FCC can best be summarized as questions:

1. What is the regulatory status of computer/communications services?

2. Are common carrier communications services and tariffs responsive to the requirements of the data processing industry?

3. Can privacy be protected against the increasing phenomenon of concentration and the exchange of data information?

These questions serve as the conceptual framework for the Mathison-Walker Study.

The first issue, the question of the future regulatory status of remote teleprocessing services, turns on the question of market entry. Are such services to be rendered by members of the communications industry? If the answer is yes, a further question follows. Should such services be tariffed or non-tariffed? If the answer is the former, then a third question remains. Does the supplying of tariffed services by regulated carriers imply that entry into this industry is limited to franchised entities only? Clearly, in a universe that posits communications and data processing as polar extremes, the answers to these questions might appear self-evident; data processing would continue to operate in an environment of market rivalry, and communication services would continue to operate under the constraints of regulation. The problem assumes a different order of magnitude with the evolution of services that embrace varying elements of both communications and data processing. It is inevitable that this mix attracts firms from both the computer and the data processsing industries.

Consider for example the common carriers. These regulated entities are moving into data processing as a natural extension of their transmission and switching services. The precise pattern varies. Some carriers are diversifying through the vehicle of a joint venture; others are forming corporate EDP affiliates, separate and apart from their regulated parent. Whatever the strategy, many carriers regard communications/EDP services as the market of the future; and their diversification efforts signal an intent to participate in that market actively.

In like manner, data processing firms are attempting to diversify into communications via the route of computer message switching. Indeed, the grafting of message switching capability to computer-based remote inquiry systems is no longer uncommon. Moreover, some EDP firms are establishing data transmission subsidiaries both to diversify into new fields and to extend the geographic coverage of their computer services into markets hitherto untapped. Again, whatever the approach, firms in the data processing industry regard teleprocessing as their logical domain and next major market, and computer-based message switching as a natural extension of their unique expertise in computer applications.

At first glance, this mutual diversification appears reciprocal. A closer look, however, suggests that regulated entities are finding it easier to move into non-regulated activities than for non-regulated firms to move into communications. Several reasons account for this imbalance. First, regulatory bodies tend to regulate services rather than firms; second, carriers may refuse to lease lines to firms they regard as poaching on communication services; third, direct entry into common carrier activities must surmount the adjudicatory process, a process noted neither for its efficiency nor its expediency. The asymmetry of market entry may thus pose as the critical variable in shaping the teleprocessing industry in the next decade. Certainly it is apt to condition the environment for computer-based services by the time public policy has sorted out its ends and means.

The second question, concerning the interface between the supplier and the user of communications channels, turns on the content of common carrier communication tariffs, customs, and practices. Many of these practices are long-standing, such as restrictions on the use of "foreign attachments," on the interconnection of private communication facilities to the common carrier network, and on the sharing of common carrier communication lines, as well as the long-standing pricing principles for communication services. The venerable telephone company practice of opposing the attachment of customer-provided devices and communication systems to the telephone network has rested upon the premise that such a policy protects the quality of the system, fosters the innovation of equipment, and identifies responsibility for repair and maintenance. However valid in the past, these practices have been challenged recently by users who seek a broader range of choice in terminal gear and equipment. Other users seek to interconnect private microwave systems to the dial network and, of course, independent equipment manufacturers seek a broader base for market participation. All of this has added pressure to establish technical interface specifications standards that facilitate the attachment of customer-owned equipment and systems to the telephone exchange network. While the FCC's Computer Inquiry addresses itself to this question, the Commission's 1968 *Carterfone* decision has been particularly instrumental in liberalizing carrier equipment practices.

Many regard relaxation of the carriers' line-sharing restrictions of importance equal to liberalization of the foreign attachment and interconnection tariff restrictions. The reason is obvious. Users are experiencing excess capacity in their leased lines and thus are searching for ways to reduce their communication costs. Until recently the carriers have insisted that line sharing is tantamount to selling communications and selling communications is equivalent to engaging in common carrier functions. Currently, sharing restrictions are being subjected to review or modification by both the carriers and the regulatory agencies. Indeed, the FCC's Common Carrier Bureau has suggested that carriers permit the reselling of bulk leased circuits as a device to check rate discrimination among users. While this suggestion is somewhat unprecedented for the communications industry, third party leasing is not an unfamiliar practice in the computer hardware market.

The pricing of communication lines has not escaped challenge either. Generally the carriers base their charges on the averaging of time, distance, bandwidth, etc. The computer industry, on the other hand, is less interested in circuit miles, but rather emphasizes the quantity of information transmitted. Perhaps discussion of precipitous rate reductions are premature at this time; but as the carriers' investment mix shifts from transmission (some 19% at present) to switching (some 58%) perhaps the promulgation of rates independent of distance is not entirely out of the question.

The question of privacy, the third issue, can be traced to the growing concentration, coalescence and interexchange of data bank information. Whether the Achilles' heel is the computer, the program, the terminal or the communication line, the question of personal privacy may well pose as the most provocative and baffling issue on the policy agenda. Only time will determine how policy addresses itself to the problem of protecting proprietary information. Some have advocated the licensing of programmers. Others have argued for cryptographic devices that raise the cost of eavesdropping on communication lines. Still others have suggested that privacy in information transmission is the responsibility of the common carrier. This request, needless to say, imposes a dual burden on the carrier; it is one thing to provide the bucket, it is quite another to protect the bucket's content.

In sum, the questions of regulatory status, of tariffs, and of privacy, constitute some of the more critical public policy issues in data communications and remote-access computer services. The FCC's Computer Inquiry can be viewed as an attempt to identify these issues. The task ahead, however, is to block out the options open to policy within the broad context of the public interest. This is both the thrust and contribution of Messrs. Mathison and Walker.

Manley R. Irwin

Durham, New Hampshire

PREFACE

During the last decade the computer and telecommunications industries have begun to converge, each becoming dependent upon the facilities and services of the other, and together giving birth to a promising set of capabilities which are significantly altering the operation and organization of industry, business, and government, and which will ultimately affect the daily life of each of us. Data communications and remote-access computing, still in their infancy, make possible worldwide data networks, conversational time-sharing computer services, centralized fast-response data banks, and a multitude of specialized computer/communication systems. The computer time-sharing services industry alone (often called the "computer utility" industry), which began in the mid 1960's, has grown at the rate of more than 100% per year and, according to a recent study by the Auerbach Corporation, will have revenues exceeding one billion dollars per year by 1973.

Great technological advances in electronic digital computers and in telecommunication systems have led to the rapid growth of data communications and remote-access computing. This growth, however, has outpaced the ability of industrial organizations to adapt to the changes, and to fully exploit the opportunities, and at the same time has outpaced the ability of government policy-making organizations to guide the evolution of this new technology.

A number of specific public policy issues have emerged over the last several years from the convergence of computers and telecommunications, and have forced themselves upon the regulatory agencies and other public institutions charged with protecting the public interest in such matters. Responding to these issues, the Federal Communications Commission began a broad public inquiry in 1966 designed to provide the basis for future policy determinations. At the time of this writing the FCC has completed the data gathering and evaluation phase of its Inquiry, and is entering a second phase focusing in more detail upon certain critical issues. During the same time period, the FCC has also faced and begun to resolve a number of related issues through its normal case-by-case adversary process of hearings and rulings.

In 1967 a Presidential Task Force on Communications Policy was created by President Lyndon Johnson, and charged with undertaking a sweeping study of national institutions, policies, and issues in telecommunications. Prominent among the subjects of interest were those relating to the interrelationship of computers and communications – those same issues being addressed independently in the FCC Inquiry. The President's Task Force reported its findings in December 1968, providing further insight for those who will attempt to resolve these issues. Viewing the issues from a national legislative standpoint, the U.S. Congress has also become increasingly concerned about the policy implications of the advancing computer and telecommunications technology, and several Congressional committees have begun to take an active role in this portion of the overall public policy arena.

This book focuses upon a number of these important issues awaiting resolution by the Federal Communications Commission and the Congress. We attempt to identify and explain the policy problems and the implications of alternative solutions, providing technical and regulatory background information where necessary. The book is intended for the professional in the computer and telecommunications fields, including management, and legal and technical readers. Graduate students in business administration, computer science, regulatory economics, and law may also find it informative.

The book is an outgrowth of a jointly-authored thesis submitted in June 1968 in partial fulfillment of the requirements for the degree of Master of Science in Management at the Alfred P. Sloan School of Management, Massachusetts Institute of Technology. Additionally, many of the thoughts presented here were developed in conjunction with work done for the Antitrust Division of the U.S. Department of Justice and for the President's Task Force on Communications Policy. Financial support for this study was also provided by Arthur D. Little, Inc.; by Project MAC, an M.I.T. research project sponsored by the Advanced Research Projects Agency, Department of Defense; by the Diebold Group, Inc.; and by the Federal Government Accountants Association.

We would like to express our special appreciation to Prof. Manley R. Irwin of the University of New Hampshire, Prof. Malcolm M. Jones of M.I.T., Prof. Martin Greenberger of The Johns Hopkins University, and Prof. Stanley M. Jacks of M.I.T., as well as to Messrs. Donald I. Baker and Harrison J. Sheppard of the U.S. Department of Justice, whose advice and criticism were invaluable.

We are indebted to many more individuals and organizations for their generous cooperation with our research efforts, than can possibly be mentioned here. Substantial assistance came from members of the staff of the Federal Communications Commission, particularly Dr. William Melody, Mr. Ernest Nash,

Dr. Boyd Nelson, Mr. Bernard Strassburg, and Mr. Robert Thorpe. Many individuals at the American Telephone & Telegraph Company were quite generous with their time, and provided valuable insights. We are also indebted to members of many other organizations including the Advanced Research Projects Agency of the Department of Defense. Bunker-Ramo Corporation, Communications Satellite Corporation, the U.S. General Services Administration, IBM Corporation, the U.S. Department of Justice, Microwave Communications, Inc., The MITRE Corporation, the President's Task Force on Communications Policy, the President's Office of Telecommunications Management, and Western Union Telegraph Company. Other individuals who were very helpful include Messrs. Paul Baran, Robert P. Bigelow, William H. Borghesani, Jr., Stephen G. Breyer, Michael A. Duggan, Anthony G. Oettinger, Lee L. Selwyn, and D. Edwin Winslow. We are also indebted to Miss Janice Bennett for her valuable assistance in preparing the manuscript.

              Stuart L. Mathison
              Philip M. Walker

CONTENTS

FOREWORD — vii

PREFACE — xi

I. BACKGROUND — 1

    A. Communications Common Carriage — 1
    B. The Federal Communications Commission — 2
    C. The Communications Industry — 4
        1. Telephone — 4
        2. Telegraph — 6
        3. International Carriers — 7
        4. Satellite Communications — 7
    D. Structure of the Computer Industry — 10
        1. Suppliers of Computer Equipment — 10
        2. Service Bureaus and Software Houses — 10
        3. Computer-Based Information Services — 11
    E. Remote Access Data Processing — 12

II. A STATEMENT OF THE PROBLEMS — 16

    A. "Computer Utilities" and the Regulation of Data Processing — 16
    B. Common Carrier Entry into Data Processing Services — 19
    C. The Provision of Message-Switching Services by Non-Carriers — 21
    D. Common Carrier Restrictions on the Use of their Communications Lines — 22
        1. The "Foreign Attachment" Limitation — 22
        2. The Interconnection Restriction — 23
        3. The Line Sharing/Resale Restriction — 23
    E. Adequacy of the Present Communications Network for Data Transmission — 24
    F. Special Service Common Carriers — 25

III.	MARKET ENTRY, PART I: COMMON CARRIER ENTRY INTO DATA PROCESSING SERVICES	26
	A. Carrier Experience and Interest in Offering Data Processing Services	26
	B. Alternative Ground Rules for Common Carrier Participation in the Data Processing Services Industry	34
	1. Option (1): Common Carriers are Permitted to Offer Data Processing Services as an Unregulated Activity, Without Restriction	35
	2. Option (2): Carriers are Permitted to Offer Data Processing Services as a Regulated, Tariffed Activity	37
	3. Option (3): Carriers are Permitted to Offer Data Processing Services Only Through a Separate Subsidiary	39
	4. Option (4): Carriers are Permitted to Offer Data Processing Services Through Separate Subsidiary Which Can Neither Sell Services to the Parent nor Lease Communication Channels from the Parent	40
	C. Conclusion	42
IV.	MARKET ENTRY, PART II: NON-CARRIERS INTO MESSAGE SWITCHING	44
	A. Description of Message Switching	44
	1. Message Switching vs. Circuit Switching	46
	2. Types of Message Switching Systems	50
	B. Regulatory Status of Message Switching	52
	1. Option (1): Maintain the Status Quo — Only Common Carriers are Permitted to Offer Complete Message Switching Services	54
	2. Option (2): Message Switching Services Remain Essentially a Common Carrier Activity; Other Organizations Can Offer Message Switching Services as Part of Remote-access Computer Services Using Shared Circuits Only if the Message Switching is "Incidental" to the Primary Non-Communications Purpose of the Services.	59

Contents   xv

       3.   Option (3): Specialized Message Switching Services Become Unregulated Activities; Non-carriers Permitted to Share and Resell Communications Capacity; Common Carriers can Offer Specialized Message Switching Services Only Through a Separate (Unregulated) Subsidiary   65
   C. Observations on Western Union's SICOM Service — A Digression   73
       1.   Nature of the Service   74
       2.   Comments on the FCC Decision   75
   D. Message Switching: Summary and Conclusions   79

V.   FOREIGN ATTACHMENTS   83

   A. The Hush-A-Phone Case   84
   B. The Carterfone Case   85
   C. Tariff Revisions   87
   D. Considerations Underlying the Foreign Attachments Rule   88
       1.   System Integrity   88
       2.   Divided Responsibility   91
       3.   Carrier Freedom to Innovate   92
       4.   Other Considerations   93
   E. Impact of the Revised Foreign Attachment Rule   96
       1.   Advances in Modem Technology   97
       2.   Tone-Generation Terminal Devices   98
       3.   Acoustic Coupling   99
   F. Summary   101

VI.   INTERCONNECTION   104

   A. Technical Issues Regarding Interconnection   106
   B. Economic Issues Regarding Interconnection   109
       1.   Background: Average-Cost Pricing   109
       2.   Cream-Skimming   111
       3.   The Validity of the Carriers' Opposition to "Cream-Skimming"   114
   C. Conclusion   120

VII. LINE SHARING/RESALE — 121

- A. "Time-Shared" Use of Communication Lines — 126
  1. Reduced-Cost Motivation — 126
  2. Message Switching Services — 128
  3. The ARPA Computer Network: An Example — 129
  4. Time-Shared Use of Lines: Conclusions — 134
- B. Channelizing — 135
  1. Bandwidth Optimization — 138
  2. Effects Upon Equipment Supply and Usage — 139
  3. Discriminatory Pricing — 142
  4. Carrier Opposition to Customer Channelizing — 145
- C. The Economic Impact of Line Sharing — 147
  1. Line Sharing: An Example — 147
  2. Price Elasticity of Demand — 150
- D. Line Sharing/Resale: Summary — 152

VIII. ADEQUACY OF COMMUNICATION SERVICE — 153

- A. Channel Bandwidths — 154
  1. Low-Speed Channels — 160
  2. Medium-Speed Channels — 161
  3. High-Speed Channels — 163
- B. Channel Reliability — 163
- C. Price Structure — 164
  1. "Short-Period" Leased Lines — 165
  2. Charging by the Bit — 166
  3. Pricing According to Channel Error Characteristics — 169
  4. Shorter Minimum Charge Times — 170
- D. A Digital Communications Network — 170
  1. Signal Transmission and Multiplexing — 170
  2. Store-and-Forward Switching — 172

IX. SPECIAL SERVICE COMMON CARRIERS — 175

- A. The Use of Microwave for Long-Haul Communications — 179
- B. Characteristics of the Proposed Services — 183
- C. Competitive Benefits of Special Service Common Carriers — 186
- D. Arguments Against the SSCC Concept — 187

	E. Competitive Response of the Existing Carriers to SSCC's	190
	F. Excerpts from the FCC's MCI Decision	192
	G. Long-Range Potential of Special Service Common Carriers	200
	H. University Computing Company's Proposed Common Carrier Data Transmission Network	203
X.	PRIVACY AND SECURITY	210
	A. Techniques for Solution	210
	B. Legal Controls	213
XI.	SUMMARY	217
APPENDIX A		221
APPENDIX B		232
APPENDIX C		241
BIBLIOGRAPHY		244
INDEX		263

LIST OF FIGURES

Figure No.

1–3	Permissible Message Switching Service Networks	53
4	The Effect of Market Structure on Average Cost-Per-Message of a Message Switching Service	70
5	Illustrative Costs for AT&T's TD-2 Microwave Relay Systems	117
6	The ARPA Network	131

LIST OF TABLES

Table No.

1	Time-Sharing Computer Centers to be Interconnected by the ARPA Network	130
2	Currently Available Common Carrier Communications Offerings Useful for Data Transmission	155
3	Comparison of Data Processing Equipment Operating Speeds with Available Transmission Line Speeds	157

CHAPTER I

BACKGROUND

A. COMMUNICATIONS COMMON CARRIAGE

Common carriage, unlike electronic data processing, is a field with a long history.[1] In its early stages, "common carriage" was exemplified by the obligations of the ferryman to serve the public generally, at reasonable prices. With the passage of time, the common carrier concept has been steadily expanded to encompass an increasing number of services thought to be fundamental in circumstances where (1) the "carrier" was in a position to charge unreasonable rates to the public and (2) the public interest required that the service be offered on a regular basis. Often, but not always, common carriers have been monopolists. In the transportation field, the concept was applied to stage coaches, ships, railroads, and more recently airlines and pipelines, as well as to privately owned toll bridges and turnpikes.

Communication by wire grew up in the nineteenth century as a common carrier service because it was treated as an ancillary part of land transportation services. The early telegraph services were generally provided by railroad companies along their rights-of-way. In fact, until 1934, both telegraph and telephone services were subject to the regulatory jurisdiction of the Interstate Commerce Commission. However, the ICC already had heavy responsibilities in the railroad field, and Congressional dissatisfaction with its inability to regulate effectively the new communications segment of the transportation industry was at least one of the factors which led to the passage of the Communications Act of 1934 and the creation of the Federal Communications Commission (FCC).[2]

---

1. The authors wish to acknowledge the assistance of Harrison Sheppard of the U.S. Department of Justice, in the preparation of this discussion and the discussion of "Legal Controls" relevant to the problems of Privacy, in Chapter X.

2. This is very clear from Senator Dill's remarks at the time:

    "The purpose of the proposed legislation [Communications Act of 1934] is to make effective the power now written into the Interstate Commerce Act of control of telephone and telegraph business in this country. The [ICC] has been so busy regulating the railroads that they have not had time to give real consideration to the problems in connection with rate regulation of telephones and telegraph, and it is only in recent years that the communications business has been big enough to demand the attention of those who use it from the standpoint of getting rate regulation..." 78 Cong. Rec. 4139 (March 10, 1934).

2 Background

Since the Commission assumed its responsibilities in 1934, the common carrier communications industry has grown to the point that it is now among the nation's largest and most indispensable industries. The Bell System alone has net assets of more than $39 billion, a sevenfold increase since 1934, and it is growing at a rate of $3.5 billion per year.[3]

The Commission has endeavored to develop a scheme of rate regulation to deal with this enormous enterprise, as well as with the other interstate common carriers. Under this scheme, called "rate base regulation," the total annual allowed revenue of the regulated firm is set equal to the sum of the annual operating expenses and a specified percentage of the book value of the invested capital (the "rate base"). The earnings of the firm are therefore proportional to the absolute size of the rate base. As a consequence of rate base regulation the behavior of common carriers is somewhat different from that of firms operating under a "competitive" reward structure.

The whole regulatory process is complex, both legally and factually, as is clear from the extended FCC investigation of AT&T costs and rates, which began in 1965 and is still in progress.[4] However, the complexities of the regulatory process should not obscure the fact that the common carriers are operating under a statute which continues to impose on them the same fundamental obligations which have historically governed common carriage generally. Today's telephone and telegraph companies are obliged to provide to any member of the public, and at reasonable rates, wire (or radio) facilities for the transmission of information.

B. THE FEDERAL COMMUNICATIONS COMMISSION

The Federal Communications Commission (FCC) was created by the Communications Act of 1934 as an independent agency to regulate interstate and foreign commerce in communications by wire and radio. The ultimate public policy embodied in the Act is—

---

3. *Statistics of the Common Carriers, for the Year Ended December 31, 1966* (Washington, D.C.: U.S. Government Printing Office, 1968), p. 25.

4. *In Re American Telephone and Telegraph Company and the Associated Bell System Companies, Charges for Interstate and Foreign Communication Service*, FCC Docket No. 16258, opened October 27, 1965.

"...to make available, so far as possible, to all the people of the United States a rapid, efficient, nationwide, and worldwide wire and radio communications service with adequate facilities at reasonable charges..."[5]

The Communications Act was passed with the intention of centralizing in one agency the task of overseeing non-federal government communications as a whole, of developing communications policies for wire and radio on an integrated basis. It was recognized this was a technical field in which Congress could not carry out the quasi-legislative process of rulemaking, or the administrative process of applying the standard of "public convenience, interest, and necessity" to numerous specific cases. Those provisions of the Communications Act, which require the Commission to study special problems and recommend legislation to cure them, explicitly reflect the intent of Congress to give the Commission special policy-forming responsibilities for telecommunications matters.

The Commission has wide regulatory powers both to compel the common carriers to conform to the broad purposes of the Act and to make any necessary inspections and investigations.

The provisions of the Act require that communications common carriers subject thereto furnish services at reasonable charges upon reasonable request. Carriers may not construct interstate lines or curtail service without Commission approval. Common carriers must file non-discriminatory tariff schedules with the Commission (or concur in the tariffs filed by other carriers) showing all charges, practices, classifications, and regulations for interstate communication services offered to the public. A tariff, when filed, automatically becomes effective unless suspended or explicitly disapproved by the Commission.

The responsibilities of the FCC for analysis and formulation of national communications policy are especially germane to the topic of this book. It appears to the authors that the FCC, as currently organized and operating, is not optimally equipped to cope with these responsibilities in a timely and effective fashion. Regarding staff and budgetary resources alone, one critic has commented:

---

5. *The Communications Act of 1934, With Amendments and Index Thereto* (Washington, D.C.: U.S. Government Printing Office, 1961), Title I, Section 1.

4   Background

"...To handle all of this [allocation of all space in the publicly-owned electromagnetic spectrum, regulation of the entire radio and television broadcasting industry, and licensing and supervision of the five million-odd transmitters using the airwaves — radio, television, marine, police, fire, industrial, transportation, amateur, citizens, and common carrier] plus regulating AT&T, a $35 billion industry, and the other common carriers, such as ITT and Western Union, plus facing the issues raised by new developments in the fast-changing communications industry, the Commission has a staff of 1500 and a budget of $17 million. This is slightly more than one third of the budget of the Bureau of Commercial Fisheries."[6]

While it may be unduly optimistic to expect any significant increases in the budget authorized for the Commission and its staff, the authors believe that such additional resources and regulatory authority could substantially improve the broadcast and common carrier communication services in the United States.

C. THE COMMUNICATIONS INDUSTRY

To put into perspective our later discussions of communications services, let us consider the structure of the communications common carrier industry in the United States. As we have seen, common carriers are firms which are licensed by the government to furnish service to the public at reasonable rates upon reasonable request. Their interstate activities are regulated by the Federal Communications Commission, and their intrastate services by the appropriate state public utility commissions. Approximately 2800 companies are recognized as communications common carriers, and they provide a variety of services including facilities for the transmission of voice, data, printed textual information, video, facsimile, telephoto, and telemetry.

1. Telephone

The largest and most important segment of the communications common carrier industry consists of the telephone companies, dominated by the Bell Telephone System, but also including over 2100 independent telephone

---

6. Elizabeth Brenner Drew, "Is the FCC Dead?", *The Atlantic Monthly*, July 1967, p. 30. Copyright, 1967, The Atlantic Monthly Company, Boston, Mass. Reprinted with permission.

companies which range in size from multi-state operations to small local firms. The telephone companies are all interconnected to form a single national network, which is used for the transmission of information in all forms, not just voice communications.

The Bell Telephone System consists of the American Telephone & Telegraph Company (AT&T), the parent organization in the Bell System; Bell Telephone Laboratories, Inc., the research and development unit; Western Electric Company, the manufacturing and supply unit; and 23 Bell Telephone operating companies. AT&T is primarily a holding company, but also operates the Long Lines Department, which supplies long distance communications. In addition, AT&T owns through Western Electric the Teletype Corporation, which manufactures teletypewriter and data communications equipment.

AT&T was incorporated in 1885, and during the early part of this century acquired and consolidated the operations of many independent telephone companies, to form the Bell System. Today, it owns more than 80% of the telephones in the United States (nearly 110 million in 1968), and serves 70% of the population. Measured in terms of assets or personnel, AT&T is the world's largest corporate enterprise, with a net plant investment of $34.8 billion in 1968, 872,000 employees, and over three million stockholders. It handles more than 117 billion telephone conversations per year (including more than 90% of the nation's long distance calls), and its annual budget usually equals "about 7.7% of the total expenditures for all businesses in this country."[7]

An important factor influencing AT&T's business activities is the constraints set forth in its 1956 consent decree with the U.S. Department of Justice. The Justice Department had brought an antitrust suit against AT&T seeking to force the divestiture of Western Electric. The suit was resolved by a consent decree wherein AT&T retained ownership of Western Electric but was no longer permitted to engage in any business activity other than the furnishing of common carrier communication services. Nor was Western Electric permitted to manufacture any equipment of a type not sold or leased to Bell operating companies for use in furnishing common carrier communication services.[8]

---

7. *AT&T Annual Report, 1968* (New York: AT&T, 1969), pp. 4, 28, 30; also "AT&T Expansion to Set New High in '68," *The New York Times,* December 29, 1967.

8. *United States v. Western Electric Co., Inc., and American Telephone and Telegraph Co.,* 13 RR 2143; CCH 1956 Trade Cases, sec. 68,246 (D.C.N.J. 1956).

6 Background

The second largest telephone organization in the United States is the General System, the telephone operating and manufacturing subsidiaries of the General Telephone and Electronics Corporation (GT&E). GT&E is a highly diversified communications and manufacturing enterprise whose operations include some 30 domestic telephone operating subsidiaries and two international subsidiaries located in British Columbia and the Dominican Republic. GT&E's domestic telephone companies constitute the nation's largest independent (non-Bell) telephone system, serving approximately 40% (over 7 million) of total independent telephones. In addition, two manufacturing subsidiaries, Automatic Electric Company and Lenkurt Electric Company, Inc., produce communications equipment for the independent telephone industry as well as commercial and military markets.

The remainder of the domestic telephone service is provided by some 2100 independent telephone companies, all of whom are interconnected with the Bell and General Systems and with each other. United Utilities, Inc., of Kansas City, Missouri, is the nation's second largest independent telephone holding company (after GT&E), with 34 operating telephone companies, plus other utility and community antenna television holdings. There are 171 independent telephone companies serving more than 10,000 telephones, and 124 which have between 5,000 and 10,000 telephones.[9] The United States Independent Telephone Association (USITA), with headquarters in Washington, D.C. represents most of the independent companies. It provides guidance to its members and coordinates their operations through committees dealing with subjects such as technical standards and accounting practices.

2. Telegraph

Second to AT&T, the communications common carrier of greatest national importance is the Western Union Telegraph Company (WU). Western Union was incorporated in 1851, soon after the birth of telegraphy, and was an important force in the development of communications in this country. Its services began with the offering of public message telegraph service, and the telegram still provides the largest, but declining, share of its business. Western Union operates communications facilities throughout the United States (although it leases a substantial portion of its transmission facilities from the Bell System), and provides a variety of communications services in addition to the telegraph Public

---

9. *Statistics of the Independent Telephone Industry*, (Washington, D.C.: USITA, 1968), p. 13.

Message Service (PMS). It furnishes custom-built private wire systems (including several large systems for the U.S. government), Telex direct-dial teleprinter exchange and Broadband Exchange services, telegraphic money orders, and a variety of leased line services including facsimile, voice, voice/data, and Telpak broadband channels. Many of these services are in competition with the telephone companies. In addition, Western Union has recently introduced two new computer-based message switching communications services, SICOM and INFO-COM, which will be the topic of later discussion.

3. International Carriers

Also of importance in the communications common carrier industry are the international carriers, which provide record and voice communications between the United States and overseas points throughout the world. The largest of the international record carriers are RCA Global Communications, Inc. (RCA Globcom), a subsidiary of the Radio Corporation of America; the twelve telecommunications operating subsidiaries of the International Telephone and Telegraph Company, who form the ITT World Communications System (ITT Worldcom); and Western Union International, Inc. (no relationship with the domestic Western Union Telegraph Company). AT&T is the exclusive international telephone carrier.

These organizations operate worldwide communications networks which provide international telegram service, Telex, private leased channels for data transmission, ship/shore services, press services, and television transmission. In addition, both RCA Globcom and ITT Worldcom have introduced computer-based message switching services.

The full range of communications media are employed in various parts of the international carriers' networks: coaxial and conventional cables, both submarine and landline, HF and UHF radio including over-the-horizon tropo-scatter, and satellite circuits.

4. Satellite Communications

Satellite communications circuits are rapidly becoming a major factor in long-haul communications, especially trans-oceanic. Under the provisions of the Communications Satellite Act of 1962,[10] a private company under joint

---

10. Public Law 87-624, H.R. 11040, 87th Congress, August 31, 1962.

8 Background

ownership of the common carriers and the public was formed to develop as expeditiously as practicable a commercial communications satellite system. This firm, the Communications Satellite Corporation (COMSAT) was incorporated on February 1, 1963, as a "carrier's carrier" to design, launch, and operate communications satellites, and sell circuit capacity to the carriers. In 1964, the International Telecommunications Satellite Consortium (INTELSAT) was formed, and today has 68 member nations. COMSAT represents the United States in INTELSAT, acts as manager on behalf of the consortium, and is the largest investor in and user of the international satellite system.[11]

Early Bird (INTELSAT I), the world's first commercial communications satellite, was launched by COMSAT in 1965, and placed in a synchronous geo-stationary orbit over the Atlantic Ocean. It proved to be quite successful with a capacity of 240 voice-grade circuits, and during 1967 three more-powerful synchronous INTELSAT II satellites were placed into service. Plans for increasing the capacity of this worldwide network call for the launching of more advanced INTELSAT III (with 1200 voice-grade circuits), and INTELSAT IV (with 3000 to 10,000 voice-grade circuits, depending on antenna configuration) satellites during the period 1968-70. All of the above satellites are for *international* communications.

The ownership and operation of a domestic satellite system is currently a controversial matter, as the Communications Satellite Act of 1962 did not clearly prescribe arrangements in domestic communications. In 1967, COMSAT proposed an experimental pilot program for U.S. domestic satellite services, a system to serve commercial and non-commercial broadcasting as well as voice and data traffic. The program would use two high-capacity synchronous satellites and a number of large and small earth stations, for a two-year test of the response of potential users of satellite communications: "what applications they require, how commercial traffic build-up takes place under a given rate structure, how the enlargement of commercial use of satellite services can sustain and benefit non-commercial use, and what the range of uses may be for educational and public broadcasting."[12] AT&T also proposed development of a general purpose

---

11. Communications Satellite Corporation, *Annual Report 1968*, (Washington, D.C.: 1969), p. 16.

12. Communications Satellite Corporation, *Report to the President and the Congress, for the Calendar Year 1967,* February 23, 1968, p. 18.

domestic satellite system, while the Ford Foundation and the American Broadcasting Companies, Inc., have proposed special purpose systems for the distribution of television programs. In its report, the President's Task Force on Communications Policy recommended a pilot program along the lines suggested by COMSAT, using advanced technology in a multi-purpose pilot program to obtain needed technical and operational data for subsequent decisions on a full-scale program. COMSAT would own the space segment of the system, as trustee, and would act as program manager. Investment in the terrestrial facilities would be open to COMSAT, the established common carriers, and the prospective users of wideband satellite transmission services (e.g., commercial broadcasters, non-commercial and educational television interests, and other public service interests).[13] If such a program is approved, it could be a significant step toward the public's obtaining the benefits of more and cheaper communications through the use of satellites.

Satellite communications are relevant in the context of the present study because they provide an alternative to terrestrial ground links for data communications and they have distinctively different cost characteristics than terrestrial communication circuits. In particular, the cost of transmission through a satellite is independent of distance.

The feasibility of using satellites for wideband data transmission between the United States and Europe was studied in 1967 in a joint effort by COMSAT and IBM. This test stemmed from the realization that there will be increasing demands for high-throughput data communications channels, with bandwidths between 48 kilocycles and 1 megacycle, and the hope that satellite technology will make them economically feasible for long-haul communications requirements. The test employed the Early Bird synchronous satellite managed by COMSAT, and the terrestrial facilities of numerous carriers, including AT&T, GT&E, French Cable Company, French Postal Telegraph and Telephone (PTT) Administration, British General Post Office, Canadian Overseas Telecommunication Corporation, and the West German Post Office. The testing program successfully demonstrated "the technical feasibility of using satellite communications networks interconnected

---

13. President's Task Force on Communications Policy, *Final Report* (Washington, D.C.: U.S. Government Printing Office, 1969), Chapter Five, "Domestic Applications of Communication Satellite Technology."

with terrestrial networks to provide high-speed data transmission over intercontinental distances."[14] Today, IBM's internal corporate data communication network incorporates a wideband satellite channel for overseas transmission.

D. STRUCTURE OF THE COMPUTER INDUSTRY

1. Suppliers of Computer Equipment

The first segment of the computer equipment industry is composed of manufacturers of main-frame computers and related peripheral devices. IBM accounts for an estimated two-thirds of the output in this basic segment of the industry, while the remaining suppliers (a dozen major ones in all) share the rest of the market.

A second segment of the computer equipment industry includes the numerous non-integrated suppliers of computer input-output devices and associated peripheral equipment. These firms supply a variety of data devices and related communications equipment ranging from typewriter and graphic display terminals to communications line multiplexers and modulation-demodulation devices.

2. Service Bureaus and Software Houses

"Service bureaus" and "software houses" are an important part of the computer industry. They offer information processing facilities and services; the service bureaus primarily provide *computer facilities* while the software houses specialize in *programming services*. However, there is considerable overlap in the activities of these organizations.

Generally, service bureaus and software houses offer some or all of five main types of services. First, they simply rent computer capacity, on an hourly basis. Second, they develop and rent generalized proprietary computer programs. Third, they offer consulting and/or systems analysis services. Fourth, they undertake contract programming for clients. And finally, they offer educational services— classes, lectures, and seminars.

---

14. IBM Corporation, *Transatlantic Data Communications Tests via Satellite,* Systems Development Division Special Report SP 11 (Raleigh, North Carolina: 1967), p. 9.

The computer service bureau and software field is a new and rapidly growing one. Its structure is at present highly fragmented, and barriers to entry are generally quite low, since computers can be rented from manufacturers rather than purchased outright; the most essential requirement for entry is the availability of skilled personnel. The service bureau industry includes over 700 independent firms, most of which are quite small.[15] In addition, a number of computer main-frame manufacturers have integrated forward into this field. Also, many large banks and other organizations, having originally obtained computers for internal purposes, have expanded into the service bureau field.

3. Computer-Based Information Services

A third major component of the computer field is what we shall call the "information service" industry. Its members are engaged in the collection, processing, and sale of information itself – an operation which today depends on computers. The information service firm differs from the service bureau in that it uses its own information rather than the customer's. Of course, both functions could be carried on by the same firm. Typical examples of information service industry offerings include credit inquiry service, stock market quotation service, and legal information retrieval services.

Because the information service field often involves relatively large fixed costs (e.g., in setting up a data base), a number of the firms in this field are larger than the average service bureau. All such information services are relatively new. Examples of the larger members of this market are Bunker-Ramo Corporation, with its nationwide Telequote III quotation service for securities and commodities brokers; Credit Data Corporation, with its on-line consumer credit reporting service (presently a regional service, and planned for nationwide operation within the next several years); and Western Union, with its Personnel Information Communication System (PICS).

Both the service bureau and information service industries are areas of rapid growth and innovation. They are particularly important to an examination of data-related communications policy because it is the service bureau and

---

15. *Response* of the Association of Data Processing Service Organizations (ADAPSO), in the FCC Inquiry into the Interdependence of Computers and Communications Services and Facilities, Docket No. 16979, pp. 5-6. (Hereinafter referenced as ADAPSO *Response,* and Responses of other organizations will be referenced in a like manner.)

12   Background

information service firms that are developing the new commercial offerings which require communications circuits for use in conjunction with computers. They thus account for a high proportion of the remote access data processing applications described in the following section.

E. REMOTE ACCESS DATA PROCESSING

Rapid advances in computer technology and in the design and programming of large computer systems have increased the commercial usefulness of "remote access data processing systems" — i.e., systems in which data is transmitted by communications links to and from a computer performing data processing functions. The trend toward remote access data processing has been largely furthered by the lower costs per unit of computation associated with large computers, by the desire to share costly system resources, by the attempt to make computing power available in a manner more convenient to the user, and by the desire to centralize the storage of related information.

Remote access computer systems can be categorized according to the speed with which they respond either to input commands or to the receipt of information. "Real-time" systems respond to inputs immediately (within seconds or minutes), while "batch processing" systems respond after a sizable time delay (hours, days, or longer). These definitions are imprecise but the terms denote useful reference points at opposite ends of a continuum.

The number and variety of remote access data processing systems, both real-time and batch processing, is already very large and rapidly growing. The following categorization of existing applications is sufficient to underscore the commercial and practical importance of the entire remote access computer industry:[16]

1. *Conversational time-sharing systems* (always real-time) involve the simultaneous sharing of a central computer among a group of users located at remote terminals and connected to the central computer by communication circuits. These systems permit a user to "interact" with a large computer, the

---

16. This categorization is based upon John G. McPherson, "Data Communication Requirements of Computer Systems," *IEEE Spectrum,* December 1967, pp. 42-45.

user working on the creative, ill-structured aspects of problem-solving and the computer rapidly performing the lengthy computations. As noted by two of the scientists at M.I.T.'s Project MAC:

> "The impetus for time-sharing first arose from professional programmers because of their constant frustration in debugging [removing mistakes from] programs at batch processing installations. Thus, the original goal was to time-share computers to allow simultaneous access by several persons... However, at Project MAC it has turned out that simultaneous access to the machine, while obviously necessary to the objective, has not been the major ensuing benefit... the most significant effect that the MAC system has had on the MIT community is seen in the achievements of persons for whom computers are tools for other objectives [than program writing]. The availability of the MAC system has not only changed the way problems are attacked, but also important research has been done that would not have been undertaken otherwise... In other words, the major goal is to provide suitable tools for what is currently being called machine-aided cognition.[17]

2. *Inquiry systems* (usually real-time) are typified by stock quotation services. In such systems, a large number of terminals are connected to a single data processing center by means of communication lines; the system enables remote users to query a frequently up-dated central store of information (e.g., the latest price bid or paid for any particular security).

3. *Data Collection Systems* (either real-time or batch) involve transmission to, and storage and processing by, a central computer, of information gathered at many remote points.

---

17. F. J. Corbató and V. A. Vyssotsky, "Introduction and Overview of the MULTICS System," *Proceedings — Fall Joint Computer Conference* (Las Vegas, Nevada, November 30, 1965), p. 186.

The global weather reporting systems operated by the U.S. Weather Bureau, the Air Force, and the Navy, are examples. Data collection systems are often combined with the "inquiry" function into a single system, having both multiple inputs to the computer and outputs to multiple users over communications links. An example of such a combined system would be an airline reservations system, which collects information (e.g., each seat reserved) from many remote terminals and permits users at remote terminals to inquire about the current status of a particular flight on the basis of information gathered from the entire system.

4. *Remote Batch Processing Systems* permit the central processing of tasks originating at and transmitted from distant locations. Processing is not done at the point of origination either because there is insufficient local volume to justify the installation of a computer there, or because the efficiencies of sharing a larger central machine outweigh the cost of transmitting the job (program and/or data) over communications facilities to the central computer.

5. *Remote Document-Production Systems* employ a central computer to transmit to remote users up-to-date reproductions of printed documents. In addition to printing business documents at remote locations, the systems can also print graphical information, such as circuit diagrams, flow charts, line drawings and specifications.

6. *Information Distribution Systems* (which could operate on either a real-time or batch basis) may very often operate like inquiry and document-production systems without the need for specific, repeated customer inquiries. That is, as information relevant to the needs of a particular subscriber is received by a central computer, the information is automatically and selectively transmitted to the subscriber via communications lines. The distribution of railroad freight traffic information to railroad traffic agents, shippers, and consignees is an example of such a system now in operation. In the future, as information is accumulated in appropriate

form, the distribution of information from machine-readable files will take a central position in automated information retrieval systems; information reflecting developments in almost any field will be distributed more directly than through the present media of publication.

CHAPTER II

A STATEMENT OF THE PROBLEMS

This book is concerned with several distinct but interrelated public policy problems involving directly both the communications common carrier industry and the data processing and information services industry. To provide the reader with an overview of these issues and the manner in which they interrelate with one another, and to set the stage for further discussion, let us briefly examine them in turn.

A. "COMPUTER UTILITIES" AND THE REGULATION OF DATA PROCESSING

Since its initial development some twenty years ago, the electronic digital computer has undergone rapid orders-of-magnitude improvements in capabilities and sophistication, and corresponding reductions in cost. The number and diversity of computer applications have followed suit, causing the computer to become an essential part of today's way of life. Several relatively recent developments in computer technology, especially time-sharing (described briefly in the previous chapter), have led to speculation by a number of noted authorities that large multi-user time-shared computer service centers serving wide geographical areas (the size of the area served being limited by the costs of communication lines) may become the pattern of the future, supplanting many of the present in-house private data processing systems and the smaller service organizations.[1]

Extrapolating this anticipated evolution of computer time-sharing, some writers envisioned the growth of general purpose "computer utilities" which would distribute computational "power" in a manner analogous to the distribution of gas and electric power by public utility firms. It was therefore questioned whether the public interest might require some form of regulation of this service activity. Interest in the data processing service business by communications common carriers also led some to speculate that perhaps communications and

---

1. See, e.g., Martin Greenberger (ed.), *Computers and the World of the Future*, (Cambridge, Mass.: M.I.T. Press, 1964); Greenberger, "The Computers of Tomorrow," *Atlantic Monthly*, May 1964; also Douglas F. Parkhill, *The Challenge of the Computer Utility*, (Reading, Mass.: Addison-Wesley, 1966).

remote-access data processing services might best be offered in combination, by communications suppliers, and might best be regulated together. It has even been proposed that supply of a service by a regulated firm might somehow make the service a candidate for regulation irregardless of the nature of the service.

Against this background the FCC has asked in its on-going inquiry into the interdependence of computers and communications (Notice of Inquiry is reproduced in Appendix A), whether it has statutory authority (under Title II of the Communications Act of 1934) to regulate data processing and information services, either stand-alone or using communication channels, and either furnished by communications common carriers or private firms.[2] In addition, the FCC asked whether such services *should* be regulated; i.e., whether changes in existing law or regulations are needed to either remove them from regulation if they are currently subject to it, or bring them under regulation if they are not subject to it at present.[3]

Most of the respondents to the FCC Inquiry addressed the question of possible regulation of data processing services. There was unanimous agreement that under existing law such services, regardless of whether using communication lines or not, and whether provided by common carriers or private firms, are not subject to regulation by the FCC; they simply are not common carriage communications. Regarding the need for additional legislation to bring such services under regulation by an appropriate governmental body, there was also agreement that this would be inappropriate. Speaking for the computer equipment manufacturing industry, the Business Equipment Manufacturers Association (BEMA) said:

> "...We believe it clear, in response to Item C, that entities providing data and information services of the type here under discussion are not now subject to regulation as common carriers under Title II of the Communications Act. The questions raised under Items D and E, however, go beyond this point; they inquire whether such services *should* be subject to Title II or similar regulation, even if this required additional legislation, or whether

---

2. FCC Notice of Inquiry, Docket No. 16979, para. 25 (C); (See Appendix A).

3. *Ibid.*, para. 25 (D,E).

they should be permitted to evolve in a free competitive market. We strongly support the position that such additional legislation would be contrary to the public interest and that the public interest would be best served by permitting — and indeed encouraging — the continued growth of these services in a vigorous competitive market context."[4]

The communications common carriers, who were at odds with the computer industry on most other issues in the Inquiry, agreed on this subject. For example, Western Union commented:

"...Western Union views 'data processing' and 'general or special information services' as being not subject to the provisions of Title II of the [Communications] Act because they do not fall within the definitional perimeters of service subject to regulation, and public policy insofar as it is presently discernible does not dictate that legislation be enacted bringing such services under regulation by an appropriate governmental authority."[5]

The Federal government respondents to the Inquiry, GSA and the Department of Justice, likewise agreed. The Justice Department said:

"It is our opinion that 'remote access data processing' is not common carrier communications and hence, is not subject to the Commission's jurisdiction under Title II of the Communications Act....While persons offering this service employ communications links, their primary business is offering data processing and the communications links are simply the means by which this is accomplished.

Moreover, as a matter of economy, we believe that it would be altogether inappropriate to extend detailed rate regulation to the remote access data processing field; and we urge the Commission to make a specific finding to this effect."[6]

---

4. BEMA *Response*, p. 67.

5. Western Union *Response*, p. CDE-47.

6. U.S. Department of Justice *Response*, pp. 64-65.

The *present* regulatory status of data and information services which employ communications facilities depends upon an interpretation of the Communications Act of 1934; and it is clear that these services fall outside the jurisdiction of that Act, and therefore of the FCC. The larger issue of the appropriate *future* status of these services — the choice between regulation and free competition — depends upon the answers to several economic and policy questions.

First, have abuses occurred in the provision of these services to an extent sufficient to warrant regulation? The responses clearly show that this is not the case; that, in fact, vigorous free competition in the markets for these services has to date produced — and will most likely continue to produce — rapid growth, innovation, and diversification, to the benefit of the consumer. Regulation would stultify the expansion and progress of the dynamic computer services industry. Secondly, the decision to regulate a business activity requires a determination of whether the activity is characterized by continually decreasing costs to scale; i.e., is it a natural monopoly which *must* be regulated? Again, numerous respondents to the FCC Inquiry have shown that data processing and information services do *not* exhibit the characteristics of a natural monopoly and are not likely to do so in the foreseeable future.

Thirdly, it has been suggested that valid public concern for the privacy and security of users' private information stored in and processed by such computer services might require some form of governmental control, including perhaps regulation of some sort. On this subject as well, there is agreement among the respondents addressing the question that this problem has not materialized to date and is not likely to do so; therefore regulation is not required. A discussion of the protection of privacy in remote access data processing systems will be found in Chapter X.

In summary, there is substantial agreement among all concerned, including the authors, that data processing and information services in any form should not be the subject of regulation by the FCC or any other governmental body. As this matter is not then at issue, further discussion of the subject is unnecessary.

B. COMMON CARRIER ENTRY INTO DATA PROCESSING SERVICES

Several common carriers, already using computers for communications functions and recognizing electronic data processing as a rapidly growing field, are actively seeking to offer computer-based services to the public. Two forms of service are envisioned: First, simply service bureau type rental of computer time, using either stand-alone or remotely accessed computer hardware; and secondly, computer-based information services.

Of the several interested carriers, Western Union has been the most aggressive. Its president, Russell W. McFall, has made a number of speeches in the past few years in which he projects Western Union as a "national information utility." Beginning to move toward this goal, the telegraph company has introduced computer-based personnel search and legal information retrieval services, and has expressed its intentions to append data processing functions to its new SICOM and INFO-COM message switching services. Other domestic and international common carriers are also entering the data processing services field. Both GT&E and ITT have formed corporate subsidiary organizations which offer such services, on a stand-alone as well as a remote-accessed basis.

AT&T might be considered the "dark horse" in the field. While it is uniquely qualified in several respects to offer such services, as will be seen later, it operates under an antitrust consent decree which restricts it to regulated common carrier communications activities. Thus, AT&T participation would require either a modification in its consent decree (quite unlikely) or an acceptance by the FCC of a tariff for certain data processing services as "common carriage communications."

The carriers' interest in providing data processing and information services poses several public policy problems. First, assuming that data processing is to remain unregulated, under what ground rules should a regulated common carrier be permitted to offer an unregulated service? Is the formulation of a wholly separate subsidiary organization a necessary prerequisite to ensure proper separation of financial accounts for regulated and unregulated activities? If the other firms which offer data processing and information services in competition with the carrier are dependent upon the carrier for their communication line facilities, how can discriminatory provision of these lines be prevented? These questions will be dealt with in Chapter III.

C. THE PROVISION OF MESSAGE-SWITCHING SERVICES BY NON-CARRIERS

Message switching is a store-and-forward method of transferring record (non-voice) information among remote stations in a communications network. In the past it has been performed by electromechanical equipment which could only perform switching functions. Today, however, it is generally performed by electronic digital computers which provide greater communications flexibility, reduce message switching costs, and may also perform data processing and information retrieval functions.

Data processing and information services as discussed above have been, and probably should continue to be, unregulated activities. However, under today's law, the practice of switching messages for others constitutes common carriage communications, and as such is an activity regulated by the FCC. Therefore, if a single organization seeks to offer a combination communications and data processing service, should the scope of FCC regulation be extended to encompass the entire service; or should the entire service be unregulated; or should the communications elements of the service be separated out, in an accounting sense, for regulatory purposes?

A controversy which arose in 1965 highlights the nature of the problem. The Bunker-Ramo Corporation offers a remote-access computer-based stock quotation service to the brokerage community. Current stock information from the various security exchanges is stored in the Bunker-Ramo computer facilities, and accessed through visual display terminals located in the brokers' offices and connected to the computers via communication lines leased from the common carriers by Bunker-Ramo. This information retrieval activity is unregulated. In 1965, Bunker-Ramo attempted to expand its service and offer to subscribers a message switching capability, using the same computers and communication lines. Brokers could then use the stock quotation system to send administrative messages among their offices and "buy and sell' orders to their representatives on the floor of the exchanges. Both AT&T and Western Union refused to lease the necessary communication lines to Bunker-Ramo, to be used in this manner, because they claimed that Bunker-Ramo, which is not a common carrier, would in effect be performing communications common carriage. Thus the new service was effectively foreclosed.[7]

22   A Statement of the Problems

As the situation now stands, many desirable services involving combinations of common carrier communications and data processing (or information services), such as proposed by Bunker-Ramo, cannot be offered. Chapter IV discusses alternative solutions to the problem, and also notes how computer technology has affected both the need for new, *pure* message switching services and the "economics" of providing such services.

D. COMMON CARRIER RESTRICTIONS ON THE USE OF THEIR COMMUNICATIONS LINES

The carriers have several long-standing restrictive rules regarding the use of their facilities, established before the advent of data communications but remaining in effect today. The rules are contained in the enabling tariffs for the respective services. It is appropriate to re-examine the rules and determine their effects in light of the nature and growth of data communications.

The major restrictions involve (1) use of "foreign" attachments on the public switched network, (2) interconnecting private communication systems with the common carrier network, and (3) sharing or reselling the use of a carrier communication line.

1. The "Foreign Attachment" Limitation

Communications common carriers have imposed several restrictions on the connection to the public switched networks[8] (both telephone and teletype) of any equipment not provided by themselves, except through the use of carrier-

---

7. For a more detailed discussion of this controversy, see Bunker-Ramo *Response*; also Manley R. Irwin, "Time-Shared Information Systems: Market Entry in Search of a Policy, *Proceedings of the 1967 Fall Joint Computer Conference,* pp. 513-520. (Washington, D.C.: Thompson Book Company, 1967.)

8. Throughout this book the terms "public exchange network," "switched network," "dial network," and "voice telephone network" will be used interchangeably. Each refers to the ordinary dial telephone system, which is used to provide local exchange telephone service, long distance (toll) service, and Wide Area Telephone Service (WATS – a flat-rate long distance billing arrangement), for both voice and data transmission. Separate low-speed switched (dial-up) networks are also used to provide AT&T's TWX and Western Union's Telex teletypewriter exchange services. Perhaps five years after the presently pending sale of TWX to Western Union is completed, the two teletypewriter networks will be consolidated.

supplied interfacing equipment. These restrictions, which were liberalized in 1969, do not however apply on private leased lines.

The most important attachment for the data communications user is the modulator/demodulator, also called a "modem" or a "data set." This device transforms data from a business machine into a form suitable for transmission over the communications line, and vice versa for data received from the line. It is important to the user because it is a limiting factor on the transmission capacity of his communications line.

The nature of the foreign attachment restrictions, their rationale and reasonableness, will be examined in Chapter V.

2. The Interconnection Restriction

The carriers have, in the past, entirely barred firms from interconnecting private communication systems (e.g., mobile radio systems, private in-house telephone switchboards, private microwave systems, etc.) with the public telephone network. This prohibition was relaxed in 1969 to allow interconnection to both the dial telephone and leased line networks, provided certain conditions were met. Chapter VI discusses the evolving common carrier rules regarding interconnection and the rationale for certain restrictions.

3. The Line Sharing/Resale Restriction

The third major common carrier restriction of concern to data communications users involves line sharing among lessees of communication channels, or resale of such channel capacity. In only limited cases is line sharing permitted. There are essentially two possible forms of line sharing: In one case a firm would lease a broadband communication channel and divide it into subchannels for simultaneous usage by several other users. In the second case several carrier subscribers would share the use of a single channel on a *time* basis; e.g., the line might be used by one firm during the day and by another user at night. The nature and validity of these restrictions are examined in Chapter VII.

E. ADEQUACY OF THE PRESENT COMMUNICATIONS NETWORK FOR DATA TRANSMISSION

The common carriers' present communications plant was developed, prior to the advent of the digital computer, for teletype and voice communication; and so channels were designed for these uses. In addition, as a response to competition from private microwave systems, the carriers have in recent years begun to offer broadband communication facilities ranging in capacity from the equivalent of 12 to 240 voice channels. However, the transmission of digital information requires communication services and facilities somewhat different from those required for voice and conventional forms of record transmission. Chapter VIII will examine these differences and suggest ways in which they might be reduced in the future.

Each type of communications channel facility has an information-carrying capacity (measured in bits of information carried per second) which is determined primarily by the frequency bandwidth of the channel. The capacities of the currently available channels are not generally well-suited to the requirements of data transmission. A data communications user may need a line with a certain capacity only to find that the available lines are "too large" or "too small" (they carry more, or less information per second than the user intends to transmit). We will examine the extent to which there is a mismatch between the channels available from the common carriers and the bandwidth requirements of the data communications users.

In addition to the capacity of a communications line, another important consideration is its reliability (its freedom from spurious noise, and its freedom from interruption or total failure), which determines the line's "error rate" at a given speed of data transmission — that is, the number of bits of information which will be lost during transmission. The exacting channel reliability requirements for data communications, which are not always satisfied today, will be discussed.

Finally, Chapter VIII will consider the concept of an entirely digital national communication network. This would be in some respects a marked departure from our previous system design concepts, would employ new technology extensively (e.g., large-scale use of store-and-forward computer switching), and would provide the means to alleviate many of the incompatibilities between the communication network and data transmission applications which are making increasing use of the network.

F. SPECIAL SERVICE COMMON CARRIERS

In 1963 a small firm, Microwave Communications, Inc. (MCI), filed an application with the FCC for a license as a "Special Service Common Carrier." MCI proposed to offer, between Chicago and St. Louis, inter-city trunk circuits furnished by microwave transmission facilities to business users for voice and data communications. These bulk private line services would, in a sense, compete with the private line services offered by the telephone companies and Western Union. The low rates and flexible use proposed by MCI attracted both wide interest in the business community, and also determined opposition from the existing carriers.

While MCI claimed that it was offering a new and different service, whose low cost was based upon the use of modern microwave technology, the telephone companies and Western Union claimed that MCI's *costs* for comparable services were no lower, in fact, than those of the existing common carriers. However, they claimed that MCI was able to offer lower *prices* because the telephone and telegraph carriers base their prices upon a national average of circuit costs, which are higher than the true costs in the dense Chicago-St. Louis corridor.

At the time of this writing the FCC has not passed upon MCI's license application. Whatever the outcome of this particular proceeding, the fundamental question remains — should the present carriers be permitted to continue their monopoly control of long-haul communications? The issue is further clouded by prospects of a domestic satellite communications system, which could be provided by a "chosen instrument" such as COMSAT, by the present land-line carriers, or by other entities not yet on the horizon. Chapter IX discusses the potential entry of new carriers into the long-haul communication services market.

CHAPTER III

MARKET ENTRY, PART I:
COMMON CARRIER ENTRY INTO DATA PROCESSING SERVICES

A. CARRIER EXPERIENCE AND INTEREST IN OFFERING DATA PROCESSING SERVICES

As explained in the previous chapter, there is a trend among several communication common carriers to employ computers not only to facilitate their role as common carriers (performing communications switching and in-house administrative data processing) but also to provide commercial data processing and information services. Western Union, for example, has made explicit its long range plans to offer data processing and information services. Russell W. McFall, Western Union's president, made the following remarks in a 1966 speech before the Industrial Communications Association:

"I referred earlier to Western Union's plans to become a national information utility. In developing this utility we plan to first merge the public message and Telex systems into a single, integrated record message system run by a national network of computer centers. Our longer range plans include the installation of large-capacity computers of advanced design at key locations throughout the country, linked by broadband facilities of the kind I have described. We plan to offer, when these are in place, the services of a national information utility."

"....the information utility will provide three broad classes of service:

1. *Message-processing services* of all kinds, including exchange services, for the transmission and delivery of data and messages received from individuals and organizations.

2. *Information processing services* for receiving and processing data, and for delivering the resulting information to the subscriber. These services will make available to business firms, for instance, real-time processing of payrolls, inventories, and sales; to professional people, billing and tax services; to individuals, portfolio analyses.

3. *Inquiry services* for disseminating certain types of information such as stock quotations; weather information; credit information and other business data; professional, scientific, and educational information; and for the interchange of information to accomplish such transactions as hotel and travel reservations, the purchase or sale of securities, mail order purchases, and so forth."[1]

Subsequently, Western Union introduced several new services in accord with its corporate strategy to become the "nation's information utility." In June 1966, it announced a computer-based legal citation service in which remote subscribers could selectively access central stores of citations to federal and state cases.[2] Shortly afterwards, Western Union announced a computer-based Personnel Information Communications System (PICS) which would "match the qualifications, salary wants, and other data on job seekers with the requirements of employers with open positions."[3]

In December 1967 Western Union obtained approval from the FCC to offer two new services, SICOM (Securities Industry Communications System) and INFO-COM (Information Communication Service), both of which are computer-based communication services. It is relatively simple, from a technical standpoint,

---

1. Russell W. McFall, "The Age of the Communicator," Address before the Industrial Communications Association, Montreal, Canada, May 2, 1966, pp. 12-13.

2. Western Union press release, June 28, 1966, p. 1.

3. Russell W. McFall, "New Partners in Progress — Communications and Computers," speech before the 21st National Conference of the Association for Computing Machinery, Los Angeles, California, August 30, 1966, p. 8.

for Western Union to expand the capabilities of SICOM and INFO-COM and offer both data processing and information services; and the company has, in fact, said that SICOM "will be augmented, in a phased program, by offerings of data processing services utilizing the data base generated by the initial service."[4] However, the FCC in its approval of the initial SICOM tariff explicitly limited the scope of the offering to communications:

> "We believe that substantial and different questions would be raised with respect to the propriety of the tariff if there should be any broadening of the SICOM offering by the addition of a fourth computer or by any other means whereby Western Union would perform or offer to perform non-communications data processing as part of the package SICOM service. If this should occur, the tariff may be subject to whatever corrective action as may be deemed necessary, either upon our own motion or upon complaint."[5]

Western Union's interest in entering the "computer utility" field led to several developments in 1968. University Computing Company of Dallas, Texas, a diversified supplier of computer time-sharing services and peripheral equipment, made a tender offer in May 1968 for some 10% of the outstanding Western Union stock. The offer was strongly opposed by Western Union's management, who apparently viewed it as the first step in a planned take-over of the company. The tender offer was headed off by Western Union's sudden announcement of plans to merge with Computer Sciences Corporation, a large supplier of computer software and related services. It was expected that the WU/CSC merger plans would receive close scrutiny, and possibly be blocked, by the FCC and by the Antitrust Division of the Justice Department, who had expressed concern regarding the entry of common carriers into the data services market. However, before the merger plans progressed to this stage, they were thwarted by another interested party, AT&T.

For years, Western Union had been negotiating with AT&T to purchase AT&T's TWX teletypewriter exchange system, which would then be integrated with Western Union's smaller Telex network. Western Union has suffered steadily

---

4. Vincent Hamill and Kenneth L. Brody, "SICOM: Securities Industry Communications System," *Western Union Technical Review,* April, 1967, p. 98.

5. *In Re Western Union Telegraph Company Tariff FCC No. 251 Applicable to SICOM Service,* Memorandum Opinion and Order, Released December 27, 1967, pp. 13-14.

declining revenues from its "backbone" business, the public telegram service, and its long-range plans for plant modernization and introduction of new services depend upon the additional revenue which acquisition of TWX would bring. In July 1968, AT&T suspended its TWX negotiations with Western Union, on the grounds that the proposed WU/CSC merger raised substantial new questions affecting the TWX sale; a few days thereafter Western Union cancelled its merger plans.

In November 1968, Western Union once more showed its desire to participate in the computer services industry. It announced plans to join Advanced Computer Utilities Corporation, of Fort Lauderdale, Florida, in the formation of a new company, Western Union Computer Utilities, Inc. WUCU would franchise computer service centers nationwide, provide training and advisory services to these centers, and lease proprietary software.

Western Union's interest in offering data processing services appears, however, to have waned somewhat in 1969. Western Union's President and Chairman of the Board, Russell W. McFall, in the 1969 Western Union Annual Report, did not mention the information utility concept. Rather, he emphasized Western Union's plan to develop a single, integrated communication system, and said:

> From the existing multiplicity of systems, we proposed to create one broad-based system. With such a system, we could take advantage of the new business opportunities before us and, at the same time, solve many of the problems in the public message business.
>
> The point is that all services would benefit from the creation of a single system. Each would be part of a communications capability more advanced, more sophisticated, and with greater capacity than could be justified for any of the individual services.
>
> ***
>
> First of all, let me explain what exactly the integrated single system is to be. It will be a system in which all the services – from public message to high-speed data handling – use the same transmission and switching facilities. It will be a system in which

compatible and interchangeable terminal equipment will be used for all the services. It will be a system with built-in flexibility in designing to handle the customer's own needs. It will be a system that permits growth as the customer's needs grow. In other words, with an integrated system, the customer can progressively upgrade — from public message to high-speed switched services and from simple switched services to more sophisticated store-and-forward systems, or to any combination of these he might need in his business.

Finally, this will be a computer-controlled system. The switching, the storing, the control and integration of all the elements are to be handled by computers.[6]

Other carriers have also shown active interest in offering data processing services, although significantly, they have taken a different organizational approach than that of Western Union. International Telephone and Telegraph Corporation operates a rapidly expanding, unregulated, data processing service bureau subsidiary called ITT Data Services. General Telephone and Electronics is also offering data processing services to the public through the mechanism of a separate subsidiary, GT&E Data Services Corporation, formed in January 1968, which also serves the telephone operating companies of the General System. The concept of the "separate subsidiary" will be important in later discussion.

In addition to GT&E, several of the other larger independent telephone companies are also entering the data services market. United Utilities, Inc., the nation's second-largest Independent (after the General System), is introducing computer time-sharing services, and Continental Telephone Corporation, in late 1968, formed a division to provide data services for the Continental System and others.

AT&T is by far the best-prepared potential common carrier entrant into the data services industry — measured in terms of technical and financial resources, existing available plant and facilities, widespread working relationships with literally millions of potential customers, and other important factors. It is currently beginning to replace its old electromechanical telephone switching

---

6. *Western Union Annual Report*, 1969, pp. 4-5.

exchanges with electronic switching systems (ESS) which are, essentially, special-purpose computers. When telephone traffic loads are light, such as at night, the ESS computers might be used for conventional data processing at little incremental cost. However, the specialized nature of these computers makes them inefficient, relative to their general-purpose commercial counterparts, for data processing.[7]

AT&T also uses commercial computers extensively for internal business data processing. It has approximately 600 such general purpose computers installed, and has established a 500-man task force to develop a comprehensive computer-based, internal Business Information System (BIS).[8]

In addition to its use of computers for communications and internal data processing functions, the Bell System also manufactures both its ESS computers and many input/output terminal devices, which can be used in conjunction with data processing services. The Teletype Corporation, owned indirectly by AT&T, manufactures teletypewriter terminals which are more widely used for data communication purposes than any competing products, as well as a variety of other data communications equipment. Also, the telephone company is currently replacing the conventional rotary-dial telephones with Touch-Tone[9] phones which can be used as both keyboard data input devices and audio output devices for computer processing. Bell is also planning the introduction of the Picturephone,[9] which is a cathode ray tube display device and which could readily be used for visual computer output.

In spite of its advantageous position, AT&T has expressly stated that it does not plan to offer data processing services. Its Chairman of the Board, in a talk before the 1968 Spring Joint Computer Conference, said:

> "Well — since we have been so concerned with data processing for these purposes of handling [telephone] calls, it should be no surprise that we also became involved in computers for other

---

7. For a full, technical description of the Electronic Switching System see AT&T Monograph #4853, "No. 1 Electronic Switching System," Papers from the *Bell System Technical Journal*, Vol. 43, September 1964, pp. 1831-2609.

8. For background on the BIS concept, see AT&T, *The Bell System's Approach to Business Information Systems*, (New York, 1965).

9. "Touch-Tone" and "Picturephone" are AT&T trademarks.

possible uses. We did in fact — using relays — build the very first electrically operated digital computer (in 1939) and up to 1950 had produced more than half of all the large ones made.

With that kind of head start, and considering also our position in transistors and solid-state technology generally, what *might* have been a surprise was that *we took ourselves completely out of the business of providing computer services.* The reason was simply that we wanted to concentrate on communications." (Latter emphasis added.)[10]

An additional reason is that AT&T operates under an antitrust consent decree which prohibits it from offering any unregulated services.[11] The other carriers are under no such constraints.

If AT&T were to file a tariff covering data processing services for hire (perhaps in conjunction with some communications service), and it were accepted by the FCC as constituting a common carriage activity subject to regulation under the Communications Act of 1934, the service would become a permissible AT&T activity in accordance with the consent decree.[12] But as noted previously,

---

10. H. I. Romnes, Address before the Spring Joint Computer Conference, Atlantic City, April 30, 1968. p. 6.

11. AT&T is confined to offering "common carrier communications services" — defined as (i) any service subject to regulation under the Communications Act of 1934 or (ii) any other service subject to direct regulation in the state(s) where AT&T offers it. See Paragraph V of Final Judgment in *United States v. Western Electric Company, Inc. and American Telephone and Telegraph Company,* 13 RR 2143; CCH 1956 Trade Cases sec. 68,246 (D.C.N.J. 1956).

12. John F. Preston, attorney for AT&T, testified as follows before a subcommittee of the House Small Business Committee:

> Mr. Potvin [subcommittee counsel]: "It would seem pertinent, too, to point out there would be some question as to the applicability of the consent decree under which you operate, as to whether the furnishing of computer services would be a proper activity for your corporation."
>
> Mr. Preston: "Yes. And that is why I initially answered as I did — that we would, if we had decided to go into this business, would have filed tariffs, and if accepted by the regulatory agencies, then this would be a common carrier communications service permissible under the final judgment."

(Hearings on Activities of Regulatory and Enforcement Agencies Relating to Small Business Before Subcommittee No. 6 of the House Select Committee on Small Business, Part 2, 89th Congress, Second Session, 1966, pp. 476-477.)

essentially all of the respondents to the FCC's Notice of Inquiry who addressed the issue (and most did), including the carriers, agreed that data processing is *not* an activity properly subject to regulation under the Communications Act and, furthermore, that the Communications Act should not be modified to bring data processing services within its (and therefore the FCC's) jurisdiction.

Carrier Provision of Private Computer Systems

In addition to public offerings of data processing and information services the carriers, especially Western Union, also supply *private* computer systems for either communications or combined communications and data processing functions — in competition with computer manufacturers and service bureaus. Mainframe computers and terminal equipment are obtained from the appropriate manufacturers. The carriers will then design, program, and in some cases, operate the entire system, which is provided to the customer on a contractual basis. According to Western Union:

> "As of December 1965, Western Union has knowledge of 57 currently operational private computerized message switching systems, some 21 of which also have electronic data processing capabilities. The 57 owners of these systems represent large corporations in diverse fields such as steel, airlines, motors, etc., as well as governmental agencies. Western Union made a proposal on 18 of these systems and was successful in six instances."[13]

The following section discusses alternative frameworks within which the communications common carriers might be allowed to operate in the provision of data processing and information services. The discussion also applies to carrier-supplied private computer systems for either communications functions or combined communications and data processing functions.

---

13. Western Union letter to the Federal Communications Commission, March 14, 1966, Re: *Computer Lease and Service Arrangements,* p. 12.

B. ALTERNATIVE GROUND RULES FOR COMMON CARRIER PARTICIPATION IN THE DATA PROCESSING SERVICES INDUSTRY

The criteria which should be used in evaluating each of the alternative ground rules under which common carriers might offer data processing services are: (1) the extent to which the carrier's participation would inhibit or encourage the growth of data processing services, and (2) the effect of such data processing service activity upon the carrier's ability to discharge its primary communications common carriage functions. In the following analysis it is assumed that data processing and information services, whether of a remote-accessed nature or not, will continue to be unregulated activities, as the authors strongly feel they should. If, at a future date, the progressive consolidation of firms offering computer-based services raises potential antitrust problems, or the services become such that the public interest requires that they be offered on a uniform and widespread basis, then the question of possible regulation of data processing services might be re-opened.

There are essentially four possible frameworks within which the communications common carriers could offer data processing and information services to the public. These are outlined below and will be discussed in the following sections.

1. The carriers would be permitted to offer data processing services as an unregulated activity with no restrictions on the facilities employed for this purpose. The costs of providing the regulated communications services and the unregulated data processing services would be separated for regulatory purposes.

2. The carriers would be permitted to offer data processing services as a regulated, tariffed activity, despite the fact that non-carriers were offering them as unregulated activities.

3. The carriers would be permitted to offer data processing services as an unregulated activity through a separate corporate subsidiary subject to strict separation rules. Neither physical facilities nor personnel could be shared by the parent and the subsidiary.

4. The carriers would be permitted to offer data processing services as an unregulated activity through a separate corporate subsidiary with strict accounting separation, and subject to the additional restrictions that the subsidiary not be permitted either to supply data processing services to the parent or lease communication lines from the parent.

Option (1): Common Carriers are Permitted to Offer Data Processing Services as an Unregulated Activity, Without Restriction

In this situation the carrier would be offering two types of services: regulated communication service in a monopoly market, and unregulated data processing service in a highly competitive market. The carrier would be using common resources—plant, equipment, and personnel—to provide both services, which would generally be more economical than using separate facilities.

Under the present regulatory framework, a carrier is entitled to earn a specified rate of return on the "book value" of the capital investment for the plant and equipment (the rate base) associated with a regulated communications service. The operating costs associated with the regulated service are recovered directly from the subscribers to the service but do not contribute to the rate base. The total revenue for a *regulated* service is therefore prescribed by the *cost* of providing the service, whereas the revenue received for *unregulated* data processing services is limited only by market considerations. Therefore, the FCC must require that the carrier separate the costs of the two types of services.

A firm which uses common resources to provide more than one product or service has two types of costs which must be allocated among the several products or services: "separable" costs which can be identified as being directly attributable to one of the firm's outputs, and "joint" costs, i.e., those which would be required to produce any one of the outputs by itself, but which, once incurred, yield several different "by-products." With respect to the joint costs, each output may be thought of as the by-product of the other outputs, and the total joint cost could be apportioned among them in several ways. Selecting any one allocation method for the joint costs is necessarily somewhat arbitrary.[14] The sum of the "separable" costs belonging to a single output and its allocated share of the joint costs constitutes the "total cost" of supplying that output. This total cost must be recovered from the sales revenues for that output.

The carrier, because of rate base regulation and the fact that it is facing both monopoly and competitive markets, has an incentive to assign a disproportionately large share of the joint costs to the regulated monopoly service. The

---

14. See, e.g., Richard Gabel, *Some Aspects of the Regulation of the Telephone Industry*, (Washington, D.C.: The Brookings Institution, 1967).

carrier will thus, by increasing the rate base of the regulated service, increase his allowable earnings from the regulated service. The fact that the price to the consumer of the regulated service increases is of small concern to the carrier because it is a monopoly market. In addition, by reducing the "book" costs of the competitive service (and therefore the revenue requirements of that service), the carrier will be able to charge the consumer in the competitive marketplace a price which is lower than that of competing firms, and which may in fact be lower than the carrier's actual cost of providing the service.

In effect, the consumers of the regulated communications service would be in part underwriting the costs of the carrier's activity in the unregulated data processing business. Competing data processing firms, whose actual costs for offering the same service may be lower than that of the carrier, may be driven from the market. For example, in a recent court case, the plaintiff charged that the defendant utility was illegally monopolizing gas appliance sales by charging all its promotional expenses on such sales to its regulated activity of supplying gas.[15]

Since the allocation of joint costs is to a large degree arbitrary, it would be very difficult for the FCC to successfully prevent improper cost allocations by the carriers, intentional or otherwise. The authors therefore disagree with the contention of the Western Union Telegraph Company concerning offerings of services involving a regulated and an unregulated component:

> "The fact of the matter is that appropriate separations and allocations procedures can be devised so as to allocate the costs properly attributable to each group of services. There is no omnipotent economic law which identifies the true distribution of these common [joint] costs. However, reasonable men who are knowledgeable about the production techniques can establish acceptable guidelines for allocating the common costs of the various services. That this can be done is a well-known fact, for separations procedures and allocations between regulated and unregulated activities have frequently taken place just as separations for jurisdictional purposes can and do take place."[16]

---

15. *Southern Blowpipe & Roofing Co.* v. *Chattanooga Gas Co.*, 360 F. 2d 79 (6th Cir. 1966). Cited in U.S. Department of Justice *Response*, pp. 78-79.

16. Western Union *Response*, p. CDE-42.

Rather, we believe that the performance of both regulated and nonregulated functions by the same entity, using common resources, is inherently likely to give rise to misallocation of costs and hence an unfair competitive advantage for the regulated carrier competing in the unregulated data processing services field.

Option (2): Carriers are Permitted to Offer Data Processing Services as a Regulated, Tariffed Activity

In Chapter II we saw that there is agreement in both the data processing and the communications common carrier industries that data services, *due to the nature of the activity*, should not be regulated, irrespective of the type of organization (carrier or non-carrier) offering the service. Exceptions to this rule should be made only if it is determined to be in the public interest for regulated common carriers to offer such services, and if their doing so on a regulated basis would avoid substantial problems or abuses. However, as the following discussion will attempt to show, such a solution might create rather than avoid difficulties.

Under the previous alternative, the common carrier would be offering two services, one regulated and the other unregulated, using common equipment, facilities, and personnel; but in the present case both services would be regulated activities. Problems associated with separation of costs might therefore be reduced; let us examine why. The strict separation of costs for individual *regulated* services is not essential, as it would be with Option (1) above, and is not currently practiced by the FCC. Rather, the present regulatory approach is to set an allowable rate of return upon the entire common carrier rate base, and only examine the costs of individual services when specific questions or abuses arise; and even when complaints arise the FCC has been reluctant to study the costs associated with individual carrier services.[17]

---

17. Although Western Union has complained for many years that AT&T has been subsidizing private line services from message toll telephone revenues, the FCC did not require AT&T to determine the costs of its individual services until the decline in revenue from Western Union's telegram service put the company's survival in jeopardy. Therefore, in 1964, in connection with an investigation of the telegraph industry the FCC requested AT&T to determine "the investment, expenses, and revenues associated with each of 7 different classes of communication service, making up the Bell System's total interstate operations, i.e., message toll telephone service, wide area telephone service (WATS), teletypewriter exchange service (TWX), private line telephone, private line telegraph TELPAK, and all other. The results of the study were presented by AT&T for the record in Docket No. 14650 on September 10, 1965. They disclosed wide variations in earnings among the various services." (AT&T rate investigation, Docket No. 16258, FCC Memorandum Opinion and Order, October 27, 1965, p. 1.)

Assuming that this practice is continued (and we do not necessarily agree that it should be continued), the cumbersome, arbitrary, and difficult process of separating the costs of individual services is minimized. However, if the costs of the individual services are "buried" in the overall capital and operating accounts, the problem of the cross-subsidization of services, raised under Option (1), is aggravated.

The case of a monopoly firm operating under rate base regulation and expanding into another regulated service in a competitive market was analyzed by Harvey Averch and Leland Johnson, research economists at the RAND Corporation. They concluded that the firm "has an incentive to expand into other regulated services, even if it operates at a (long run) loss in these markets; therefore, it may drive out other firms, or discourage their entry into these markets, even though the competing firms may be lower cost producers."[18]

There are, in fact, several illustrations of the expansion of a regulated carrier into new services in competitive markets, to the detriment of the competing firms. AT&T, for example, has attempted to offer (subject to the approval of state regulatory commissions) private mobile radio services, and paging and wired music systems. It has also filed tariffs for apartment, hospital, and farm intercom systems. AT&T's entry was vigorously opposed by the existing firms offering these services.[19]

Therefore, we believe that if the carriers offered data processing services as regulated activities they would have opportunities and incentives similar to those present under Option (1), for supporting these competitive services from revenues obtained from monopoly communication services, and would thereby gain unfair competitive advantage.

---

18. Averch, Harvey, and Leland Johnson, "Behavior of the Firm Under Regulatory Constraint," *American Economic Review*, Vol. 52, No. 5, December 1962, p. 1052.

19. See Kenneth Madson, "Antitrust: Consent Decree — The History and Effect of Western Electric Company v. United States 1956 Trade Case," 45 *Cornell Law Quarterly* (1959), p. 45. See also U.S.D.C. Southern District of New York (sic), *New Jersey Communications Corporation, Action Systems Co., Division of Connecticut Consolidated Industries, Inc., L. J. Loeffler Inc., TN Telephone Sales Corporation, The Delta Communications Corp., and Private Communications Inc. v. American Telephone and Telegraph and New York Telephone Co.*, p. 2. (Cited in Manley R. Irwin, *Vertical Integration in the Communication Equipment Industry: A Public Policy Critique,* report to the President's Task Force on Communications Policy, 1968; Chapter VI, p. 13.)

Option (3): Carriers are Permitted to Offer Data Processing Services Only Through a Separate Subsidiary

The inherent problems with the previous two alternatives lie in the difficult and arbitrary process of separating the cost of providing those services, whether regulated or unregulated, which employ common plant, equipment and personnel. To avoid this artificial separation of costs, many have suggested that a separate carrier-owned subsidiary be formed for data processing services; i.e., the carriers would be barred from using common facilities for both communication offerings and data processing services. Several carriers (e.g., ITT and GT&E) have already adopted this organizational approach as a politically acceptable way to enter the data processing services industry.

Although cost separation problems would be alleviated, other problems of preferential or discriminatory treatment remain. For example, the parent corporation could provide the subsidiary with superior communication channels, better line maintenance, earlier delivery of communications equipment, and advance notice of changes in the price or availability of communication services. The parent corporation could also, through its many relationships with firms which lease communication lines, and its requirements that such firms disclose the purpose for which the communication lines are to be used, obtain proprietary information about the activities and plans of companies competing with the parent carrier's data processing subsidiary.[20] The parent would also be inclined to take into account the needs of the subsidiary when considering which new services to offer, and when scheduling the introduction of these services in different areas of the country.

Another inherent problem is the possibility that the subsidiary, which has no ceiling on its rate of return, could sell data processing services (systems design, programming, machine-time rental, etc.) to the parent corporation at inflated prices. The parent would properly treat such expenses as operating costs which would be added to its revenue requirement and passed on to the subscribing public. Since the market value of many data services, such as systems design, are not readily measurable, it would be very difficult to detect the overpricing. The

---

20. For example, AT&T's interstate Private Line Tariff states, "the purpose or purposes for which the private line service is to be used must be made known to the Telephone Company prior to such use." (AT&T Tariff FCC No. 260, para. 2.2.4. (A)).

subsidiary could earn substantial guaranteed profits from sales to the parent and could, therefore, undercut competing data service organizations which had no assured high-return market, in the offering of services to the public.[21]

The possibility of the subsidiary selling substantial data services to the parent corporation is not a remote one. GT&E's subsidiary, GT&E Data Services Corporation, formed in January 1968, "will mainly provide data processing services to the General System of more than 30 telephone operating companies, the nation's second largest system." The subsidiary will also offer data processing services to business, industry, and government.[22]

The problems of unfair competitive advantage arising from the close relationship between a regulated parent common carrier organization and an unregulated data processing subsidiary lead the authors to conclude that the unconstrained subsidiary is a less than satisfactory approach to carrier provision of data services.

Option (4): Carriers are Permitted to Offer Data Processing Services Through Separate Subsidiary Which Can Neither Sell Services to the Parent nor Lease Communication Channels from the Parent

To correct the defect in the previous alternative arising from the possibility of a carrier affiliate intentionally overpricing data services provided to the parent carrier, it has been suggested that sales of data processing services (including systems design, consulting, programming, etc.), by the subsidiary to the parent be forbidden. Perhaps certain services such as the rental of computer time, whose value is roughly determinant, might be exempted from this prohibition, subject to full public disclosure by the carrier of the transactions between it and the subsidiary. Alternatively, the subsidiary might be allowed to sell data services to the parent but only after open competitive bidding.

The problem of the carrier giving preferential treatment in the provision of communication lines still remains; but because of the generally uniform "quality" of available communication channels, this may be of minor importance. More

---

21. To the authors' best knowledge this problem was first recognized by Professor Manley R. Irwin of the University of New Hampshire.

22. "GT&E Unit to Process Data for Parent Firm and Others," *Wall Street Journal* (New York), January 11, 1968.

serious are the opportunities for unfair competitive advantage in the offering of remote access computer services arising from the subsidiary's ability to obtain advance inside information regarding the carrier's plans for new services and price changes, and the carrier's incentive to take into account the needs and competitive position of the subsidiary when planning the introduction of new communication services. This is an unavoidable problem whenever the subsidiary obtains communication services from the parent carrier.

It would not be inconsistent with established practices in the common carriage field to prohibit a carrier from providing common carriage service to its subsidiary organization. As noted by the U.S. Department of Justice, the commodities clause of the Interstate Commerce Act (49 U.S.C. 1 (8)) prohibits a railroad common carrier from transporting on its own lines any goods which it owns or in which it has an interest.[23] The purpose of this provision is to prevent a carrier from discriminating "in favor of itself against other shippers in the rate charged, the facility furnished, or the quality of the service rendered."[24] Similar forms of discrimination would be avoided by prohibiting a common carrier from leasing communication lines to its data processing subsidiary.

Another, perhaps more serious, problem which remains under the present alternative results from the carrier's tariff provisions requiring advance disclosure of the intended use of communication channels by lessees. The parent carrier may make available to its subsidiary privileged information regarding the activities and plans of competitors in the remote-access computer services business. The original intent of the disclosure requirement in the tariffs was to prevent customer usage of communication facilities in ways contrary to other provisions of the tariffs; e.g., line sharing or resale. As will be discussed subsequently, the authors believe that several of the restrictions imposed by the carriers on the use of their communication lines are unjustified in their present form. Perhaps, with the relaxation of these restrictions, the requirement for the customer to disclose his business activities and plans to the carrier may be unnecessary.

---

23. U.S. Department of Justice *Response*, p. 81.

24. *Delaware etc. R. Co.* v. *United States*, 231 U.S. 363, 370 (1913); Cited in U.S. Department of Justice *Response*, p. 82.

Without the removal of this tariff provision it does not seem that a fair basis for competition between a carrier subsidiary and other firms offering remote access computer services can exist. Of course, if the carrier subsidiary were to restrict itself to the offering of stand-alone computer services not employing communication lines, this problem would not occur; but the rapid growth of remote-accessed services, and the business potential in this market, make that situation highly unlikely.

C. CONCLUSION

We have seen that the entry of common carriers into the unregulated data services market would be attendant with opportunities for them to compete on an unfair basis in that market by discrimination in the provision of communication services (in favor of their remote-access computer service subsidiary vis-à-vis its competitors), by covert subsidization of their data services from their communications services, and by improper use of non-public information. A requirement that data services be furnished by a carrier only through a separate affiliate organization would remove most of these problems, especially if the affiliate is barred from either obtaining communication services from or selling data processing services to the parent carrier.

Another, alternative, probably unwarranted, which should be mentioned is to completely bar communications common carriers from offering data processing or information services to the public. The only argument for such a complete exclusion of common carriers from the data services market seems to be the possible adverse affects of such activity upon the carriers' ability or motivation to adequately fulfill their proper role as communications carriers. The responses filed in the FCC Inquiry indicate a substantial difference of opinion between the carriers and their customers as to the adequacy of present carrier services and the carriers' responsiveness to user demand. For example, while AT&T said that its "facilities are capable of carrying information at transmission rates which meet today's data communication needs from the very low to the very high speeds,"[25] the various Federal agencies, in a jointly-filed response, stated that:

> "A number of definitive needs and problems have emerged from agency experiences. Among the more important of these are:

---

25. AT&T, *Response*, p. 2.

Common Carrier Provision of EDP Services 43

1. More reliable digital transmission circuits are needed...

2. Additional tariff offerings at higher transmission rates are required...

3. Common carrier dial-up service to include speeds at least up to 40,800 baud should be offered...

4. A high order of priority should be given to the conversion of communication facilities to truly digital transmission systems...

5. Existing technical interface standards and operational standards should be reviewed."[26]

Although there is controversy regarding the degree to which the carriers are satisfactorily fulfilling their obligations to serve the public, it appears that their inertia is a defect inherent in the industry structure, and would not be substantially affected by allowing them to provide data services. Therefore, we believe that communications common carriers should be permitted to offer data processing and information services to the public — through a separate subsidiary which would be barred from either selling data services to the parent carrier or leasing communication channels from the parent.

---

26. General Services Administration, *Response*, pp. iv-v.

CHAPTER IV

MARKET ENTRY, PART II: NON-CARRIERS INTO MESSAGE SWITCHING

Store-and-forward message switching has existed as a concept for many years. With the advent of the computer, however, it has taken on new significance, due to the computer's ability to efficiently perform message switching, both as a stand-alone service and as part of a remote-access data processing system.

A. DESCRIPTION OF MESSAGE SWITCHING

In its simplest implementation, a modern message switching system consists of a number of teletypewriter terminals connected via communications channels to a central computer.[1] Any terminal can originate a message and transmit it to the central computer. The message will be stored at the central computer until it is possible to forward it to the destination terminal(s). In older message switching systems, punched paper tape was used as the storage medium instead of a computer. These included manual "torn tape," semi-automatic, and automatic paper tape systems.

A more sophisticated implementation would consist of a larger number of terminals and several message switching computers. The terminals would each be connected, via low speed communication lines, to the nearest message switching computer. The several computers would be tied together via a network of high speed trunk lines. The overall configuration is not unlike the way telephones are connected to central exchanges by wire pairs and the central exchanges are interconnected via a network of broadband channels. The use of several message switching computers is called a *distributed* network. Figures 1 and 2 (on page 53) illustrate single-node message switching systems; Figure 3 illustrates a simple distributed message switching network. Variations on the above theme are possible, such as the use of multiplexing units,[2] or line concentration devices,[3] but conceptually these diagrams suffice for purposes of discussion.

---

1. In most message switching systems the communications channels are leased, dedicated lines. However, in low volume applications it is possible to design a message switching system that uses dial-up communication lines. For example, Western Union offers to its Telex subscribers a computer message switching service called Telex Computer Communications Service (TCCS), which may be used for such applications as sending "broadcast" messages to a number of receiving stations.

Non-Carrier Message Switching Services 45

In prior years, when the central message switching mechanism was manually or electro-mechanically handled paper tape, the only possible form of message switching was *pure* message switching; i.e., the central facilities were not capable of performing other functions. However, when a computer is used to perform the switching function, it is also possible to use it for data processing tasks — such as computation, information retrieval, report preparation, and data storage. Many in-house corporate message switching systems, today, configure their systems to handle multiple activities. For example, a manufacturing firm may use its central computer system and remote terminals to collect and store information on the status of the inventory in the several company warehouses. Any location can then enter a query, through its terminal, to determine the current level of inventory. Simultaneously, the same central computer, communication network, and terminals may be used to transfer messages from one location to another. This dual purpose system might be termed a "hybrid" system.

Both pure and hybrid systems are in use today. Many corporations operate their own internal message switching systems (provided either by the carriers or the computer manufacturers). Western Union has also introduced two pure message switching networks which can be "shared" by a number of different firms: SICOM (Securities Industry Communications System) and INFO-COM (Information Communications Service). With these systems, each subscribing firm obtains a dedicated subsystem and cannot use the shared transmission and

---

2. A multiplexer (also called channelizing equipment or channelizer) is a device for allocating a large-capacity communication channel to a number of smaller channels. Frequency-division multiplexing (FDM) subdivides a large-bandwidth channel into a number of smaller bandwidth channels. Time-division multiplexing (TDM) interleaves data from a number of low-speed channels into a single high-speed channel. FDM is the traditional telephone multiplexing technique, but TDM is beginning to be used for voice telephone channels (see discussion of Digital Transmission in Chapter IX) as well as for data channels.

3. A line concentrator for data transmission is a device which uses the statistically distributed properties of data terminal demands for service to allocate a few "trunk" channels from a computer to many terminal channels. The device is typically used in locations remote from the computer to reduce the communication cost for serving remote terminals. Terminals may contend for access to the trunk channels, or buffer storage may be provided; in either case, the transmission of a message may be temporarily delayed if a trunk channel is not available. The terms "concentrator" and "multiplexer" are often incorrectly used interchangeably, since data concentrators often employ TDM "front ends" to interleave data from several low-speed terminal channels into a single high-speed trunk channel.

switching facilities to communicate with the other subscribing firms.[4] However, INFO-COM subscribers can transmit messages to Telex and TWX subscribers, and to Western Union offices and agencies for delivery as telegrams. Further discussion of these services will be found later in this chapter.

The only "hybrid" systems in use today are internal corporate systems. As will be discussed on the following pages, common carrier tariff restrictions prevent private organizations which offer commercial remote access computer services (including the provision of communication lines) from adding message switching features. And the FCC has, so far, not permitted a common carrier to offer data processing in conjunction with a tariffed communications service.[5]

1. Message Switching vs. Circuit Switching

Message switching, as distinguished from circuit switching, is characterized by the temporary storage or delay of the message at the switch (or several switches in sequence), before retransmission to the ultimate destination. No direct electrical connection between terminal points is established, as is the case with circuit switching. As a result, message traffic for an outgoing line can be queued at the switch, if that line is busy. Also, and perhaps more importantly, a *distributed* message switching network permits the trunk lines between switches to be "time-shared" among many users on a per-message basis; a given line is only dedicated to a particular user for the time required to transmit his message, and is then available for the transmission of another user's message.

---

4. For example, "Western Union's Phase I Securities Industry Communications (SICOM) System provides the service of shared transmission and computer-directed switching of orders and administrative message traffic for members of the brokerage community. Each subscribing firm ... utilizes a portion of the common transmission and switching facility to derive a service equivalent to that obtainable by employing dedicated computer and transmission facilities. *No Communication is provided between the subscribing firms.*" (Emphasis supplied) Vincent Hamill and Kenneth L. Brody, "SICOM — Securities Industry Communication System," *Western Union Technical Review,* Vol. 21 (1967), p. 98.

5. For example, in its SICOM decision in 1967, the FCC said, "the principal question to be resolved is whether the package SICOM Service offering includes any significant non-common carrier, non-communications services. If the SICOM offering includes the offer of any significant non-common carrier, non-communications services then the tariff should be rejected under Section 203 of the Act which requires tariffs to be filed only for interstate or foreign common carrier communications services." *In Re Western Union Telegraph Company Tariff FCC No. 251 Applicable to SICOM Service,* Memorandum Opinion and Order, released December 27, 1967, p. 9.

When two users, or a user and a conversational time-sharing computer, engage in a dialogue in which there are significant gaps in time between messages, the "idle" line time can be made available to another user. Higher line utilization and correspondingly lower costs can thereby be attained. With the conventional circuit-switched network, users in such a situation must either "hang up" and re-dial for each message in the dialogue (which is impractical for many applications because of the long "connect" time required to establish a connection), or hold the line and pay for the idle time between messages.

For many data communications applications, message switching has a number of advantages over the conventional circuit-switching network philosophy:

1. By making more efficient use of the communication lines in a common-user network, as discussed above, fewer lines are required for a given volume of communications.

2. By centrally storing and forwarding messages, it is possible for a sending station to transmit a message even if the receiving terminal is unavailable (busy or inoperative) at that time. Messages for inoperative terminals are intercepted and either stored or rerouted to alternate terminals.

3. Use of a central computer for switching permits communication between terminal stations with different operating speeds and different character code formats (e.g., ASCII, EBCDIC, Baudot).

4. A message transmitted once from its origination terminal can be "broadcast" by the switching computer to a selected set of destination points.

5. Extensive message processing functions can be performed by the switching computer:

    * Error detection and correction

    * Format control and other types of message editing

    * Verification of destination address; or automatic routing based upon message content

* Numbering and date-stamping of messages, to protect against loss in transit

6. Messages can be stored ("logged") for backup and later retrieval if necessary (for maintaining an audit trail, for example).

7. Traffic priorities can be established, and high-priority messages expedited.

8. Messages can be "batched" for subsequent transmission over high-speed trunk lines — yielding greater cost/effectiveness.

9. Traffic statistics can be collected, facilitating network optimization.

Also, and often most significantly, since the central network control unit may be a general purpose computer, it is possible to perform data processing functions with the same facilities that perform the communication functions. This permits a much higher level of utilization of the equipment (therefore distributing its costs over a broader base), than if it could be used only for switching, as is the case with conventional circuit-switching equipment. Since communications traffic volumes typically fluctuate greatly during the 24-hour day (with traffic peaks in midafternoon, and lulls during the late evening and early morning hours) this ability to utilize otherwise-wasted processing capacity is of great importance.[6]

In addition, the reliability requirements of many message-switched communication networks are often met by the installation of "duplexed" computer switching equipment — a duplicate backup equipment configuration at each switching center, which picks up the switching workload in the event of failure of the primary switching computer. During normal operation of the primary switching computer, the backup equipment is available for data processing tasks, thus minimizing the idle time and consequent cost burden imposed upon the

---

6. For example, Western Union has estimated that in its Phase II Information Services Computer System (ISCS), to be implemented in the early 1970's, the UNIVAC 1108 message switching computers—configured to handle the daily message traffic peaks—will have approximately 70% of their capacity over a 24-hour period available for data processing.

communications network.[7] In a circuit-switching system, on the other hand, this cost burden for reliability cannot be avoided. Either added expense must be incurred to make the switching equipment very reliable, or duplexed equipment must be used and the backup equipment (which generally has no other possible use) allowed to lie idle during normal operation of the primary system.

On the other side of the coin, circuit switching has characteristics which make it advantageous for use with some data communications applications:

1. The circuit switching apparatus itself is generally less expensive than a third-generation computer adequate to handle an equivalent traffic load on a message switching basis.

2. The circuit switching apparatus (not the line) is characterized by a high fixed call set-up cost, and a very low cost variable with message length or line holding time. With message switching, which must bring into buffer storage and retransmit each character sent, the opposite is true. Circuit switching therefore becomes more favorable when a large amount of data is transmitted in each message.

3. After the desired connection is established, data passes through the circuit switch instantaneously, whereas significant delay may be encountered in the store-and-forward message switch. Real time data applications have therefore favored use of circuit switching or of dedicated (non-switched) lines rather than distributed message switching networks; however, current developments in rapid-throughput computer message switching systems may reverse this trend.

4. Circuit switching equipment is completely transparent to the signals passing through it, is insensitive to data code and format, and will handle any speed up to its design limits.

---

7. James Martin, *Teleprocessing Network Organization* (Englewood Cliffs, N.J.: Prentice-Hall, Inc., 1969), Chapter 14, "Message Switching." Additional discussion of the characteristics of computer message switching systems may be found here.

5. Likewise, a circuit switch will pass either digital or analog signals. Message switching, however, can be used only for information in digital form; thus, voice signals cannot be sent through today's message switches. To do so would require digital encoding (e.g., pulse code modulation, or PCM) of the voice, and the use of very fast-throughput switches.

The advent of interactive computing has brought about a demand for communications service charged on a "per bit"-transmitted basis, which the use of a distributed message switching network would allow. This topic will be discussed further in Chapter VII on Line Sharing. At present, the only publicly-available services of this nature are Western Union's SICOM and INFO-COM. These services, which were the subject of controversy after their announcement in 1967, contain some of the features which will be desirable in more advanced systems of this type, whether supplied by a common carrier or an independent enterprise. They are not, however, designed to provide the "charged by the bit" communications, with message response times of 1-2 seconds, desired for applications such as remote conversational computing.

2. Types of Message Switching Systems

For purposes of discussion, we categorize communication systems according to the set of parties with whom a subscriber of a given system can communicate. A *public system* is one to which anyone may subscribe, and may then communicate with all other subscribers. The TWX and Telex teletypewriter exchange services are examples of public *circuit switching* systems for record communications, as is the dial telephone network for both voice and record communications. Western Union telegrams are transmitted by means of a network of paper tape *message switching* centers, which might be thought of as a public message switching system. At the time of this writing there are no other public message switching systems in operation in the U.S.

A *private message switching system* is one in which a firm's terminal station can communicate only with other stations belonging to that corporate organization. There are, as mentioned previously, a large number of private corporate message switching systems in operation today. Some employ electro-mechanical store-and-forward devices (paper tape) for switching, while others employ computer equipment. Most private systems are operated by the user firms themselves. In addition, in 1968 Western Union introduced two systems which it operates as tariffed communications services — INFO-COM and SICOM. These are large message switching systems which are "shared" by several subscribing firms, each of whom leases terminal stations which, by system design limitations, can communicate only with other stations leased by that subscriber. The system is categorized here as "private" because each subscribing firm effectively obtains a private subsystem (both communication channels and switches) from Western Union.[8]

Another type of "private" system is AIRCON, supplied on a non-tariffed basis by RCA Global Communications, Inc. AIRCON consists of a store-and-forward computer operated by RCA, to which a firm can attach its own leased, multidrop (i.e., with several terminal stations attached) communications lines. Again, each firm obtains a "private" subsystem.

A third category of message switching systems is termed *industry-oriented;* subscription to such a system is limited to members of a given industry, who may then use the system for communications with one another. Such a system might well be of a hybrid nature, offering certain data processing features (e.g., access to industry-wide data bases maintained by the system operator) tailored to the requirements of its class of subscribers. At present, common carrier tariff restrictions on resale of communication channels (explained in the following section, and discussed in detail in Chapter VII) have barred non-carriers from offering complete "package" (i.e., lines plus switch), industry-oriented message switching systems. An exception is the message switching service which the FCC has permitted Aeronautical Radio, Inc. (an unregulated private firm) to offer to members of the commercial airlines industry. Due to a number of factors, such as lack of familiarity with the business operations and consequent specialized data communications requirements of various industries, and a traditional degree of insensitivity to the demands of communications submarkets, the common carriers presently offer no industry-oriented message switching services as we have defined them. Thus, the market for such services remains an untapped business opportunity.

---

8. The INFO-COM network is linked to Western Union's Telex Computer Communications System (TCCS), so that INFO-COM subscribers can also send messages to Telex and TWX subscribers and to Western Union offices and agencies for delivery as telegrams. This feature does not change the categorization of INFO-COM as a "private" system, however; its use remains primarily for intra-company communications.

B. REGULATORY STATUS OF MESSAGE SWITCHING

Currently, any organization, not necessarily a common carrier, can perform message switching for hire *provided* that its customers individually or jointly supply their own communication lines (i.e., normally, lease such lines from a carrier) to the message switching computer.[9] (See Figures 1 and 2.) The lines cannot be provided by the message-switching entrepreneur, nor can a shared distributed network be used, except by a common carrier, because the existing carrier tariffs prohibit any organization (with certain minor exceptions) from reselling leased or dial-up communication channels to any other organization (i.e., providing a communication service for hire, traditionally a common carrier activity).[10] (See Figure 3.)

From a policy point of view there are two questions: (1) Should the "switching" of messages, where the operator of the switch does not provide the lines, be an unregulated or a regulated (tariffed) activity? (2) If the operator of the switch is unregulated, should the resale of communications channel capacity be permitted, to enable non-carrier operators to provide complete message-switching services? To permit non-carriers to switch messages and to forbid them from reselling the use of communication lines effectively excludes the non-carriers from the business of offering message switching services, except for very specialized switching services (e.g., Bunker-Ramo's TOPS service, noted in footnote 9 *supra*). Therefore, the fundamental question is whether or not the nature of complete message switching services (lines *and* computer switch) is such that they

---

9. For example, the Bunker-Ramo Corporation provides the securities brokerage community with a store-and-forward computer service called TOPS (Tele-center Omni Processing System). However, Bunker-Ramo does not lease the private access lines; the lines are leased from the carriers by the subscriber. TOPS switches "buy and sell" orders for stock brokers and also edits messages, verifying each order against the current trading price of the stock. Since this service provides intra-company communications to its subscribers, rather than communications between different brokerage firms, it is a "private system" under the definitions suggested above.

10. AT&T's private line tariff prohibits a lessee from using a private line "...for any purpose for which a payment or other compensation shall be received by either the customer [lessee] or any authorized or joint user, or in the collection, transmission, or delivery of any communications for others..." (AT&T Tariff FCC No. 260, Section 2.2.3.) This restriction will be discussed in **Chapter VII**, on **Line Sharing/Resale**.

PERMISSIBLE MESSAGE SWITCHING SERVICE NETWORKS

A and B are subscribers to a message switching service offered by a non-carrier, X. $\boxed{X}$ is a switching computer operated by X. Communication channels are leased from a common carrier by the organizations as shown.

FIGURE 1

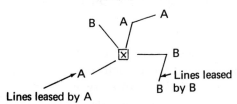

This is permitted today; A can communicate with either A or B.

FIGURE 2

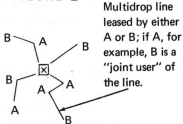

Multidrop line leased by either A or B; if A, for example, B is a "joint user" of the line.

This became permissible in February 1969, when multi-user sharing of private leased lines was first permitted. A can communicate with either A or B.

FIGURE 3
(Distributed Message Switching System)

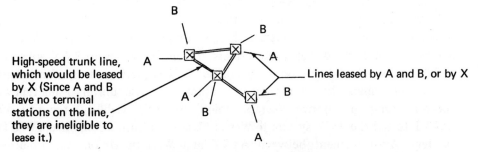

High-speed trunk line, which would be leased by X (Since A and B have no terminal stations on the line, they are ineligible to lease it.)

Lines leased by A and B, or by X

This is not permitted today, unless X is a common carrier, because transmission of traffic for A & B over X's trunk lines constitutes "resale" of the lines, which is forbidden to non-carriers by the carrier tariffs.

should be offered exclusively by regulated common carriers. There are three possible alternatives:

1. Maintain the status quo; only common carriers are permitted to offer complete message switching services.

2. Message switching services remain essentially a common carrier activity; other organizations can operate "hybrid" remote-access computer systems and offer message switching services using shared circuits only if the message switching is "incidental" to the primary non-communications purpose of the remote-access service.

3. Special purpose message switching becomes an unregulated activity (including the freedom of non-carriers to resell communication line capacity); common carriers can offer such message switching services only through separate unregulated corporate subsidiaries.

Option (1): Maintain the Status Quo — Only Common Carriers are Permitted to Offer Complete Message Switching Services.

The primary argument in favor of treating message switching services as regulated activities is the public utility nature of these services — i.e., they are essential to the community and are, to some extent, natural monopolies.

Industry requires a widely available service for the rapid and inexpensive transmission of written messages. AT&T's TWX system, with approximately 60,000 subscribers, and Western Union's Telex system, with approximately 25,000 subscribers, have fulfilled this function to date. These two systems previously competed with one another, although Western Union made arrangements which enabled Telex subscribers to send messages to TWX subscribers. AT&T, however, did not similarly enable TWX subscribers to send messages to Telex subscribers. In an attempt to obtain a single universal teletypewriter network (and to improve Western Union's financial viability) the FCC urged AT&T to sell the TWX system to Western Union to form a combined TWX/Telex system. An agreement between AT&T and Western Union was finalized in January 1969 (negotiations began approximately 25 years ago) under which AT&T agreed to sell its TWX terminal equipment to Western Union and to operate, for five years, all TWX switching equipment that is an integral part of AT&T's voice telephone switching facilities. As a result of the merger, the capability for a subscriber of either system to communicate with any other subscriber is likely to be provided. Ultimately, these public circuit-switched systems will be replaced by one or more public message-switched systems which are more flexible and cost-effective.

The advantages of a *single* public message switching system should be clear: the ability (1) to reach the widest number of subscribers; (2) to realize whatever economies of scale exist in message switching communication systems (discussed in more detail later); (3) to employ average-cost pricing for the service, thus ensuring its availability both in densely populated areas where the cost of providing the service is low and in sparsely populated areas where the cost of providing the service is high; (4) to employ "value-of-service" pricing;[11] and (5) to "optimize" the overall system.

---

11. "Value of Service" pricing is the term used in the communications industry for the widespread practice of price discrimination in the supply of common carrier services. Successful application of this concept, which is explained in more detail in Chapter VII on Line Sharing/Resale, requires that the seller divide his total market into segments, each segment consisting of buyers who are willing to pay the same amount for a particular service. In economic terms each segment consists of buyers with the same "price elasticity of demand" schedule. For example, business organizations represent an "inelastic" market for telephone service—the telephone is essential to their operations (its "value" is high) and they are insensitive to the fixed monthly station charge (but not necessarily to variable toll charges) for its use. Residential households constitute a more "elastic" market for telephone service—the household consumer is more sensitive to price changes. The telephone company therefore charges business organizations a higher rate than residential households for the same telephone service, and loses few business customers, but gains large numbers of household customers, who are attracted by lower rates. The overall social merit of such price discrimination may be debated, but it has one undeniable result: it maximizes the total number of subscribers to the service.

Effective price discrimination requires a protected monopoly market. Discrimination is difficult if not impossible to accomplish in a competitive market because other firms would be attracted to undercut a service which is priced appreciably above its true cost. Thus, a common carrier offering a monopoly message switching service might, by employing price discrimination, attract a larger number of subscribers than could a number of competing non-carriers offering a similar service.

The significance of an increased number of subscribers lies in the fact that (1) greater economies of scale may be thereby realized, and (2) the utility of the service to a single user may be proportional to the number of other users with whom he may communicate, i.e., the total number of subscribers to the service.

The desirability of providing widespread services suggests that the provider of the service should be *obliged* to make the service available everywhere. In return for the promise to serve all comers, the provider should be given a privileged franchise to the region he has agreed to service; i.e., he should be a common carrier.

Additionally, if one of the existing national communications common carriers, such as AT&T or Western Union, were to provide the store-and-forward message service, the carrier could employ existing facilities at little incremental cost, and could plan the implementation of the new service so as to minimize the impact upon the existing teletypewriter services (e.g., by gradual phasing out of the TWX/Telex services, and the continued utilization of the same terminal equipment).

The possibility that the carriers, such as AT&T or Western Union, may be able to use existing personnel, equipment, and office facilities for message switching at little incremental cost perhaps during off-peak hours, is not merely hypothetical. AT&T's Electronic Switching System (ESS), for example, might be expanded to include a message switching service. In fact, Bell Laboratories has already designed a modified ESS which has message-switching capability. It is called No. 1 ESS-ADF (Arranged with Data Features), and became operational on an experimental basis during 1968 for the internal use of the Bell System Companies. If this experiment is successful, AT&T might make a public offering of a common user store-and-forward message service.

However, although Bell's Electronic Switching System is capable of offering message switching services, it is not likely to be as cost-effective as equipment designed for message switching purposes. The Bell ESS is built around a specialized computer whose most critical design requirement is dependability.[12]

---

12. "Certainly a new system must at least be comparable to existing offices. This means outages of no more than a few minutes in 40 years... In fact, the dependability objective represented one of the major challenges of No. 1 ESS development. Since No. 1 ESS is a large digital information processor, it is a cousin to the general-purpose digital computer. However, *the dependability requirements requires that No. 1 ESS be a very different kind of system with a much higher level of redundancy."* (Emphasis added) (W. Keister, R.W. Ketchledge, and H.E. Vaughn,"No. 1 ESS: System Organization and Objectives," *The Bell System Technical Journal*, Vol. 43, September 1964, p. 1841. Copyright, 1964, The American Telephone and Telegraph Co., reprinted with permission.)

Therefore, ESS is constructed with costly components and extensive sub-system redundancy, using well-understood second generation, discrete component transistorized hardware with proven reliability. (General-purpose computers are now in their third generation, using integrated circuitry techniques, and fourth generation machines will be available in the early 1970's.) As the computational requirements of a circuit switching exchange are low, the internal processing speed of ESS is, by design, several times slower than that found in many of today's general-purpose computers.

ESS hardware, with a design reliability and a projected service lifetime many times that of commercial data processing equipment, is well suited for use in a circuit switching mode, but, because of its slow processing speed, is not well suited for use as a store-and-forward message switching processor. The performance of a message switching processor is measured by its "throughput," i.e., the number of characters it can pass per second, which is proportional to the processor's speed of operation. ESS therefore would have a low throughput when operating in a message switching mode. Reliability of components is important, of course, but not to the degree found in the telephone network. The distributed network concept, under which a large multi-computer message switching service would most likely operate, is specifically designed to accommodate greatly reduced reliability of the individual switching centers and communication links between centers without overall system degradation.[13] Today's commercial computers have hardware failure rates well within tolerable limits, and exhibit superior performance for a distributed store-and-forward application.

Furthermore, although the rate at which ESS exchanges are being introduced into the telephone system is relatively rapid (it should be one exchange per day by the early 1970's) the installation process will not be complete until the year 2000. At the end of 1968 there were only some 40 ESS exchanges in service, with 1969 plans for the installation of 30 more, out of the 12,000 Bell System central exchanges in the nation. Although the software used by ESS is modifiable and will be refined and improved, the ESS hardware technology is "frozen" by the long equipment design lifetimes and is, in certain respects, already obsolete.

---

13. For example, Paul Baran of the RAND Corporation, father of the digital distributed network concept, says:

"...But, *even though we will use unreliable equipment, we expect to achieve better reliability* [with the "Distributed Adaptive Message-Block Network"] *than we are accustomed to today.* Systems such as this can be built only if one appreciates that unit reliability and systems reliability are two separate things, and that in the properly designed system, system reliability can be greater than unit reliability, unlike some systems that have been built." (Emphasis supplied)

A Briefing on the Distributed Adaptive Message-Block Network, RAND Corporation paper, P-3127, April 1965, p. 27.

If the primary social goal is to obtain a message switching network covering the entire nation, it can be argued that permitting competitive systems would produce this result more rapidly than would a monopoly service. If message switching services were found to exhibit economies of scale, "survival of the fittest" would lead to one or several large systems. If these services do not exhibit economies of scale, then perhaps it would be wiser to forego the advantages of a *single,* national message switching service in exchange for a number of smaller competing firms catering to the specialized requirements of different user groups. If a need for inter-system communication develops, the many firms could readily standardize interface parameters and interconnect their facilities, as it would often be in their interest to do.

In its evaluation of the issues in the FCC Inquiry, Stanford Research Institute recommended to the FCC that unregulated message switching be permitted, but that:

> "In cases of this sort [message switching services which are desired on an industry-wide basis, and thus would be more valuable if interconnected] and in the case of services like Telex-TWX, it appears that the Commission could make an important contribution by encouraging the establishment of technical standards which would allow the interconnection of systems designed by different equipment manufacturers. Providers of teleprocessing services in such a market would be able to compete both in price and in auxiliary features, but all would have to meet the same standards of signaling, message format, and technical performance. In each type of service in which interconnection appeared to be an important issue, a careful study of the particular industry or industries affected would have to be made before any particular approach to standardization could be acopted."[14]

The development of technical standards and the encouragement of interconnection, as suggested by SRI, would surely enhance the overall utility of certain competitively-offered message switching services, but in some cases this may not be sufficient. Where competing message switching services are reluctant to standardize and/or to interconnect, and the public interest would be served by

---

14. Stanford Research Institute, *Policy Issues Presented by the Interdependence of Computer and Communications Services,* Vol. 1 of Report to the FCC, Docket 16979 (Menlo Park, Calif.: Feb. 1969), p. 22.

such standardization/interconnection, the FCC should be empowered to require it as a condition of service.

Option (2): Message Switching Services Remain Essentially a Common Carrier Activity; Other Organizations Can Offer Message Switching Services as Part of Remote-access Computer Services Using Shared Circuits Only if the Message Switching is "Incidental" to the Primary Non-Communications Purpose of the Services.

While there may well be a demand for a standardized widespread message switching service, such as described above, it seems more likely that many of the message switching systems of the future will be of the hybrid (i.e., message switching integrated with data processing of some sort) variety found internally in many organizations today. Hybrid systems are growing in number largely because it is often cost-effective to combine related communications and data processing activities into a single facility for multiple, even unrelated, functions.

For example, a typical hybrid system is currently being implemented by a major insurance company. The branch offices of the company are connected with the main office, and with each other, via a computer-based message-switching network. The same system is also designed to answer inquiries, keyed into the teletypewriter terminals at the branch offices, such as "What is the cash value of life insurance policy number XXXXX ?".

Another example of a hybrid service, drawn from the authors' experience working with the Project MAC time-sharing system at M.I.T. (the Compatible Time Sharing System, or CTSS), illustrates that a message switching capability is often an integral and necessary part of a remote-access data processing system. CTSS offers a feature called "the mailbox."[15] If one user wants to communicate with another user (who may or may not be on-line at that moment) he can enter a message at his keyboard to be put into the other user's "mailbox," which is simply storage space in the central computer's mass memory facilities. When the addressee logs into the system, the computer indicates to him that there is a message stored in his "mailbox." He can then have the message printed at his terminal if he wishes.

---

15. P.A. Crisman, Ed., *The Compatible Time-Sharing System, A Programmer's Guide*, (2nd Ed.), (Cambridge, Mass.: the MIT Press, 1965), Section AH.9.05.

A variant of this concept, called the "WRITE" command, permits a CTSS user to communicate directly with another user who is logged-in at the time.[16] After giving the command and identifying the desired recipient, the sender types his message; whenever he hits the "carriage return," the line he has typed is immediately sent by the computer to its destination. Lines of text sent may be intermixed with lines received from the other party. A conversational dialog may thus be established between distant users — quite useful when lack of an adjacent telephone prevents convenient voice communication, when such voice communication would incur costly additional long distance toll charges, or when a written record of the conversation is required.

Typically, the "mailbox" concept is useful when a team of users are working on the same problem, either at different hours of the day or at physically distant terminals. This valuable capability must be permitted in order to obtain the full benefits of future time-sharing systems. The potential of such systems was noted by two of the designers of the new MULTICS time-sharing system currently being implemented at Project MAC:

> "We are optimistic about technological progress, and can envision computer systems that permit communication (voice and other) interspersed with data processing. On a 'conference telephone call' the third party would be a computer. Such a system would enhance, by orders of magnitude, the ability of people to interact and cooperate with one another in a manner both convenient and meaningful to each of the individuals concerned."[17]

In fact, this type of message switching has been important in the implementation of the MULTICS system itself — which was a large cooperative effort between teams at M.I.T.'s Project MAC in Cambridge, Massachusetts; Bell Telephone Laboratories in Murray Hill, New Jersey; and General Electric Company in Phoenix, Arizona. The existing CTSS at Project MAC was used extensively as a store-and-forward message switch for communication between the MULTICS designers located hundreds of miles apart by means of the "mailbox" feature and shared user files, which was legal in this situation only because there was no service offering for hire, and no "third party" usage of communication lines. This

---

16. *Ibid,* Section AJ .2.01.1.a.

17. E. E. David, Jr., (Bell Telephone Laboratories, Inc.), and R.M. Fano (M.I.T.), "Some Thoughts About the Social Implications of Accessible Computing," *Proceedings of the Fall Joint Computer Conference* (Las Vegas, Nevada, November 30, 1965), p. 245.

store-and-forward message switching is credited by one of the MULTICS project leaders as an indispensable factor in the success of this complex joint effort.[18]

These forms of message switching point to another problem which occurs if message switching in a remotely accessed data processing service continues to be prohibited. Namely, as time-sharing systems become more widespread and as their capabilities increase, enforcement of the prohibition will become increasingly difficult. Users of time-sharing systems, even if the "mailbox" capability is not available, could communicate with each other by entering messages which are stored as program or data files, along with their computer programs, in the on-line mass storage of the system. Then, when a subsequent user wanted to read the message, he would call for the appropriate file from storage, which would be a message rather than a sequence of computer instructions. The only way to prevent this would be not to provide users with the ability to access common data files or to share the computer programs which they wrote. This is a very large price to pay in order to prevent clandestine message switching.

Assuming that there is a need for hybrid services, either because of the interrelated nature of the switching and processing activities or because of the economies of scale and smoothed workload associated with the combining of different services, one then asks who should be permitted or encouraged to offer the hybrid service. Earlier we concluded that data processing services should not be regulated. Perhaps, however, when offered as an inextricably intertwined part of a hybrid service, data processing should be regulated and the entire service offered only by a common carrier. This is the argument of Western Union.[19]

Or perhaps the message switching portion of a hybrid service should be considered unregulated and the entire service offered by a non-carrier. This is the position of many non-carriers who responded to the FCC Inquiry. The authors are in agreement with this view, as a policy which would be in the greatest general interest.

Hybrid services, such as a stock quotation service, tend to be narrowly designed for a particular industry or class of subscribers. It is unlikely that the common carriers have sufficient familiarity with specialized industry requirements

---

18. Talk by Dr. F. J. Corbato' at the Sloan School of Management, M.I.T., May 9, 1968.

19. See e.g., Western Union *Response,* p. CDE-43.

to offer many of these services. Also, it is undesirable for the carriers to spread their talent, energy, and capital thin across new activities which may detract from proper fulfillment of their communications responsibilities.[20]

Real-time computer systems are each unique and their performance in particular applications is dependent upon the degree of sophistication of the computer technology. Consider the relative rates of application of new technology in the computer and communications industries. For example, Bell Telephone Laboratories developed the transistor some twenty years ago. Yet, how many transistors are actually found in telephone company equipment today? Very few, relative to the number of opportunities for their use. Meanwhile, in the computer industry, the transistor is becoming obsolete as solid logic technology and large scale integration are introduced. The FCC has recognized this characteristic of the communications industry relative to its competitive brethren.[21]

---

20. Common carrier diversification may have directly detrimental effects upon the carrier's primary communications services. For example, Western Union has for many years used revenues from its Public Message Service (conventional telegrams) to support its ventures into new business areas. In its 1966 Domestic Telegraph Investigation, the FCC concluded:

"...the user of the message telegraph service was being requested, on the Western Union ratemaking theory, to pay for the increased wage bill for *all* services of the company.

"It appears that Western Union burdened the message telegraph service to support the company's diversification program. Since the middle 1950's, when the emphasis was placed upon diversification, Western Union's new investment in plant and equipment has been primarily in the non-message telegraph areas, principally in the private wire field." (Emphasis in original text; *Report of the Telephone and Telegraph Committees of the Federal Communications Commission in the Domestic Telegraph Investigation,* Docket No. 14650, 1966, pp. 102-103.)

21. See the Commission's findings *In the Matter of AT&T and Associated Bell System Companies' Charges for Interstate and Foreign Communication Service* (Docket Nos. 16258 and 15011):

"Unlike many manufacturers, Respondents rarely, if ever, junk obsolescent plant because of competitive impetus. Costs are averaged and plant is continued in service until replaced in a schedule proposed by them. For example, Respondents now propose to take several decades to convert fully the nationwide switching system to electronic switching, thus permitting a gradual and controlled phase-out of electro-magnetic switching systems now in use. If Respondents were operating in a competitive field, they might very well be required to accelerate plant retirements even though the plant may not have been fully amortized. This schedule has been taken into account in the setting of Respondents' depreciation rates so that the customers will fully pay for the retired plant. We agree with Respondents' witness, Nathan, that the Bell System is in a more favorable position to control the rate at which innovations are introduced."

Of course, one cannot criticize the telephone companies too harshly for their lack of innovation, since their primary objective is to provide adequate and reliable service at reasonable cost. However, innovation among the common carriers tends to be retarded relative to free enterprise, such that it may be unwise to give the carriers the exclusive right to provide message switching services, especially of the hybrid variety.

If the problem is resolved by allowing non-carriers to offer hybrid services but not pure communication services, then the operational problem arises as to where to draw the line between pure communication services and hybrid services. The "test" suggested initially by IBM in an open letter to the FCC and subsequently supported by many members of the computer industry — manufacturers, service bureaus, and users — is the "Primary Business Test."[22] Essentially, under the primary business test, message switching by a firm providing a data processing or information service, which is incidental or related to the provision and use of the data processing or information service, would be permitted and outside the purview of FCC regulation. Precedent for this procedure is found in a case before the FCC in which Aeronautical Radio, Inc. (ARINC), a non-common carrier, sought to obtain permission to provide a service to the airlines which involved the transmission of messages. ARINC's service provided information to the airlines on weather, navigation, traffic, and reservations. The FCC held that "safety of life and property in the air [is] the principal purpose for which ARINC was organized and is operating."[23] The Commission ruled that ARINC would be allowed to offer its message switching service because the messages it proposed to send to the airlines were sent "in carrying out purposes of its own distinct from those of a communication carrier for hire."[24] This rule, as applied to hybrid services, would ask whether the unregulated firm proposing to offer a particular hybrid service would be performing communications transmission as its primary business. If not, then the firm would be permitted to offer the hybrid service.

---

22. Burke Marshall, then IBM Vice President and General Counsel, to Ben F. Waple, Secretary of the FCC, February 15, 1966, p. 4.

23. 4 FCC 155, at 162 (1937).

24. *Ibid.* at 164.

Difficulties Inherent in the Primary Business Test

Adoption of the primary business test would require a case-by-case examination of each situation to determine whether message switching by the non-carrier was in fact an adjunct to his primary business of data processing or whether message switching itself was the primary business. The difficulties of a case-by-case examination are clear: First, it would increase the risk of entering such businesses and would thus discourage the growth of new enterprises; second, the costly delays and evidence gathering required for the hearings would prove more burdensome to the *small* businesses with limited sources of funds; third, it would be socially costly to burden a court or administrative agency with examination of questionable services; fourth, because of the arbitrariness of the determination (explained below), the primary business test may be an invitation to discrimination and actual or apparent distortions of services; and fifth, the service or revenue mix of a business might fluctuate or shift over time, requiring new determinations.[25]

One way to lessen these difficulties would be to require approval of only those services to which the common carriers objected. However, this would give the carriers an opportunity to discourage, through delaying tactics, the entry of smaller firms who could not afford the time and cost of a lengthy adjudication process. Greater equality of power could be given to the entrepreneur by shifting the burden of proof to the carrier, who would be required to furnish the

---

25. Stated differently,

"Communication and computer firms will obviously employ the [primary business] test to their own ends, however, those ends are defined. That such a test borders on the arbitrary can be seen in its application. In the eyes of Western Union, flower and data processing service are "incidental" to communications. Yet the communication industry becomes less tolerant when the identical test is invoked by the data processing industry, to wit, the Bunker-Ramo case.

Thus, the primary business test cuts both ways; it can be employed to rationalize a movement from communication to data processing; or from data processing to communications. And if such a test is resurrected as a policy guide, it takes little imagination to visualize a backlog of adjudicatory proceedings in which the adversaries seek the magic number of 49%."

(Manley R. Irwin, "Time-shared Information Systems: Market Entry in Search of a Policy," *Proceedings of the 1967 Fall Joint Computer Conference,* [Washington, D.C.: Thompson Book Co., 1967], p. 518.)

requested communication channels until such time as he proved that the message-switching service in question was not proper under the regulations of the Commission.

Another difficulty in allowing the courts or an administrative agency to use "primary business" guidelines is in deciding what criteria should be used for measurement. The variety of different services offered would make the determination an arbitrary one in many cases. One possible criterion might be the volume of message traffic as compared with the volume of data processing traffic in the system under consideration. However, volume can be measured in several ways: number of messages, number of characters, amount of processor time used, etc. The system might be artificially and inefficiently designed to meet whichever measurement criteria were selected.

Another possible criterion might be a court's subjective determination of the primary business of the firm. This would require firms to disclose their future business plans and would tend to lead them to "disguise" the actual nature of their service. If a firm wanted to improve an existing service by adding new features, a reassessment would have to be made to determine whether the nature of the service was thereby changed. Innovation might be inhibited.

For the above reasons we believe that use of the primary business test would be unsatisfactory, although it is preferable to the complete exclusion of non-carriers from the business of message switching.

Option (3): Specialized Message Switching Services Become Unregulated Activities; Non-carriers Permitted to Share and Resell Communications Capacity; Common Carriers can Offer Specialized Message Switching Services Only Through a Separate (Unregulated) Subsidiary.

There are, as mentioned earlier, three major classes of message services in the U.S.: public message services, private message services, and specialized systems. Public message services are exemplified by Western Union's telegram service, aptly termed the Public Message Service (PMS), and the TWX and Telex teletypewriter exchange services. These services, employed largely by business and industrial firms, provide the nation with a rapid, low-cost method of transferring "hard-copy" messages. Such services are a basic component in the nation's "backbone" communications plant. They, like the highways, railroads, power lines, etc., are part of the "infrastructure" of the economy.

Since AT&T agreed in January 1969 to sell the TWX service to Western Union, all three of the above mentioned message services become the responsibility of the telegraph company. There is substantial merit in having one rather than several organizations provide the nation's public message services. (In many countries a single government agency provides public telegram and telegraph services.) The goal of a public communication system is to enable all citizens to subscribe to the service and to allow each to communicate with any other subscriber. A single communication system, supplied by one firm, can achieve this total access capability at lower cost than could many interconnected communication networks. Additionally, if several services are involved, they are likely to be provided at lowest cost if common physical plant is used where technologically appropriate. Toward this end Western Union is planning to integrate the TWX, Telex, PMS, and several other communication systems and services into a single nationwide communication system, the Information Services Computer System (ISCS). ISCS will initially employ four computer switching centers interconnected by high-speed broadband communication channels provided by Western Union's transcontinental microwave system.

Where services to the public, such as the message services described above, are required on a widespread and regular basis, and where the supplier is in a position to charge unreasonable rates for the service (by virtue of his being sole supplier, a direct consequence of the "natural monopoly" nature of the service), the public interest requires that the supplier be regulated.

Private message systems are distinguished from the public systems described above by the class of users that can be contacted through the system. Private communications systems, message or circuit switched, are usually owned (or leased) and operated by a corporate or government organization for the purpose of sending messages between its various remotely located offices, branches, plants, warehouses, and outlets. These systems cannot be used to send messages to, or receive messages from, stations other than those operated by that organization. Corporations generally have private systems for large-volume communications among their own offices (these systems are cheaper than public systems and/or are designed with special features), and use public services (they may lease TWX or Telex terminals if their traffic volume warrants, or use the ordinary telegram) for messages to all others. Private communication systems, which are not, by definition, offered to the public, and upon which the public does not rely, do not require regulatory control (an exception occurs where the private communications system uses a scarce "public" resource such as a portion of the electromagnetic frequency spectrum).

The third major class of message service, the specialized or industry-oriented type, is neither opened to the general public, nor restricted to use by a single firm. These services are usually designed to meet the specialized requirements of various industry groups. For example, as mentioned earlier, Aeronautical Radio, Inc., provides the several major airline companies with a message communications service also providing weather, flight, and safety information. Additional examples of specialized message systems may be found in the educational and research community. Seventy-one colleges and universities have formed the Interuniversity Communications Council (EDUCOM) for the purpose of establishing an interscholastic communications network. Also, the Advanced Research Projects Agency (ARPA) of the Department of Defense is sponsoring the development of a high-speed message-switched communication network for interconnecting the time-sharing computers at nineteen universities and firms performing research under ARPA contracts. The "ARPA network" is discussed in more detail in Chapter VII on line sharing.

Systems of the type described above would, in many cases, be "hybrid" systems, i.e., they would be performing functions other than purely communications. Some, however, would simply be custom-designed communication systems which, for example, would employ sophisticated error detecting and correcting codes to ensure message accuracy, or send only strictly formatted messages to minimize costs, or use encrypted messages for security reasons, or would be specialized along some other dimension important to a class of subscribers but not to the general public.[26]

---

26. Different classes of communicators have widely diverse functional requirements for message and data communications services. They may have messages or blocks of data of widely varying lengths, and with different formats, traffic volumes, and priorities. They may each require communication systems with different message transit times (response times), error detection and correction capabilities, message retrieval capability, system reliability, incidental processing (such as message time and date stamping, or sequence numbering of messages), etc. They may each employ terminal stations which differ in transmission speed, character set, error detection capabilities, local storage (buffering) facilities, method of output (visual or hard copy), and terminal accessories (paper tape reader/punch, card reader/punch, line printer, etc.).

The carriers have argued that the store-and-forward component of a hybrid message switching/data processing service for hire should be regulated, and that all pure store-and-forward services for hire should be regulated. However, there is substantial doubt as to whether all store-and-forward services have those characteristics exhibited by traditional common-user voice and message networks which led to the regulation of these services. Perhaps specialized message switching services lack these characteristics and are therefore inappropriate candidates for regulation.

The important difference between ordinary circuit switching communication systems, and message switching systems, is the greater ability of the latter to be customized to the varying functional requirements of different classes of communicators.

There are many specialized communication services which can be satisfactorily provided by circuit-switched systems. These services can be consolidated and provided over large common-user communication systems (the national dial telephone network is the best example), reducing user costs by achieving substantial economies of scale. Certain other specialized communication services, however, are best offered by a message-switched system, possibly because of the need for tight error control, frequent transmission of broadcast messages, incidental processing, on-line retrieval capability, message formatting, message priorities, use of "canned" messages, collection of traffic statistics, etc. These specialized message-switched communication services may be provided through the use of separate systems or a single multi-service system. Whereas the consolidation of circuit-switched service requirements into a single large system has produced economies of scale, consolidation of disparate message-switched systems into a single multi-service message-switching system might result in *diseconomies* of scale. To a certain extent, of course, some diverse message switching services could be efficiently accommodated in a single common-user system; however, many could not. For example, the Department of Defense's AUTODIN system, and Western Union's INFO-COM network are common-user message switching systems. AUTODIN allows several different types of terminals (keyboard, paper-tape, punched cards, etc.), several security and priority levels, and a high degree of error checking. INFO-COM provides different classes of service to subscribers depending upon their individual message volumes. But both systems have restrictive message formats, limited terminal types, rigid procedures, and other necessary limitations upon the individual user's flexibility of use.

Very clearly, certain communication services which are most naturally provided via a message-switched system, could *not* be easily or economically accommodated by an all-purpose message switching net. Of course, as the state-of-the-art in message switching progresses, it may then be possible to accommodate a larger variety of functional requirements in a common-user message-switching system than is possible today. Western Union's long-term growth is based, in part, upon this premise. However, the state-of-the-art is far from allowing this now, and rather than await possible technical developments, it is probably advisable to proceed with the specialized communications systems as they are required. This approach is likely to speed the development of store-and-forward equipment and techniques.

The carriers could be designated as sole suppliers of specialized communication systems. Alternatively, the rules delineating the communication services which must be offered only by licensed common carriers could be liberalized so that specialized communication systems could be operated and offered for hire by non-carriers. A possible arrangement would allow industry groups to develop their own communication systems, or contract with a private firm (or carrier) to provide the system. An example of the type of relationship which we feel should be encouraged is that between the airlines industry and Aeronautical Radio, Inc. (ARINC) which, as has been mentioned earlier, operates a specialized message switching communication system for the subscribing airline companies.

One of the strongest arguments *against* a multiplicity of communication services, such as would obtain in the unregulated environment proposed here, is the existence of substantial economies of scale in communications. While it is certainly true that potential economies of scale do exist in message-switching services at any point in time (the "static" case), subject to the technical inefficiencies of serving diverse requirements with a single system as discussed above, it is also true that these static economies may be outweighed by other "dynamic" factors over a period of time. Conceptually, this is illustrated as follows:

Consider the behavior over time of the average cost curve of a store-and-forward service. (See Figure 4.) This curve shows the relationship between the average cost-per-message and the total volume of messages. Naturally because of technological improvements, the curve will shift downward over time, but the critical question is at what rate. A single firm offering message-switching services would operate at the highest volume and lowest cost point on the average cost curve. Multiple firms, each of which would handle a smaller volume and,

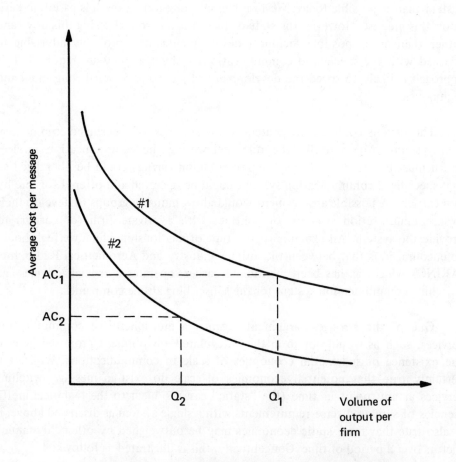

FIGURE 4 THE EFFECT OF MARKET STRUCTURE ON AVERAGE COST-PER-MESSAGE OF A MESSAGE SWITCHING SERVICE

therefore, operate at a higher average cost point on the curve may, because of competition and specialization, introduce innovations and technological developments that cause a more rapid downward shift of the curve. This may be visualized in Figure 4. The former, monopolistic case gives AC curve #1, which for a total firm (and industry) output of Q_1 will result in a cost of AC_1. With, for example, two firms operating in the market, increased innovation might result in AC curve #2 per firm. On this curve, the same level of total industry output, or Q_2 per firm, may be obtained at a lesser cost $-AC_2$.

The above is the classic argument for encouraging a multiplicity of firms rather than only a single firm to offer a given product or service. However, with regard to communications services there is an additional dimension which suggests a multiplicity of unregulated firms is preferred, where possible. The key concept is that of *risk aversion* on the part of corporate organizations. Firms are willing to assume different degrees of risk depending, largely, upon the potential reward. In industry, reward is measured by the rate of return of the prospective venture on the risk capital invested. Where the likelihood of success in a given venture is low, the investor requires a very high potential rate of return. Where the likelihood of success is almost certain, the investor requires a lower rate of return from the ensuing enterprise. This is a well established capital budgeting principle.

The overall rate of return on capital investment of the larger regulated communications common carriers is limited by the regulatory agencies (to approximately 6-8%). The constraints upon the rates of return for individual carrier services are less strict but there is considerable pressure upon the carriers to maintain each rate of return at a value close to the required overall rate of return. The carriers, behaving rationally given these pressures, avoid undertaking risky ventures. Consequently, products or services based upon unproven technologies, or opening new markets, are likely to be introduced only very slowly by the carriers, if at all.

The social cost of restricting the provision of all multi-firm communication services to regulated and therefore risk-averse common carriers can be substantial. As an instance of the social need for, and benefit of, firms willing to undertake risky ventures, consider IBM's decision in 1962 to develop what has become, today, the world's most widely used family of computers — the System/360. Regarding IBM's decision, *Fortune* magazine said the following:

> The decision by the management of the International Business Machines Corp. to produce a new family of computers, which it

calls the System/360, has emerged as the most crucial and portentous — as well as perhaps the riskiest — business judgment of recent times.

* * *

... No company had ever introduced, in one swoop, six computer models of totally new design, in a technology never tested in the marketplace, and with programming abilities of the greatest complexity.

* * *

... IBM was staking its treasure (some $5 billion over four years), its reputation, and its position of leadership in the computer field on its decision to go ahead with the System/360.

* * *

... The head of one competing computer manufacturing company acknowledges that at the time of the System/360 announcement he regarded the IBM decision as sheer folly.[27]

Although the results are not wholly in, it is generally conceded that IBM's System/360 has made a major contribution to the productivity of the U.S. business and industry. Had the company been unwilling in 1962 to undertake the development of the System/360 (which it probably would have been, had its potential profits from the venture been limited by regulatory control), these social benefits would have been lost.

In sum, we believe that it is inadvisable to continue the present scheme of rate and profit regulation of all *inter-company* communication systems. In the particular case of store-and-forward communication systems, which portend to have an important future both as stand-alone services and as components of

---

27. T.A. Wise, "I.B.M.'s $5 Billion Gamble," *Fortune*, LXXIV, No. 4 (1966), pp. 118, 120; continued in Part II, "The Rocky Road to the Marketplace," *Fortune*, LXXIV, No. 5 (1966), p. 138.

remote-access data processing services, we feel that significant steps forward would result from the establishment of numerous specialized, or industry-oriented, store-and-forward communication systems. It is unlikely that the present communication common carriers could, or should, assume sole responsibility for the diverse store-and-forward communication services which new technology has made possible and which industry requires.

C. OBSERVATIONS ON WESTERN UNION'S SICOM SERVICE – A DIGRESSION

In early 1967, Western Union announced two computer-based message switching services which incorporated several features not previously found in common carrier offerings, and which became the subject of considerable controversy. SICOM (Securities Industry Communications System) is a service offering exclusively to the stock brokerage community, which provides store-and-forward message switching among several offices of a subscribing brokerage firm. It uses three Western Union computers (UNIVAC 418's) located in the New York City area, and low-cost communication lines to link the customer's remote offices to the computer center. INFO-COM (Information Communication Service) is an expanded version of SICOM, but it is not restricted to brokerage firms; we will not discuss INFO-COM specifically but most of the following comment applies to it as well.

Both the Bunker-Ramo Corporation and Scantlin Electronics, Inc., who offer computer-based stock quotation services with which the proposed ultimate SICOM service (with both message switching and data processing functions) would compete, petitioned the FCC to order the suspension and investigation of SICOM. The Business Equipment Manufacturers Association (BEMA) joined these organizations in opposition to INFO-COM as well. The FCC denied both petitions for suspension.[28]

While a complete discussion of the nature of the Western Union service is beyond the scope and intent of this book, the authors are of the opinion that the FCC erred in its analysis of the SICOM service, and failed to recognize the antitrust implications of this service, which fall under its enforcement jurisdiction

---

28. *In Re Western Union Telegraph Company Tariff FCC No. 251 Applicable to SICOM Service,* FCC Memorandum Opinion and Order, released December 20, 1967.

as regulator of common carrier communications. We should note that the FCC has been given specific statutory authority to enforce the antitrust laws of the United States "where applicable to common carriers engaged in wire or radio communications."[29]

1. Nature of the Service

The SICOM service offering consists of several distinct segments, sold only as a "package." Each of the customer's participating offices has a teletypewriter (or up to six of them) leased from Western Union, connected by a normal Western Union Private Wire Service (PWS) low-speed line[30] to a "Dalcode" time-division multiplexer in the nearest of fifteen SICOM terminal cities. From the terminal city to Western Union's New York computer, transmission is via a low-speed channel which is derived on a time-division multiplexed (TDM) basis from a voice-grade line. Twenty-six such TDM channels are derived from a single voice-grade line, at a unit cost much lower than the cost of the frequency-division multiplexed (FDM) lines employed in Western Union's Private Wire Service (or AT&T's similar Series 1000 private line service).

This cost differential may be seen in a comparison of SICOM network charges (the charge for the TDM lines only) with PWS circuit charges. SICOM per-mile circuit charges are from *one-sixth* to *one-twelfth* (depending upon distance, with greater differentials at shorter distances) of the corresponding PWS charges between the same points.[31] This cost differential is crucial to a proper

---

29. Section 11 of the Clayton Act, 15 U.S.C.A. 21; see also Section 314 of the Communications Act, 47 U.S.C.A. 314, and *United States* v. *Radio Corporation of America,* 358 U.S. 334 at 348 (1959).

30. For transmission distances greater than 10-20 miles, these channels are derived from a larger baseband channel by means of frequency-division multiplexing (FDM), as are normal voice-grade lines. For the discussion which follows, it is important to realize that these FDM channels are considerably more costly than their newer TDM counterparts, such as used in the long-haul portions of the SICOM network.

31. From San Francisco, California to New York City, the SICOM per-mile circuit charge is $0.083, compared with the PWS rate of $0.499 — cheaper by a factor of six. For shorter-haul links, the discrepancy doubles: Springfield, Massachusetts to New York City is $0.120 per mile, compared to $1.452 for PWS; and Pittsburgh, Pennsylvania to New York City costs $0.105 per mile, compared with $1.339 for PWS. (From Horace J. DePodwin Associates, Inc., *Major Economic Issues in Data Processing/Data Communications Services,* [*BEMA Response,* Attachment 2] February 1968, p. 153.)

understanding of SICOM, since the customer must buy the entire SICOM package in order to obtain these savings in line costs.

The last part of the SICOM package is the message switching computer, which is shared by many SICOM subscribers, and which performs several processing functions auxiliary to the principal message switching task; in addition, Western Union proposed to ultimately expand SICOM to include data processing services.[32] The SICOM customer obtains the dedicated use of both terminal equipment on his premises, and low-speed lines to the computer (FDM lines from the customer to the Dalcode multiplex units, and TDM lines from this point to the computer), and shared use of the message switching computer itself.

2. Comments on the FCC Decision

In its December 20, 1967 decision to permit the SICOM service offering, the FCC seems to have misunderstood the fact that the SICOM service "package" consists of three distinct components, as described above: teletypewriter terminals, *dedicated* low-speed communication lines, and the message-switching computer. Each of these *could* be offered separately, and in only one, the low-speed communication lines, does Western Union enjoy a privileged position. Both the provision of terminals and the message switching computer service constitute competitive markets which Western Union is attempting to enter by tying the leasing of these items to the leasing of a product (communication lines, sold on a basis very advantageous to the customer) over which it has economic power.

On this point the U.S. Department of Justice warned that one danger is:

". . . that a monopolist entering a competitive market involving a related product will use its monopoly position as a means of illegally tying the sale of other products to those over which it has a monopoly power; that is, of course, illegal. See *International Salt Co. v. United States*, 332 U.S. 392 (1947); *Mercoid Corp. v. Mid-Continent Investment Co.*, 320 U.S. 661 (1944)."

---

32. "This shared transmission and computer-directed-switching service will be augmented, in a phased program, by offerings of data processing services utilizing the data base generated by the initial service. These additional offerings will be oriented to the particular needs of the brokerage firms and may include day order matching, open order file maintenance, research reports, retrieval functions, and back office data processing applications." (Vincent Hamill and Kenneth L. Brody, "SICOM – Securities Industry Communication System," *Western Union Technical Review,* Vol. 21 [1967], p. 98.)

"Western Union's SICOM and INFO-COM services offer a good example. An ordinary data processor cannot permit its customers to share lines used in remote access data processing (or message switching) services, because of the common carriers tariff restrictions.... On the other hand, Western Union, as a carrier, is able to offer a computer-based service which allows customers to share circuits [i.e., use economical TDM channels derived from a larger common channel, which are not obtainable as a stand-alone carrier offering or as a service provided by a non-carrier] and thereby secure economies not otherwise available to non-carriers. In realistic terms this means that Western Union can be regarded ... as ... tying the sale of one service (message switching and/or data processing) to a second service (circuits on a line-shared basis) over which the carrier has economic power."[33]

Essentially, the Commission did not seem to realize that the nature of the components of SICOM is such that they *could* be offered separately. Western Union makes no showing to the contrary. Rather, Western Union says,

"A stock broker may come to Western Union and request only a part of the communications service being offered [terminals, lines, and switch] indicating that he either has available a computer of his own or may be able to obtain computer services from a service bureau operation. In such an event, *Western Union would offer a Private Wire Service to the stock broker* so that he could obtain the necessary communication facilities to combine with the customer furnished computer functions. *Since there would be no refusal to provide the communications part of the package offer, there would be no violation of the prohibitions against tie-in sales.*" (Emphasis added.)[34]

However, there *is* refusal to provide the communications part of the package offer! The PWS that Western Union proposes to offer to the stock broker above is not an equivalent substitute for the TDM derived subchannels employed in the

---

33. U.S. Department of Justice *Response*, pp. 72-73.

34. Reply of Western Union to Petitions for Suspension and Investigation, November 20, 1967, para. 34.

SICOM and INFO-COM services. The FCC, in summarizing the Western Union position, states that "the 75-baud circuits derived from the shared 2400 baud trunk circuits through the use of line concentrators are not comparable to dedicated 75-baud circuits furnished under the ordinary private line services."[35] Western Union, in fact, argued that because lines obtained with SICOM are not "like" other 75-baud channels offered by Western Union, there is a valid basis for the greatly reduced SICOM line charges.

However, even if we were to view SICOM and PWS channels as functionally similar (as we will argue below that we should), the significantly higher charges for PWS channels makes them non-equivalent. A monopoly communications service (the TDM subchannels used in SICOM) is thus, in effect, tied to the use of Western Union computer service and terminal equipment. Western Union should not, therefore, be permitted to reserve the much cheaper TDM channels for sale with the complete SICOM service.

A second aspect of SICOM which was misunderstood by the Commission is the functional nature of the long-haul communications channels which it uses. As mentioned earlier, low-speed SICOM teletypewriter channels are derived from 2400 baud (voice grade) lines on a time-division multiplexed (TDM) basis, using "Dalcode" multiplexers located in the fifteen SICOM terminal cities. Western Union makes much of the fact that the SICOM customer "shares" a line with other users, and thus his lower cost of communication (as compared with usage of Private Wire Service) is justified.[36] Accepting this concept, the FCC says:

> "However, in the SICOM service Western Union is proposing to make it both practical and economical for *different users to share in the 2400 baud channel facility* through the use of line concentrators that utilize time-division multiplexing principles and to *construct the charges therefor on the principle accepted in exchange types of service of distributing the revenue requirements for such shared facilities on the basis of relative use by those sharing therein.*" (Emphasis added.)[37]

---

35. FCC Memorandum Opinion and Order, para. 31. The term "baud" is a measure of data signaling rate, and for binary pulses (used here) is equivalent to "bits per second."

36. Reply of Western Union to Petition for Suspension and Investigation, paras. 14 and 16.

37. FCC Memorandum Opinion and Order, para. 38.

Two points must be mentioned regarding this conclusion: first, Western Union charges for the SICOM communication lines on a flat rate basis, *not* "relative use." The customer pays a fixed network charge, based upon distance, and fixed station equipment charges. The usage charge of 18 cents per 1000 characters transmitted covers only the cost of line termination and message switching at the computer center. "It has been estimated by computer service companies that, based on a projected monthly volume of traffic, Western Union's usage charge is comparable to that which any computer center would have to charge for a similar service."[38] Thus, the FCC has erred in saying that the revenue requirements for the "shared" communication channels are distributed on the basis of relative use by those sharing therein.

Furthermore, there would be no logic in Western Union's charging for the communication lines on the basis of volume of the customer's usage, for in the SICOM "shared" line there is no common pool of resources which, if one customer is not using, remain for the use of another. It is this common pooling of system resources which permits telephone exchange service to be charged on the basis of use, and it is this concept to which the FCC refers. But the FCC was apparently unaware that the SICOM customer does not, in fact, "share" the use of a channel in this sense. The TDM technology used for SICOM lines and the older frequency-division multiplexing (FDM) used to derive PWS lines produce functionally identical results. In either case, the customer gets a dedicated channel of 75 bit-per-second transmission capacity. With the use of TDM technology, line "sharing" occurs only in the sense that bit streams from the several low-speed lines are interleaved on a millisecond basis, and transmitted over the higher-speed baseband. This process is transparent to the user at the terminal, who is unaware of the fact that he is occupying a repetitive time slice on a high-speed line rather than a dedicated low-speed "copper connection," just as his FDM counterpart is unaware that he is sharing a carrier frequency spectrum with other simultaneous users. From the user's viewpoint, the SICOM channel is different only in that it is much less expensive, and (as Western Union was quick to point out) that the TDM concentrator is designed to function with only particular types of terminal equipment and code conventions.

---

38. DePodwin, *op. cit.*, p. 151.

Additionally, the FCC notes the claim of Western Union that these economical TDM lines could not be provided for general-purpose PWS usage,

> "*because they operate only at specified fixed and unvarying speeds.* This use, Western Union states, is suitable for SICOM but is not suitable in the ordinary private line service where customers want a choice of a range of low and high speed services and also want to be able to operate their stations in a particular service at varying speeds subject only to a maximum speed." (Emphasis added.)[39]

Let us consider whether this is in fact the case. First, while it is true that users "want a range of low and high speed services," terminal manufacturers and users would certainly welcome the availability of economical fixed-bit-rate TDM lines; and terminals suited to the available line speeds would be produced and used. (Indeed, for 75-baud lines as at issue here, tens of thousands of teletypewriter terminals are already in use.) This is an insufficient reason for not making TDM channels available on a PWS basis. Second, however, could a given terminal (e.g., a 75-baud teletypewriter) be operated "at varying speeds subject only to a maximum speed," using a TDM line? The answer is "yes." TDM channels derived by means of multiplexers such as used in SICOM are indeed designed to operate at varying keying speeds; although the 75-baud teletypewriter terminal is an asynchronous device, it transmits a stream of pulses (bits) at a fixed speed of 75-baud for each character struck on the keyboard or read from paper tape; there are merely varying amounts of idle time between such 75-baud bursts of bits when the user types slowly.

D. MESSAGE SWITCHING: SUMMARY AND CONCLUSIONS

The purpose of this chapter was twofold: to place into perspective store-and-forward message switching as a communication technique of increasing importance, and to determine through an examination of the technical and economic nature of this service whether, for the greatest general benefit, it should be a regulated or competitive activity.

---

39. FCC Memorandum Opinion and Order, para. 38.

The chapter began with a description of message switching as a store-and-forward technique of information transmission, as distinguished from circuit switching which is merely the establishment of a "copper connection" between two points. It is logical to charge the user, in a circuit switching system, for the duration of his connection, regardless of the amount of information transmitted during that time. However, in a message switching system, in which the user does not have a dedicated line but rather shares with others the lines that he uses, it is possible to charge on the basis of the amount of information actually transmitted. In addition, the controlled line utilization often possible in a message switching system, and the system's flexibility for special operations (such as the ability to store messages until a destination terminal becomes available) makes such systems more cost-effective for many data communications applications.

There are many pure message switching systems and hybrid systems — performing both switching and data processing functions — in use today. These are in-house systems; i.e., they are operated for private purposes. Under present common carrier tariff restrictions non-carriers cannot lease lines and set up commercial for-hire message switching services, either pure or hybrid.

We discussed the need for publicly-available message switching services and indicated the three alternative sets of ground rules for the provision of such services:

1. Maintain the status quo; i.e., only carriers can offer public message switching services.

2. Adopt the "primary business test;" i.e., an organization can offer a hybrid service if message switching is only an incidental part of the service.

3. Make special-purpose message switching an unregulated activity which non-carriers can offer to the public and carriers can offer only through separate unregulated subsidiaries.

It was concluded that alternative (3) is preferable because message switching services, as presently implemented (superimposed upon a circuit switching network) are not generally continually-declining unit-cost activities (natural monopolies), involve substantial opportunity for innovation, and are employed in many dissimilar applications each of which requires a specialized system and vendor. This solution is consistent with a well-established policy of prohibiting communications or transportation common carriers from controlling newly-emergent forms of carriage, particularly when based upon new technology.[40]

It should be noted that this chapter has dealt with message switching in the near future, the next ten to fifteen years. In the longer range, the authors believe that serious consideration should be given to the conversion of the present common carrier physical plant, which is based upon analog transmission and a circuit switching philosophy, to a network using exclusively digital transmission, multiplexing, and switching, with extensive use of store-and-forward message switching. The rapid implementation of such systems seems much more likely for the military, in the Defense Communications System,[41] and perhaps other advanced "dedicated" system applications, than for publicly-available common carrier offerings. (Project Mallard, for example, a joint U.S.–U.K.–Canada–Australia program, is currently developing an all-digital voice and data tactical military field communication system, using advanced digital technology for both transmission and switching.) This may be due to a number of factors, including our present system of carrier incentives, which suggests the need for study of the adequacy of these regulatory and incentive structures, and modification of them to encourage innovation and the implementation of advanced technology without undue delay.

Procedurally we would urge that an increasingly large portion of new common carrier plant construction should be devoted to digital transmission, multiplexing, and switching equipment to be used for data communications; this would yield the highest payoff and the lowest cost of implementation. Ulti-

---

40. The authors do not wholly subscribe to the *general* principle of prohibiting carriers from controlling newly-emergent forms of carriage; however, as applied to message switching we believe this policy is sound.

41. Electronic Industries Association, *Trends in Communications Research and Development*, (Washington, D.C., February 27, 1968), pp. 11-12.

mately, voice communications would be digitized, for reasons of transmission economy, and integrated into the growing digital network.[42] This could occur on a large scale in the 1980's and beyond.

The present chapter also discussed two commercial computer-based message switching services recently announced by Western Union: SICOM and INFO-COM. We noted that technologically these new services are in the public interest since they will employ highly cost-effective time-division multiplexing (TDM) and store-and-forward digital techniques. The dramatic communications cost savings allowed by these services are evidence of the fact that TDM should be used by the carriers on a wide scale in the derivation of low-speed data channels, in addition to the older frequency-division multiplexing techniques currently employed.

The analysis of the new Western Union message switching services attempted to show: (1) that they consist of three components – *dedicated* low-speed communication lines, switching computers, and terminals – which are separable but are sold only as a package; and (2) that Western Union, by enabling SICOM and INFO-COM subscribers to use economical TDM channels, without offering such channels to its other communications customers, has economic control over part of the package offering. Therefore, it appears that both these services, although technologically admirable, constitute illegal tie-in's contrary to the antitrust laws of the United States. The recommended solution is to require that Western Union provide similar TDM communication channels to the public as a general offering independent of the computer and terminal equipment components of the SICOM and INFO-COM services.

---

42. At present, digital transmission facilities, used for voice, visual, and data communications, account for some 1/1000th of the nation's common carrier interexchange plant. (The Bell System has 200,000 circuit miles of T1 digital carrier installed, out of more than 215 million circuit miles of interexchange facilities nationwide; AT&T *Response*, pp. 38 and 45.)

CHAPTER V

FOREIGN ATTACHMENTS

Common carriers have traditionally emphasized that they provide a communications "service" as opposed to simply communications facilities; that they are responsible for providing end-to-end communications capability and, therefore, in order to ensure the continued quality and reliability of the service, they must have end-to-end control of the facilities. If a subscriber were to provide a portion of the communication facility the carriers would no longer be able to control the entire communications network for which they are responsible. Therefore, the carriers have included provisions in their tariffs giving subscribers only limited permission to attach non-carrier provided devices, termed "foreign attachments," directly to the dial telephone network. For reasons explained later, the foreign attachment restrictions do not apply to communication lines *leased* from the carriers.

Restrictions on the use of customer-provided attachments have long occurred in both the interstate and intrastate common carrier tariffs.[1] Until January 1969, telephone company tariffs prohibited the use of customer-provided voice telephone equipment entirely and, much more important to the data communications community, required that the customer who would use this computer or data terminal for digital transmission on the dial telephone network do so only through the use of a carrier-supplied interface device for modulation/demodulation called a modem or data set.[2] The modem converts the direct

---

1. The provisions of the Communications Act of 1934 require that communications common carriers file tariff schedules, which must be nondiscriminatory, with the Federal Communications Commission showing all charges, practices, classifications, and regulations for interstate communication services offered to the public. Similarly, tariff schedules for intrastate services are filed with the state regulatory commissions. A new tariff automatically becomes effective within a specified period of time unless rejected or temporarily suspended by the Commission.

2. "No equipment, apparatus, circuit or device not furnished by the Telephone Company shall be attached to or connected with the facilities furnished by the Telephone Company, whether physically, by induction or otherwise, except as provided in 2.6.2 through 2.6.12 following (which, among other exceptions, provides for the attachment of customer-provided data transmitting and receiving equipment through use of a Telephone Company DATA-PHONE data set). In case any such unauthorized attachment or connection is made, the Telephone Company shall have the right to remove or disconnect the same; or to suspend the service during the continuance of said attachment or connection; or to terminate the service." (AT&T Tariff FCC No. 263, Sec. 2.6.1.; cancelled and replaced January 1, 1969.)

current electrical pulses originating from a business machine into a series of audible tones suitable for transmission over the voice telephone channel (modulation), and it performs the inverse function (demodulation) for data received from the channel.

In January 1969, the foreign attachments tariff restrictions were significantly relaxed, as will be discussed in some detail later in this chapter. This action by the telephone companies, initiated by AT&T and followed by the independents, was ordered by the FCC after years of controversy on the subject. Suits against AT&T had been brought by two small manufacturers of foreign attachments, the Hush-a-Phone Corporation and the Carter Electronics Corporation. These cases, resolved in 1956 and 1968, respectively, served as catalysts for FCC action.

A. THE HUSH-A-PHONE CASE

The legality of the blanket foreign attachments restriction was first addressed in 1956 by the U.S. Court of Appeals, in *Hush-a-Phone Corporation* v. *United States*.[3] In that case the court held that a prior version of the present AT&T Tariff FCC No. 263 (the interstate long distance tariff) was illegal under Section 201(b) of the Communications Act of 1934.[4] The foreign attachment in question was a rubber cup-like device to be attached to the microphone portion of the telephone handset to provide privacy in conversation. The court held specifically that the existing tariffs "... are an unwarranted interference with the telephone subscriber's right to use his telephone in ways which are privately beneficial without being publicly detrimental."[5] The general prohibitions of the tariffs were found to be more restrictive than necessary to ensure the preservation of the quality of telephone service, which could be achieved by tariffs containing *minimum technical specifications* for equipment to be attached to the telephone network. AT&T was ordered to revise its tariff to permit use of the Hush-a-Phone device. This was done, but the general foreign attachment prohibition remained in effect. Shortly thereafter the Carterfone episode began.

---

3. 99 U.S. App. D.C. 190; 238 F. 2d 266 (D.C. Cir. 1956).

4. 47 U.S.C. Sec. 201(b), which requires that all regulations be "just and reasonable."

5. 238 F. 2nd 266, at 269.

B. THE CARTERFONE CASE

The Carter Electronics Corporation of Dallas, Texas, developed and sold an acoustic/inductive coupling device for interconnection of the base station of a mobile radio system (or other private communications system) with the dial telephone network. From 1959 through 1966, approximately 3500 "Carterfones" were sold in the United States and overseas. Customer reaction was favorable in spite of considerable pressure to discontinue use of the Carterfone exerted upon Carter's actual and potential customers by the Bell System and the General Telephone System. The telephone companies warned that use of the Carterfone violated their tariffs and that customers who persisted in using the Carterfone would have their telephone service disconnected. In 1966, Carter brought an antitrust suit against the Bell System and General Telephone Company of the Southwest, which was referred by the U.S. District Court in Texas to the Federal Communications Commission because of the Commission's primary jurisdiction in communication matters.

In testimony before the FCC, AT&T and GT&E opposed use of the Carterfone on the grounds that it violated both the "foreign attachment" and "interconnection" (discussed in Chapter VI) tariff provisions for public exchange telephone service. Several technical arguments were given to support the position that "telephone system integrity" necessitated the use of only carrier-supplied attachments, whether acoustically coupled or directly wired to the telephone network. Overruling these objections, an FCC Hearing Examiner, in August 1967, approved the Carterfone for use on the dial telephone network, and ordered the carriers to modify their tariffs to specifically allow its use.[6] Upon appeal, the full Commission in a unanimous opinion issued on June 26, 1968, upheld this decision and broadened it to include all harmless customer-provided attachments, finding that the tariff provisions prohibiting foreign attachments "are, and have since their inception been, unreasonable, unlawful, and unreasonably discriminatory under the Communications Act of 1934." The Commission stated further:

> "...Our conclusion here is that a customer desiring to use an interconnecting device to improve the utility to him of both the telephone system and a private radio system should be able to do so, so long as the interconnection does not adversely affect the

---

6. *Carter Electronics Corporation* v. *AT&T et al*, Docket Nos. 16942 and 17073, Initial Decision of Hearing Examiner Chester F. Naumowitz, August 30, 1967.

telephone company's operations or the telephone system's utility for others. A tariff which prevents this is unreasonable. It is also unduly discriminatory where, as here, the telephone company's own interconnecting equipment is approved for use. The vice of the present tariff... is that it prohibits the use of harmless, as well as harmful devices."

* * *

"In view of the unlawfulness of the tariff, there would be no point in declaring it invalid as applied to the Carterfone and permitting it to continue in operation as to other interconnection devices. This would also put a clearly improper burden upon the manufacturers and users of other devices. The appropriate remedy is to strike the tariff and permit the carriers if they so desire, to propose new tariffs which will protect the telephone system against harmful devices, and they may specify technical standards if they wish."[7]

The Bell System, General Telephone System, and U.S. Independent Telephone Association initially responded to the FCC's ruling by filing petitions for reconsideration of the decision. Calling the telephone network "an intricate and delicately balanced mechanism," AT&T claimed that "the cumulative effect, especially over an extended period of time, of the proliferation of (customer-provided) devices, each of which in and of itself might be found to be 'harmless,' can be very injurious to the performance of the network." Moreover, AT&T stated, "a particular device designed and manufactured to be 'harmless' can become 'harmful' because of the way in which it is used or because it is not properly maintained...."[8]

In September 1968, the Commission announced its decision not to reconsider its original order, forcing the carriers to go to the courts if they wished to object further.[9] Shortly thereafter, both the Bell and General Systems appealed

---

7. *In the Matter of Use of the Carterfone Device in Message Toll Telephone Service*, Docket Nos. 16942 and 17073, FCC Memorandum Opinion and Order, adopted June 26, 1968; 13 FCC 2d 420.

8. *Telecommunications Reports*, XXXIV, No. 33, August 5, 1968, p. 4.

9. 14 FCC 2d 571, adopted September 11, 1968.

the decision to the U.S. Court of Appeals for the Second Circuit, in New York City. The Chairman of the Board of AT&T had previously stated that an appeal would be taken with respect to the FCC's finding of past unlawfulness of the tariff (which, if unchanged, would permit Carter Electronics Corporation to collect triple damages in its then pending antitrust suit against AT&T), but not with respect to the FCC's order that the tariff restrictions be liberalized. The appeal, however, was later withdrawn and AT&T and Carter Electronics Corporation settled out of court. After the FCC's refusal to reconsider its decision. AT&T filed with the FCC revised interstate long distance and WATS tariffs, which specifically permitted use of customer-provided equipment heretofore barred. These became effective in January 1969.

C. TARIFF REVISIONS

The January 1969 tariff revisions allowed customers to attach their own data modems and voice telephone equipment to the switched telephone network, with essentially three restrictions: First, such attached devices must limit maximum total power output and maximum energy distribution through the audio spectrum (including certain frequency limitations to avoid interference with the proper operation of the telephone companies' automatic switching equipment). Second, data or voice equipment may be attached by direct electrical connection only through an appropriate "protective connecting arrangement" supplied at nominal cost by the telephone company (discussed in more detail below). Subject to the limitations on signal output described above, acoustic or inductive coupling of customer equipment is permitted without a "connecting arrangement." Third, the tariffs require that all "network control signaling" functions, i.e., conventional rotary or Touch-Tone dialing, be performed by an AT&T-furnished unit.

These tariff filings also incorporated AT&T's first provisions allowing the interconnection of private communications systems with the public switched network subject to certain restrictions. The impact of these revised interconnection rules is discussed in the following chapter.

We now consider, first, several factors which have conditioned the relaxation of the foreign attachments rule; and, second, the significance of that relaxation for the data communications user.

D. CONSIDERATIONS UNDERLYING THE FOREIGN ATTACHMENTS RULE

The carriers have given three specific reasons for the necessity of the foreign attachments prohibition. First, the prohibition has been necessary to preserve the "system integrity" of the national communications network; i.e., to avoid impairment of the communications service of other subscribers using the public network, and to protect the safety of customers and telephone company personnel. Secondly, in order to avoid "divided responsibility" for the proper operation and maintenance of the national communication network, it is necessary for the carrier to retain control over all equipment attached to the public network. Thirdly, the use of customer supplied attachments could "hamper" carrier innovation — it could reduce the telephone company's freedom to introduce modifications into the national network.

Before reviewing these considerations in detail, we should observe that the foreign attachment restriction grew up in another area: the earlier days of the telephone system, when it was necessary to protect the network from the attachment of all manner of largely homemade "Rube Goldberg" devices (protection against harmful devices is, obviously, still necessary) and before the growth of data communications created a real need for the attachment of non-carrier supplied equipment to the network. The arguments of the carriers, outlined above, were largely valid under those circumstances. However, recent changes in the communications environment and in user requirements have led to a reevaluation of the significance of these dangers, and thus to changes in the foreign attachment regulations.

1. System Integrity

The first and primary argument urged by the carriers in defense of restrictions on the use of foreign attachments is "system integrity." There are basically two ways in which a user's attached equipment could affect the integrity of the telephone system. First, excessive signal levels from a user's attached equipment can cause "crosstalk," which interferes with the telephone service to other subscribers. Secondly, a user's improperly functioning attachment may interfere with the carrier's automatic switching, signaling and charging equipment.

Crosstalk occurs when signals from one communications channel are inadvertently introduced into a parallel channel. For transmission over distances greater than 10 or 20 miles, channels are "stacked" together onto a composite higher frequency carrier signal; although the individual voice channels are often thought of as separate "pipes," the carriers prefer to think of the subchannels as "troughs," such that if the signal level of a particular subchannel is too high, it will "wash over" into an adjacent "trough" and degrade the quality of another subscriber's connection. This phenomenon is called intermodulation distortion. Interference can also occur in local transmission, by electrical induction between parallel wire pairs. The background voices that are sometimes heard during telephone conversations are examples of crosstalk. The telephone companies argue that improperly designed or poorly maintained devices attached to the communication network may produce excessive signal levels and cause crosstalk.

The revised AT&T tariffs minimize this danger by specifying the maximum signal output levels of customer-provided equipment, and by requiring that such equipment be attached via the "protective connecting arrangement" mentioned above, supplied by the telephone company. AT&T uses a protective connecting device called a Data Access Arrangement (DAA) for this purpose, installed on the subscriber's premises, at nominal monthly cost.[10] The DAA contains a power sensing device, a variolosser, an audio filter and a transformer. The power-sensing device limits in-band signal power; the variolosser protects the station against high voltage; the filter eliminates out-of-band signals; and the transformer provides low-frequency and direct current isolation between customer-provided equipment and the line. The filter and power-limiting device minimize crosstalk, noise, or distortion in shared transmission elements of the network, which could interfere with other users; the variolosser and transformer prevent interference with network signaling, or harm to people or equipment, due to possible improper voltages.

The second way in which user-provided equipment might adversely affect the communication system is by interfering with the automatic switching, signaling and charging equipment associated with dial telephone service. Faulty or fraudently designed devices attached to the public switched network could result in (1) frequent dialing of wrong numbers, (2) inadvertent disconnection of subscribers using this equipment, (3) tying up of expensive shared central exchange equipment (e.g., switching registers), and (4) defeating the automatic billing equipment used for toll calls.

---

10. The FCC's *Carterfone* decision suggested that technical standards would be required for attached devices. AT&T chose to require the use of a protective device, in lieu of specifying comprehensive equipment standards and attempting to enforce these standards. Several equipment manufacturers and others protested this move, claiming that a more satisfactory approach could be worked out, including the development of more comprehensive standards. (See *Telecommunications Reports*, October 21, 1968, pp. 1-5, 16-22.) At the time of this writing the issue remains unresolved.

AT&T reacted to this danger by exempting "network control signaling" (i.e., dialing) apparatus from the relaxation of the foreign attachments rule. The January 1969 tariff provides that the subscriber must continue to use a network control signaling unit (NCSU) rented from the telephone company. This rotary dial or Touch-Tone keyboard unit is installed at or near the customer-provided modem and data terminal equipment, or voice telephone equipment. For the data user, this results in a certain degree of inconvenience and added cost relative to the forbidden alternative of a dialing unit incorporated into the terminal/modem itself. With this restriction the user could not employ his own tone-generation keyboard for both dialing and transmitting data. In addition, the use of customer-provided voice telephone equipment is for all practical purposes excluded, except for special-purpose applications, because of the diseconomy and physical inconvenience associated with the requirement that today's telephone, if provided by the customer, be "split" into two parts — the dialing apparatus controlled by the telephone company, and the voice transmission equipment (handset, hybrid coil, etc.) supplied by the subscriber.

The carriers have a legitimate fear that switching, signaling and charging difficulties would occur if unrestricted attachments of customer-provided equipment were permitted, particularly in light of consumer tendencies to compromise in the quality of their equipment purchases and their maintenance procedures. On the other hand, AT&T's approach to the problem, prohibiting customer-provided network control signaling equipment altogether, imposes added cost and inconvenience upon the user. As noted by most of the respondents in the FCC Computer Inquiry, and the FCC itself in its *Carterfone* decision, properly set and enforced equipment standards could minimize most of these potential problems. There has been little evidence to show that customer-provided equipment would, with any significant frequency, cause the above-mentioned difficulties. AT&T has informally admitted that all of the dangers described above could be "defended against" by the replacement or addition of certain central office equipment. However, AT&T also claims that the cost of such measures would probably outweigh the possible benefits to the subscriber. Unfortunately, no cost data is available to substantiate this claim.

Some of the possible switching, signaling and charging difficulties associated with the use of customer-provided equipment are the result of the present use of certain central exchange signaling conventions employing "in-band" tone signals. These are audible tones which are transmitted on the subscriber's telephone circuit between central offices to indicate the initiation and termination of a call and billing period. These signaling conventions were designed under the assumption that non-carrier attachments to the network would not be allowed. The widespread use of customer-provided attached equipment could make it advantageous for the telephone companies to convert to the use of invulnerable "out-of-band" signaling conventions. Such new techniques are being phased into the telephone network at present, and the new AT&T Electronic Switching Systems are designed to be impregnable to the difficulties described above.

The question of customer-provided network control signaling equipment can be resolved in a rational way only if answers are provided to these critical questions: What is the probability of occurrence of each of the difficulties which the carriers fear? And, what would be the cost of modifying certain carrier signaling conventions and central exchange subsystems, and/or of installing subscriber line test equipment, to defend against faulty or fraudulent customer equipment?

As we noted above, data sufficient to provide definite answers to these questions is not yet available. However, it is our general belief that the already large number of successfully interconnected devices and networks in the nationwide and worldwide common carrier systems is persuasive evidence that customer-provided network control signaling equipment would not cause sufficient interference with the carriers' switching, signaling, and charging procedures to offset the benefits which would accrue if these devices were allowed. The issue is not a simple one, however, and additional considerations and possible alternative approaches to a controlled liberalization of the rule are discussed in Section 4 below.

2. Divided Responsibility

The second argument in support of the foreign attachment rule is that relaxation of the rule results in "divided responsibility" for the operation of the national communication system. That is, the carrier would no longer control all the equipment used in the customer's end-to-end communications path; equipment manufactured and maintained by another organization would perform the communications interface and network control signaling functions at the end of the line. However, some customers regard the carrier as having total responsibility for the end-to-end communication line, including attached devices, with the result that the carrier may be blamed for any difficulties, from whatever source. This could have two undesirable effects upon the carrier: a poor public image, and the cost which is incurred when the customer calls carrier maintenance personnel for a difficulty which turns out to be caused by a non-carrier attachment. The practical problem of costs incurred in "false alarm" service calls can, however, be solved fairly easily. For example, a flat charge can be levied should a telephone repairman called to a customer's premises discover the trouble to be in customer-owned equipment rather than telephone company facilities. At the time of this writing, such a policy has been implemented by a number of telephone operating companies.

With respect to data communications, which is the chief concern of this book, "divided responsibility" has always existed whenever the carrier's customer elected to use terminal or computer equipment not supplied by the carrier (as is often the case). Use of a carrier-supplied data set does not eliminate this division of responsibility; it merely shifts the carrier/non-carrier interface from the end of the communications line itself to the business machine side of the data set. Division of responsibility has also existed with private leased lines, on which the user has long been free to attach independently-supplied and maintained equipment (both voice and data). The fact that this flexibility has been workable on leased lines without undue hardship to either the carrier or the user is evidence that divided responsibility will not be a significant problem on the switched network.

3. Carrier Freedom to Innovate

The third argument which has been used in support of the foreign attachments prohibition is that it is necessary to protect the carrier's freedom to innovate and to introduce improvements into the network. Customer-provided network interface equipment (for modulation and/or dialing functions) might be adversely affected by a carrier-initiated change in the network, and modification or replacement of the customer attachments might be required. Customers would oppose such network modifications.

On the other hand, very few of the equipment changes and upgrades which the carrier makes to the network are likely to have this effect upon customer equipment. Such changes would also obsolete much of the carrier's own equipment which was designed, similar to the customer's attachments, to be compatible with the prior system. Also, change comes very slowly to the telephone network, more slowly than in the technology of the devices attached to the network, because of (1) the huge capital investments required for major conversions,

(2) the carrier's necessity to maintain compatibility with his own existing equipment, and (3) the absence of a competitive spur in the telephone industry. The telephone subscriber may, in most cases, use his attached equipment for many years, probably to the limit of its economic lifetime, without difficulties caused by network change affecting him. Finally, network changes are introduced gradually, in one geographic area at a time, so that customers can be warned years in advance of forthcoming changes. The customer may then make equipment purchase decisions accordingly, and/or plan for the modification of existing equipment.

4. Other Considerations

It is interesting to note that the foreign attachment rule has prohibited the attachment of customer-provided devices to the public switched telephone network, whereas their use has long been permitted on private leased lines. Such lines are in fact an integral part of the same common physical plant with the lines used for the public switched network. The result is that the use of poorly designed or malfunctioning data communications equipment on a leased line causes as much crosstalk (the primary carrier argument against foreign attachments) or other degradation of the common carrier system (including adverse effects upon both other leased and dial-up lines) as if this equipment were used on a public switched line. Thus, the public is very definitely affected by the correct or incorrect selection, operation and maintenance of "foreign" data equipment for use on a leased line.

This would imply that the carriers' apprehensions that non-carrier attachments will produce crosstalk should apply equally to their use on leased as well as switched lines. However, the users of leased lines are typically large corporations and government organizations to whom system performance and reliability is just as important as to the carriers themselves. Thus, such users can presumably be relied upon to employ "quality" equipment, to operate it correctly, and to keep it well maintained. Such assumptions regarding the general use of the dial network can be made less easily. In addition, the "policing" problem for the carrier may be much less severe on leased lines than on the general public network. This is because customer violations of the equipment or maintenance standards, which would require carrier action, would presumably be less likely to occur. Also, the number of leased lines currently in service constitutes a small percentage of the total carrier plant, and the "routing" of a leased line is fixed, whereas the routing of a dial-up line (on successive connections between the same two parties) varies with the pattern of network loading. Therefore, difficulties arising from the use of leased lines might be more readily detected and located.

In the light of these factors, it seems that the telephone company may have been pursuing a cautious, but responsive policy — that is, permitting the use of independently-supplied attachments where such use is reasonably certain not to damage the network. Or it may be that certain classes of customers have more economic power and thus exert considerably more influence upon telephone company policies. This cannot be overlooked as a major reason for the permission of foreign attachments only on leased lines prior to the *Carterfone* ruling.

Pursuing a policy of maximum liberalization of restrictions wherever possible without incurring undue danger of damage to the network, perhaps limited permission of customer-provided network control signaling units (NCSUs) can be granted on the dial network. The objective should be to maximize economic benefit to the user, while minimizing the risk to the telephone network. The use of non-carrier NCSUs is *desirable* for many data communications applications, and is *essential* for others; in general, it is *essential* if non-carrier voice telephone instruments are to be economically employed. On the other hand, carrier fears of system difficulties caused by the permission of non-carrier NCSUs seem to be more conditional upon *who* is permitted to use these attachments rather than for *which* applications (data and/or voice communications) they are permitted. Thus, we see that there are at least two ways in which the pie can be sliced, to achieve the goal stated above.

As a first alternative, non-carrier NCSUs could be permitted only for use with data communications terminal devices; customer-provided voice telephone equipment would continue to be barred from performing dialing functions, thus requiring use of a carrier-supplied NCSU with such voice equipment. As a result, customers would most likely find that savings through the use of non-carrier voice telephone equipment would be small or non-existent; and such savings being their primary motivation for the use of such independently supplied voice equipment, they would largely avoid its use. Such a tariff modification has the advantage that it would meet the needs of those data communications applications requiring the use of customer-provided NCSUs, while at the same time eliminating many of the dangers which unrestricted NCSU proliferation could cause.

One of the principal fears of the carriers is in regard to the residential consumer as a user of non-carrier attachments — because of his perhaps dangerous tendencies toward shopping for bargains and disregarding proper preventive maintenance, and because of the policing problems associated with detecting inferior or defective equipment in use and either having the trouble corrected or the equipment removed from the network. There is little residential consumer demand for data communications at present, and such demand will begin only when economically attractive "computer utility" services for the residential consumer are offered. And even with the advent of such services, the consumer demand for data equipment will be extremely small as compared with the size of the market for voice telephone equipment. Therefore, removing the prohibition on customer-provided NCSUs for use with data communications terminal equipment while continuing to prohibit their use with voice telephone equipment, might be a compromise solution which would satisfy most of the requirements for these devices, while eliminating one of the largest potential dangers which the present rule guards against. There is, however, at least one significant difficulty with this solution: many existing attachment devices can be used for both data and voice communication — examples are the modem with necessary voice capability for transmission coordination purposes, and the Touch-Tone telephone instrument which may be used as a keyboard data input device as well as for normal voice communications. Presumably the former of these "combination" devices would be considered a "data" attachment and thus a non-carrier NCSU for it would be permitted, but the latter, with its various possible permutations and adaptations, could pose problems of definition.

A second approach to the goal of limited permission of customer-provided dialing apparatus might be to permit the use of these devices — with either data or voice equipment — by business telephone subscribers, but continue to prohibit their use by the residential subscriber. This is feasible because there are in general separate local exchange tariff provisions for business and for residential telephone service, with the businessman generally charged considerably more for the same service. Tariff provisions for business service could be modified to permit use of non-carrier NCSUs, thus making it feasible for the business subscriber to obtain both data and voice telephone equipment from independent suppliers. This might be justified to irate consumers on the grounds that since the businessman is charged higher rates for telephone service (often several times the residential rate), he is entitled to special privileges. In the future, if the unrestricted use of foreign attachments by businesses establishes the advantages of such use (as the authors are confident will be the case), the residential restriction could be liberalized as well. Beginning with the more limited business market would provide a testing and proving ground, under more controlled conditions.

E. IMPACT OF THE REVISED FOREIGN ATTACHMENT RULE

The impact of the relaxed foreign attachment rule upon the data communications user falls primarily upon the modem, which, prior to January 1969, had to be obtained from the communications common carriers if used on the dial telephone network or the TWX or Telex teletypewriter networks.

Modems are produced by manufacturing subsidiaries of the communications common carriers, principally Western Electric Company, by computer manufacturers, and by a number of independent electronics manufacturers; previously the non-carrier firms could sell modems only for use on private, leased lines. Many varieties of modems exist, differing in transmission speed, modulation method, mode of transmission (serial or parallel), automatic or manual line equalization capability, channel bandwidth and conditioning requirements, error rate, design lifetime and price. Although Western Electric supplies a number of these different types, it tends to restrict itself to those modems for which there is a large and assured market, and Bell generally prices these modems, which it will only lease, at levels substantially higher than units with similar performance characteristics offered by the non-carrier suppliers.

The cost of the modulation/demodulation function often may be reduced by incorporation of the modem circuitry into the terminal or computer with which it is used, rather than forcing its separate implementation. For example, modems are incorporated as an integral part of certain of the Teletype terminals leased by the carriers to their customers. Since the relaxed foreign attachment provisions permit the use of customer-provided modems, independent manufacturers may integrate modem and terminal or computer equipment into a single unit. This would eliminate redundant circuitry (e.g., by sharing logical components, power supply, and clocks), permit reduced packaging size and costs, and improve the appearance and physical layout of computer centers, especially multiple-line installations, such as remote-accessed computer centers, which typically have several racks filled with modems. Since Western Electric does not manufacture multiple-line modem configurations (whereas such could be conveniently incorporated with the computer's multiplexed input/output channel), the user has been at a particular disadvantage in this case.

The availability of "special purpose" modems for use on the switched telephone network was also promoted by the relaxation of the foreign attachment rule. Western Electric data sets have generally been designed for "typical" applications and mass markets; submarkets have usually been ignored. The relaxation of the rule allows independent modem manufacturers to respond to market requirements for modems of varying and unusual transmission speeds, error rates, reliability, and costs.

1. Advances in Modem Technology

The carriers' application of advances in modem design has suffered not from lack of technical expertise but from lack of incentive to innovate, as the foreign attachment rule on the dial network gave the Telephone Company a protected monopoly market for modems. Application of laboratory developments may have been retarded by the fact that modem design improvements which effectively increase line capacity tend to reduce the customer's required line holding time (and therefore reduce the carrier's revenues) for a given volume of communications — a fact which, under the present telephone pricing structure, is contrary to the carrier's interests.

In addition, telephone company policy is to design equipment for long lifetimes and to avoid the use of accelerated depreciation methods (which would encourage more rapid equipment replacement). Digital technology is changing at such a rapid rate that the expected technological lifetime of a new modem is only a few years. Therefore, it is not economical to design it for a longer physical life, both because of the added manufacturing expense and because in a monopoly market it may tend to retard the introduction of future design improvements. The relaxation of the foreign attachment rule, allowing unrestricted equipment procurement by users, will stimulate the carriers and the computer industry alike to adjust equipment design lifetimes and depreciation policies to more closely correspond to the technological obsolescence rate, thus promoting innovation and efficiency.

In discussing innovation and the communications common carriers, an interesting aside concerns the history of the modem itself. Punched paper tape has been used in conjunction with teletypewriter equipment since the 1920's, and was initially used for data transmission over the teletype network, by the military in World War II. However, this method was slow, and required data conversion from tabulating cards to paper tape and back to cards again. 1954 marked the birth of data communications as we know it today. In that year, IBM introduced a card transceiver which provided direct card-to-card transmission over leased telephone lines. The transceiver included its own modem, since none was available from the carriers. However, since this modem was a "foreign attachment," the transceiver could not be used on the dial network; thus, use of this equipment was impractical for many applications and many potential users. Finally, in 1958 AT&T recognized that data communications was here to stay, and introduced its first data set; its function was exactly the same as the ones provided previously by IBM, but this one was legal for use on the public dial network. The IBM modems previously built into the card transceivers were then replaced with interfacing circuits to connect the transceiver to the AT&T data set.[11] This historical incident may provide insight into the motivations of users who have pressed for a relaxation of the foreign attachment restrictions, hoping to thus take advantage of the full innovative potential of the competitive marketplace.

In the past, communications engineering expertise was much more restricted to the telephone companies than it is today. This strengthened the argument that the telephone company was best able to provide the equipment to be used with its lines. During the past twenty-five years, however, the electronics industry has developed from its infancy and, aided by massive experience with military contract work, has come to rival the carriers' competence in almost all aspects of communications technology. This is especially true as applied to digital equipment; the digital computer is the best-known example of this technology, and modem design is closely linked. Digital technology is not the primary business of the carriers (although this will gradually become less true, as the use of electronic switching systems, time-division multiplexing equipment, and digital carrier systems becomes more widespread), whereas for many members of the independent electronics industry it is the stock in trade.

2. Tone-Generation Terminal Devices

Under the revised tariff it is still impossible to employ a customer-owned tone-generation keyboard, functionally similar to a Touch-Tone keyboard but with additional keys as determined by the specific application, requiring no modem and therefore possibly very inexpensive, both for dialing and for transmitting data once the connection was established. The conventional keyboard terminal generates a unique digital bit pattern for each depression of a key, which is converted by the modem into an audio signal for transmission over the telephone network. An alternative approach is to generate audio tones directly

---

11. Charles R. Doty, Sr., *Data Communications ... A Detailed Review ... and a Glimpse at the Future ... of the Development and Application of Equipment, Systems and Technology.* (Poughkeepsie, N.Y.; IBM Corporation, 1964), p. 49.

with each key depression, as is done by the Touch-Tone telephone. This method, which does not require the use of a modem at the transmitting end, is an economical alternative for the design of low-cost or special purpose terminal devices. A terminal designed for use with the switched network could employ the numeric keys on a tone-generation keyboard for the dual purposes of dialing and data entry. The original foreign attachments rule prohibited *all* such direct attachments to the dial network, and as discussed earlier in this chapter, the revised tariff prohibits customer-provided network control (i.e., dialing) equipment on the switched network.

A low-cost tone-generation terminal, perhaps in conjunction with audio answerback for output information, might for example serve as the basis for a future household terminal. Special purpose designs could accommodate such remote-access computer services as keyboard calculators for the home, instructional terminals for on-line testing and tutoring of children and adults (e.g., career advancement) and various information services such as stock market trading prices, bank account information, etc. Innovation in terminal design for these consumer services which will require low-cost terminals is inhibited by present and revised foreign attachment regulations.

3. Acoustic Coupling

Acoustic coupling is the process by which electrical signals (from e.g., a modem connected to a terminal device) are converted to audible sounds which are then picked up by the microphone in the ordinary telephone handset, reconverted to electrical signals and finally transmitted over the telephone network. At the receiving end of the telephone channel, the process is inverted. (Reception of the signal from the telephone network may be by means of inductive instead of acoustic coupling, but two-way coupling devices which employ acoustic coupling *to* the telephone network, and inductive coupling *from* the network are often referred to as "acoustic couplers.")

The acoustic coupler permits voice or data signals from subscriber-provided equipment to be introduced into the telephone network without there being a direct electrical connection between this equipment and the telephone line or telephone instrument itself. These devices have therefore been employed for data communications applications for two reasons: First, acoustic coupling was an expedient to permit the use on the switched network of an inexpensive or special-purpose modem supplied by an independent electronics manufacturer. Secondly, the use of an acoustically coupled modem is necessary today if portability of a data terminal is desired. This may be important for computer applications such as remote conversational time-sharing, for which some users have "portable" teletypewriter terminals. They use the terminal in many unanticipated locations, such as different engineers' offices in a company, and therefore need the ability to connect to the telephone network from a variety of access points.

Even though acoustic coupling is an expedient to avoid direct electrical connection to the telephone network, its use has not been without considerable carrier opposition. The telephone companies have been historically opposed to any sort of coupling or interconnection to or from the telephone network, except as done by them. For many years, they prevented the use of any sort of couplers which would permit tape recording of both sides of a telephone conversation. This restriction was relaxed by FCC order in 1948.[12] Radio amateurs have for many years used home-built "phone patches" for coupling the telephone to their radio transmitters and receivers, to permit conversation between a distant amateur radio station and a telephone subscriber. Since these "relaying" activities have proven useful in a public service context — e.g., establishing a radio and telephone connection between an overseas serviceman and his family in the U.S. — their abridgement of tariff provisions has been "winked at" by the telephone company. However, pressure from the telephone company has largely prevented commercial marketing of such interconnection devices for amateur radio use.

With respect to the use of acoustic coupling devices to connect data communications equipment to the telephone network, the carriers have been somewhat less adamant. Although the foreign attachment provisions of the domestic carriers' public network tariffs have prohibited the use of non-carrier supplied equipment, whether attached to the carrier network by direct electrical connection or by acoustic or inductive coupling, in recent years the rule has not been generally enforced as applied to the acoustic coupling of data terminal devices. Prior to the tariff liberalizations in January 1969, a number of independently-manufactured "data acoustic couplers" (i.e., modem and acoustic coupler built into a single unit) were available on the market. The revised common carrier tariffs explicitly allow acoustic/inductive coupling of customer equipment, provided a telephone company supplied network control signaling (dialing) unit is used.

---

12. *In the Matter of the Use of Recording Devices in Connection with Telephone Service,* FCC Docket No. 6787, Commission order effective June 30, 1948.

Since the foreign attachment prohibition has been lifted, however, acoustic coupling to permit use of customer-supplied data modems is no longer necessary. Such equipment can now be attached by direct electrical connection. Acoustic coupling is therefore useful today only if one requires the second feature which it offers: data terminal portability. It is less than an ideal solution for this requirement, however. As a means of interfacing with the telephone network, acoustic coupling is undesirable because it adds an unnecessary stage of signal energy conversion (electrical to acoustical and back to electrical); it is costly; and it introduces noise into the telephone channel, therefore increasing the probability of transmission errors and reducing the channel's transmission capacity.

Given the relaxation of the foreign attachment rule to permit use of independently-supplied equipment on the dial network, it would not be difficult for the telephone company to provide plug-in jacks on the telephones, which would permit temporary direct electrical connection of a data modem (or other customer attachments, such as a tape recorder), and eliminate the need for acoustic coupling to provide portability of a terminal. Such jacks could be an integral part of future telephones (e.g., the new Touch-Tone telephone which is expected to be widely used as a data input device, and to which one might wish to attach a low-cost visual output device), or they could be made available upon customer request.

F. SUMMARY

This chapter began with an explanation of the nature of the common carriers' long-standing "foreign attachments" rule, which has prohibited subscribers from attaching non-carrier supplied devices to the dial telephone network. The legality of the rule was first tested in 1956 in the *Hush-a-Phone* case, in which a Federal court concluded that a subscriber may use his telephone in ways which are privately beneficial and are not publicly detrimental. However, the carriers' broad foreign attachment restrictions remained in effect until the FCC's landmark *Carterfone* decision in 1968 found these restrictions to be illegal, and ordered their removal. The carriers then revised their tariffs to permit the attachment of customer-provided devices, subject to the conditions that (1) limitations on signal power and bandwidth are met, (2) carrier-supplied protective interface devices are used, and (3) only carrier-supplied dialing and hook-switch devices (termed "network control signaling units" by AT&T) are used. These conditions — the latter two of which have generated considerable controversy — were imposed to prevent customer-supplied attachments from introducing noise or crosstalk into the telephone system, thereby degrading service to other subscribers, and to prevent such attachments from interfering with the proper operation of the carriers' automatic switching, signaling, and charging equipment on the dial network.

It has been suggested that the carriers are reluctant to allow greater liberalization of the attachment rules because of fears that widespread replacement of their telephone instruments with "foreign" telephone equipment in residential households might, even with strict technical interface standards, lead to equipment maintenance problems and necessitate costly "defensive" measures at the central switching exchanges. However, the actual need for liberalized attachment rules comes from the business and industrial community, and it is unlikely that these organizations would use improper attachments, or fail to maintain them properly. Therefore, two alternative mechanisms which would permit controlled liberalization of the present attachment restrictions are discussed: First, subscribers could be permitted to install their own network signaling units *only* for use with data communication terminal devices (residential households will not, for several years at least, make appreciable use of data terminals); or second, only business and other organizational subscribers could be granted additional attachment privileges, relying upon the assumption that these organizations are inherently more trustworthy and responsible than are the millions of residential subscribers.

For the data processing industry, the impact of the revised tariff provisions falls largely upon the modem, which is necessary in order to connect data transmitting and receiving equipment to a communications channel. The revised rules allow dial telephone network users to employ modems obtained from independent equipment suppliers. The increasing number of such vendors, and the resulting competition, can be expected to reduce the prices for this equipment, speed up the rate of innovation, produce a wider assortment of modems, and generally stimulate the data communications industry.

The revised tariffs also explicitly allow the acoustic coupling of portable terminal devices to the telephone network, which was previously forbidden but in actual practice has largely been condoned by the telephone company. To the best knowledge of the authors, however, the telephone companies have no plans to introduce telephone instruments equipped with "jacks" or "terminal outlets" which could be used for the direct electrical attachment of portable data terminal units, thus avoiding the inefficient expedient of acoustic coupling.

While certain questions remain with respect to the necessity for the use of protective connecting devices, and the requirement that only carrier-supplied network signaling equipment be used, the 1969 tariff revisions will clearly prove beneficial to the data processing community, and are unlikely to create additional problems for the carriers in their provision of high-quality telecommunications facilities and services for voice and data communications.

CHAPTER VI

INTERCONNECTION

"Interconnection," as used in this book, is the direct connection of a subscriber-provided communication network (e.g., a private mobile radio system, a microwave system, an intercom system, or a telephone switchboard), or a specialized common carrier network, to facilities provided by a telephone or telegraph common carrier. Until January 1969, interconnection with both dial telephone[1] and leased line facilities[2] was prohibited. Exceptions were allowed for railroads, pipelines, and other right-of-way companies, as well as for NASA and, in certain circumstances, the military.

Since 1969, AT&T has permitted customers to interconnect private communication systems with the public dial telephone network and with leased private lines of voice-grade bandwidth or less, subject to certain restrictions discussed below.

---

1. For example, AT&T Tariff FCC No. 263, for interstate and foreign message toll telephone service (binding on the associated companies of the Bell System, and also, through concurrences, on the telephone operating companies of the General System and most other independents) provided:

    "No equipment, apparatus, circuit or device not furnished by the Telephone Company shall be attached to or connected with the facilities furnished by the Telephone Company, whether physically, by induction or otherwise, except as provided in 2.6.2 through 2.6.12 following." (Para. 2.6.1.)

    * * *

    ". . . Except as provided in 2.6.5(A)(1) preceding, nothing herein shall be construed to permit the use of a device for the recording of two-way telephone conversations, or of a device to *interconnect any line or channel of the Telephone Company with any other communications line or channel of the Company or of any other person.*" (Para. 2.6.9; emphasis added.)

2. For example, AT&T Tariff FCC No. 260, for interstate private line service, provided:

    "No line, instrument, appliance or apparatus not furnished by the Telephone Company or by the Telephone Company and its other participating carriers shall be connected with, attached to or used in connection with the lines, equipment, apparatus or service furnished by the Telephone Company except where such connection, attachment or use is made in accordance with 2.1.4 and 2.2 preceding or 2.6.4, 2.6.5, or 2.6.6 following, or in accordance with the tariffs of this Company's other participating carriers governing the furnishing of private line service." (Para. 2.6.1(A).)

Private communication systems which are interconnected with the dial telephone network must comply with three conditions, basically the same as the restrictions under which the attachment of customer-supplied devices is allowed, as discussed in the previous chapter. First, the total output power and the energy distribution over the audio frequency spectrum of signals from interconnected systems must fall within prescribed limits. Second, the interconnected system must be attached to the carrier's dial telephone network through a protective connecting device supplied by the carrier. Third, all network control signaling (i.e., dialing) over the dial telephone network must be performed by carrier equipment, rather than by the private system, at the point of interconnection.[3]

Private communication systems may be interconnected with channels leased from the carriers provided three important conditions are met: First, the interconnection must be made at a point on the user's premises where he has a "regular and continuing requirement for origination or termination of communications" over his private communication network. Second, the bulk of the traffic over the interconnected private network must either originate or terminate at the point of interconnection (e.g., a business office or a manufacturing facility), rather than continuing onto the leased common carrier channel. That is, points of interconnection cannot be established primarily for the purpose of passing "through" traffic over a "pieced-out" system consisting of both leased common carrier links and privately-operated facilities. And thirdly, only interconnection with leased channels of voice-grade or less will be allowed, except in certain specified circumstances.

Restrictive conditions are imposed upon interconnected private systems for both technical and economic reasons. The restrictions associated with interconnection to the dial telephone network are largely technical, while those for interconnection with leased channels are primarily economic. The underlying reasons for these restrictions are discussed in the remainder of this chapter.

---

3. While the recent granting of permission to interconnect private communication systems with the dial telephone network was a significant departure from the modus operandi of earlier years, its impact was substantially reduced by the telephone companies' requirement that carrier-supplied network control signaling units and protective interface devices be used. This restriction may remain an active point of contention between the carriers and the users and equipment suppliers for some time to come.

A. TECHNICAL ISSUES REGARDING INTERCONNECTION

Three technical arguments have, historically, been given by the carriers in defense of their prohibitions against the interconnection of private communication systems. To a large extent, the recent change of attitude on the part of the carriers towards interconnection with both the dial telephone network and private leased lines suggests that these technical problems associated with interconnection are not insurmountable, although they must be recognized and considered.

The first difficulty is that signals from interconnected private networks might adversely affect subscribers using the public switched carrier network. Secondly, there would be divided responsibility for the total communications network. And thirdly, interconnection, if permitted, would hamper innovation by the carriers. The reader will note that these "technical" arguments are similar to the arguments long given by the carriers in defense of the foreign attachment prohibition.

The first of these arguments is illustrated by the following excerpt from the 1968 Carterfone case. Here, AT&T argued that owners of private communication systems:

> ". . . will naturally be primarily interested in the operation of their own system, rather than in the telephone service being furnished to the general public.
>
> * * *
>
> ". . . If private mobile systems were connected with the message toll network, they would not only be receiving from it, but would be putting something into the network. As a result, the private mobile radio system, when in use, would actually be functioning as part of the network. Therefore, its characteristics would be as vital to good quality service as that of the public telephone equipment and facilities, the common carrier facilities.
>
> *Improper input levels or noise levels would have as adverse an effect on the quality of the telephone service as if they originated in the public telephone equipment.*" (Emphasis added.)[4]

---
4. *In Re Use of the Carterfone Device in Message Toll Telephone Service, et al.*, FCC Docket No. 16942, Transcript of the Oral Argument before the Commission en banc, April 22, 1968 (hereinafter called Carterfone Oral Argument): Wayne Babler, AT&T, p. 1220.

As mentioned in the previous chapter, the carriers have established interface standards to ensure that interconnected customer-provided networks and devices will have no adverse effects upon other subscribers to the dial telephone network. As noted above, protective interface devices are also employed at the point of interconnection, to protect the common carrier network against damage in the event of malfunction or substandard performance of the customer-owned communication system. While these initial standards are somewhat experimental in nature, and several years experience will be required to confirm their adequacy for protection of the telephone network and to "iron out" any difficulties which may remain, past experience in related areas indicates that the effort will be successful. For example, standards for interconnection of the Bell System with the nation's more than 2100 independent telephone companies, and with certain privileged user organizations (e.g., the government) have existed for many years, and have proven workable. Despite the fact that some of the independent telephone companies use equipment manufactured by firms other than Western Electric, the carriers have been able to avoid detrimental effects upon the public network. Many of the independents have even integrated their telephone systems into Bell's direct distance dialing system, using interconnection standards published in an AT&T handbook entitled *Notes on Distance Dialing*. The success with which interconnection has been accomplished has been noted by F.J. Singer, an engineer at the Bell Telephone Laboratories:

> "The task of interconnecting many telephone systems and subsystems to provide the present direct distance dialing network has been accomplished despite the problems of overcoming inherent incompatibilities."[5]

In addition, there are over 8000 farmer or "service" lines (private lines, owned and operated by farmers in remote areas) which are interconnected with the public network.[6] Standards have been successfully worked out for these interconnecting service lines.

The carriers have given a second argument in defense of interconnection restrictions; one which, again, is also raised in justification of their views towards the use of "foreign attachments." They argue that interconnection leads to

---

5. F. J. Singer, *Electrical Engineering,* Feb., 1962.

6. Carterfone Oral Argument: K. E. Griffith, FCC Common Carrier Bureau, p. 1298.

"divided responsibility in the operation of the nationwide telephone network."[7] This is certainly true; today the responsibility for the national network is divided among the more than 2100 interconnected independent telephone carriers.[8] However, this has not impaired the telephone companies' ability to provide service unexcelled elsewhere in the world.

As is true with regard to the use of non-carrier attachments, divided responsibility seems to be a small price to pay in order to obtain the benefits of diversity, flexibility, and specialization of organizations providing service to the public, and competition among them. These considerations apply both to private communication systems and to facilities provided by specialized common carriers, whose service franchise may or may not overlap with other carriers, depending upon the nature of the service being provided. As will be discussed in more detail in Chapter IX on special service common carriers, a number of such firms presently operate in several communications submarkets (e.g., mobile radiotelephone service), and license applications have been filed by firms seeking to enter other submarkets (e.g., the Microwave Communications, Inc. authorization to provide microwave channels on a "private line" basis for voice and data transmission).[9]

The arguments by the carriers that interconnection, if permitted, would impair their freedom to innovate and introduce improvements into the network, were discussed in the previous chapter on foreign attachments. The points to be considered pro and con are essentially the same with regard to interconnection, and will not be repeated here.

In summary, it appears to the authors that from a *technical* standpoint, full, free interconnection absent of today's artificial restrictions is feasible without harm to the common carrier communications network or to its subscribers. The

---

7. Carterfone Oral Argument: Wayne Babler, AT&T, p. 1220.

8. However, it is recognized that the Bell System dominates the multitude of smaller carriers. Indeed, perhaps the reason why effective standards have been set is the dominance and centralized control provided by the Bell System. If that is the case then surely the Bell System will exercise similar control over interconnected private communication networks.

9. See *In Re Applications of Microwave Communications, Inc., for construction permits to establish new facilities in the Domestic Public Point-to-Point Microwave Radio Service between Chicago, Illinois, and St. Louis, Missouri,* FCC Docket Nos. 16509-16519, Decision of the Commission, Issued August 13, 1969.

standards-making process is not unfamiliar to the communications industry, and adequate equipment specification standards and operating regulations can be derived and enforced. After studying this subject, and taking testimony from the carriers and others, the FCC likewise concluded in its 1959 *Above 890* decision that, "It is clear from the record that, from a technical standpoint, interconnection of private systems with the common carrier systems is feasible where compatible and adequate transmission standards are maintained."[10]

B. ECONOMIC ISSUES REGARDING INTERCONNECTION

The established common carriers impose a number of restrictions upon the interconnection of their leased line channels with competing or private communication systems, on the grounds that unrestricted interconnection would encourage "cream-skimming." In the following sections we explain the cream-skimming concept and discuss the validity of the carriers' opposition to it. First, as essential background, average-cost pricing is explained.

1. Background: Average-Cost Pricing

Ideally, according to economic theory, to obtain efficient allocation of resources the prices of goods and services should reflect the true cost of providing them. However, utility pricing in some cases departs from this principle, charging similar prices for services with dissimilar actual costs, by averaging the costs of these services. Average-cost pricing offers a simpler rate structure and provides a subsidy to subscribers for whom the actual cost of providing service is high. This is sometimes, but not always, socially desirable.

Geographic Averaging

In general there are two ways in which cost averaging can occur in communications utility pricing. First, there is the geographic averaging of costs which permits uniform pricing of interstate services, though the costs of communications facilities in sparsely-populated regions are higher than in densely-populated regions or in regions with favorable terrain.

---

10. 27 FCC 359, at 397 (1959).

The uniform average price which is charged permits the low-cost areas to subsidize the high-cost areas, and thus insures widespread availability of the service. Such widespread use is to the benefit of all subscribers, for the value of telephone service in the message toll network is directly proportional to the number of other parties with whom contact can be made. An additional advantage of such an averaging method is the simplicity of calculating charges for use of the system; a uniform mileage charge for interstate toll and private line services applies everywhere. The communications common carriers do attempt to make prices reflect costs, somewhat, in that the individual operating companies determine the prices in their own local areas for intrastate services. However, by and large, it is claimed that it would be very difficult and costly to have the prices accurately reflect the costs in each area.[11]

Technological Averaging

The second form of averaging is what we will refer to as *technological* averaging. Within a given segment of the total network there may exist several varieties of equipment in service. For example, on a single route between two cities, there may be old open-wire circuits, more modern cable circuits, and still newer microwave links. Hardware of different ages, different degrees of technological advancement, and with different operating costs will be in service side-by-side, and obviously the real costs of furnishing the same service will vary depending upon which is used. The assumption is made that it would be undesirable for the rate charged the user to reflect these physical cost differences, and so they are averaged and the user pays a uniform price regardless of which segment of the total plant carries his traffic.

Technological averaging was probably a useful and necessary concept in the past, but recent rapid advances in communications technology may force us to rethink the underlying assumptions here. Methods such as satellites and low-cost microwave provide communications at a physical cost many times below that of the older more conventional techniques, but the carriers' cost averaging does not encourage the carriers to use the lower-cost facilities since the demand for all facilities is independent of their cost.[12]

---

11. However, this historical difficulty may be reduced by the availability of modern data processing equipment, which will permit and make economically feasible more accurate detailed cost breakdowns, service usage measurement, and billing.
12. Congress has recognized the inadvisability of giving the carriers control of the new satellite technology and, in the Communications Satellite Act of 1962, which formed the Communications Satellite Corporation (COMSAT), limited carrier ownership of COMSAT to 50%.

One approach to consider is that represented by several "special service common carrier" organizations, discussed in Chapter IX, in their attempts to secure licenses to provide specialized cut-rate microwave systems which will not only be less expensive than existing carrier microwave facilities, but will pass this cost advantage on to the public without dilution by averaging with older more expensive equipment. But is this not unfair competition with the established carriers who are saddled with their previous investment in older equipment? Perhaps, but who should bear the added cost of that portion of the carriers' transmission plant which today remains undepreciated but has become technologically obsolete — the carriers themselves (their stockholders) or their subscribers? Historically, the subscribers have borne this burden, due to the lack of pressures upon the carriers to the contrary. A new system of incentives is needed which will encourage the rapid amortization of old common carrier equipment, and its replacement by more efficient facilities. Permitting market entry by new carriers may serve this function. Speeded-up write-offs might in the short run lead to increased communications costs, which would be undesirable; in the long run, however, costs are likely to fall, both because of the use of newer and more efficient technology and because of the demand effect of this technology. Reduced costs would lead to increased communications usage, which in turn could lead to further reductions in cost.

2. Cream-Skimming

As discussed above, the cost of providing communications channels varies across the nation with the density of the population, the character of the terrain, and the age of the installed facilities. The prices charged represent an average of the high-cost and low-cost facilities.

If small new firms were permitted to provide communications channels in localized regions in competition with the large carriers, these local firms would tend to select areas for service where the cost of providing the communications channels is lowest. Then they would be able to price their channels below the "average" prices charged by the larger carriers. Ultimately, the large carriers would serve only the high-cost areas and would be forced to increase the rates charged to their subscribers. Thus the argument runs. Activity of this sort by small firms is called "cream-skimming," and is opposed by the established carriers because they would lose lucrative markets.

The established carriers have also used the argument of "cream-skimming," unsuccessfully, to oppose the entry of private microwave systems in 1959:

> "...the carriers justified a policy of entry restriction based on the carrier pricing system. These rates were determined on the basis of averaging costs. Some services provided by the industry were supplied over facilities that were high-cost or low-profit routes. Other services were provided the subscriber over low-cost routes. The final charge levied to the public was a flat rate derived as an average of both high and low cost facilities. AT&T submitted that owners of private microwave would be tempted to provide their own communication requirements over low-cost routes but then would turn to the carriers and lease requirements over the more expensive routes. Private microwave, would, in short, preempt the low-cost routes in a form of cream-skimming which would consign the expensive runs to the common carriers. The net effect of this cream-skimming would be to lift the average cost of communication services for all customers and hence increase the general price to the consuming public. In the name of lower communication prices, then, AT&T concluded that the public interest was best served by a continued policy of restricting the use of subscriber-owned radio systems."[13]

Once private microwave systems were sanctioned by the FCC, the carriers' next line of defense against the "cream-skimmers" was to prohibit such systems from interconnecting with any common carrier facilities. As noted earlier in this chapter, the interconnection prohibition was relaxed in 1969, although such connection is largely limited to channels (dial-up or leased) of voice-grade bandwidth or less. The policy of the established carriers towards would-be small specialized common carrier entrants with regard to interconnection is the same as it was initially towards private microwave systems; and in this instance it is unlikely that the established carriers will willingly make any concessions allowing interconnection with either the dial telephone network or their private line services. This will be discussed in more detail at the end of this chapter.

---

13. Manley R. Irwin, "The Communications Industry and the Policy of Competition," *Buffalo Law Review,* Vol. 14 (1965), p. 261.

Refusal to interconnect has been, in fact, the major economic weapon of the carriers in restricting the development of competing forms of communication — both private communication systems and public services offered by new carriers. The development of microwave technology has added significantly to the potential of each of these. But, as one commentator has observed,

> "The overall question remains whether the de facto monopoly condition prevailing in microwave and the private line markets can be reversed and restructured under a regulatory environment assuring effective free market competition. *And the critical immediate issue centers around the interconnection question.*
>
> "...the [Federal Communications] Commission has *permitted AT&T to use interconnection to defeat common carrier competition and to restrict private microwave competition,* thereby limiting the utility and services of competing microwave systems." (Emphasis added.)[14]

The independent telephone companies with whom interconnection is permitted are not in competition with Bell. Rather, they largely serve undesirable rural areas, and do not overlap with the Bell System. The argument of "cream-skimming" can be turned around and applied against the Bell System, for it has historically chosen to serve the densely populated, profitable areas, while leaving the "chaff" to the small independents.[15]

---

14. Donald C. Beelar, "Cables in the Sky and the Struggle for Their Control," *Federal Communications Bar Journal,* Vol. 21, No. 1 (1967), p. 40.

15. Although the Bell System operates approximately 84% of the nation's telephones, its franchise covers less than half of the U.S. geographic area. Bell telephone exchanges, predominantly located in the large metropolitan centers, each service an average of 21,800 telephones. Independent exchanges, located in the smaller cities and suburban and rural areas, each service an average of only 1,570 telephones, and thus cannot obtain similar efficiencies of scale. (Calculated from U.S. Independent Telephone Association, *Independent Telephone Statistics,* [Washington, D.C.: 1968], Vol. 1, pp. 2, 6.)

3. The Validity of the Carriers' Opposition to "Cream-Skimming"

As discussed in the previous section, cream-skimming can occur when the prices charged for long haul and local communication services are generally uniform over a large geographical area and over equipment of varying vintages. A "cream-skimming" specialized carrier[16] could offer a selected communications service, and/or offer it in a selected local area, in competition with a "national" carrier (which must serve all areas within its franchise) and undercut the national carrier's prices.

The specialized carrier could underprice the national carrier for two possible reasons: First, the cost of providing the service by the specialized carrier may actually be lower than that of the national carrier because of the ability of the small carrier to use the most modern equipment. Second, although the national carrier's equipment costs may actually be lower because of economies of scale, the specialized carrier may be able to undercut the national carrier's uniform price which, because of geographical cost averaging, may be well above his (the national carrier's) local costs.

The established carriers argue that "cream-skimming" is bad *per se*. From the carrier's point of view that may well be true. However, from a public interest standpoint, geographic cream-skimming is undesirable only if the national interest requires uniform prices for uniform communication services irrespective of the cost of providing the service in each locality. Clearly, if geographic cream-skimming is permitted, carriers would be pressured to bring their prices in local areas closer to their true costs, thereby defeating the uniform pricing policy. If the national interest would be served by pricing services according to true costs, then geographic cream-skimming would be one mechanism to accomplish this result.

---

16. We use the term "specialized carrier" here as a catch-all for the various types of organizations which could practice some form of cream-skimming. There are at least three important possibilities: First, a truly specialized common carrier could offer a particular communications service, probably in a somewhat restricted regional area, using equipment especially well-suited to the particular service being offered. This is the "Special Service Common Carrier" concept, discussed at length in Chapter IX. Second, a small "local" common carrier could offer a service or number of services in an attractive localized geographic area, in competition with larger carriers also serving that area — for example, a mobile radio service. And third, a private firm could provide communications service for its own use (or perhaps shared use with others), employing economical modern equipment for the task — for example, a private branch exchange (PBX) or a private microwave system.

Rapid Technological Change

We believe that, in general, the averaging of the costs of a new technology representing an incremental network improvement, with the costs of the previous technology employed in the common carrier network, is a sound policy.

It would be inappropriate for subscribers to be quarreling over where new technology, with its attendant cost reductions, should be introduced first. However, in instances of major technological developments it may be advisable to encourage "technological cream-skimming" as a means to pressure the carriers to introduce new technology. If, for example, COMSAT were competing with the carriers in transoceanic communications rather than acting as a "carrier's carrier" (leasing the satellite channels directly to the carriers) then, because of the lower cost to the public of the satellite channels, the carriers would be forced to introduce their own satellites rather than continuing to install expensive undersea cable. This not being the case today, the four major international carriers are planning to lay an additional submarine cable to Europe with a capacity of 720 voice circuits, even though such capacity could probably be more economically provided by communications satellite.[17]

The present "rate base" reward structure encourages the carriers to retain costly equipment such as submarine cables, in spite of the availability of more cost-effective equipment. The FCC has recognized this, and in an earlier investigation of the telephone industry made the following observations:

> "Experts testified that it was their opinion, deduced from records of the American Co. [AT&T] that developments capable of improving the service and decreasing its cost have been withheld for considerable periods; that the practice of limiting associated (telephone operating) company purchases of apparatus and equipment to those furnished by the Western Electric Co., has prevented the use of improvements in the art developed outside the Bell System for considerable periods after the improvements were available;

---

17. See, e.g., dissenting statement of Commissioner Nicholas Johnson in FCC approval of TAT-5, *Telecommunication Reports*, Vol. 34, No. 11 (March 4, 1968), pp. 20-21.

and that the American Co., has persisted in the installation of certain types of equipment which its studies indicated to be unduly expensive, and uneconomical to operate."[18]

In this report, the FCC then gave examples of investment decisions (or rather, decisions not to invest) made by the telephone company that were not in the public interest, although they were in the interest of the stockholders of the telephone company. The FCC further noted that:

"The Bell System organization, with the American Co., control the profits from both telephone manufacturing and telephone operation, *tends to limit the incentive to reduce the costs of telephone plant and telephone service* so long as revenues are close to the maximum fair return on the value of the property devoted to the public use allowed by the regulatory bodies and courts." (Emphasis added.)[19]

The authors, therefore, feel that limited "technological cream-skimming" should be allowed by the FCC to "prod" the carriers. Although it might adversely affect the revenues of the large carriers, it would encourage them to rapidly introduce the most cost-effective technology.

Economies of Scale in Long Haul Facilities

If a point-to-point communication service exhibits continual economies of scale and involves slowly changing technology then, in general, it should be provided solely on a common carrier basis. For example, Figure 5 illustrates the economies of long-haul microwave transmission.[20] However, in certain cases where there are economies of scale in the provision of long-haul services there may be offsetting reasons to permit either non-carriers or new carriers to provide the long-haul service. In its *Above 890* decision the FCC came to this conclusion

---

18. Federal Communications Commission, *Investigation of the Telephone Industry in the United States.* (Washington, D.C.: U.S. Government Printing Office, 1939), p. 585.

19. *Ibid.*

20. From FCC Docket No. 16258, Bell Exhibit 24 (Witness: A.M. Froggatt, Attachment C).

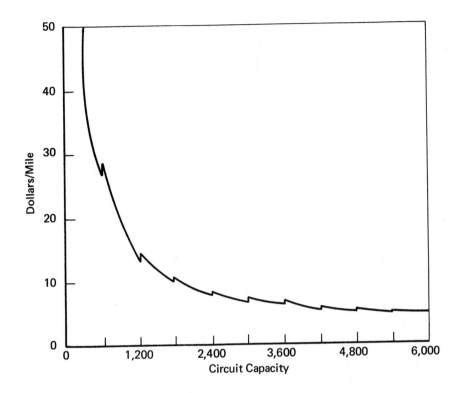

FIGURE 5   ILLUSTRATIVE COSTS FOR AT&T's
TD-2 MICROWAVE RELAY SYSTEMS
Book Costs Per Voice Circuit Route Mile

and allocated radio frequencies (above 890 MHz) for the use of private microwave communication systems.[21] Although the carriers could offer a specific service at a lower cost than a private firm because of the economies of scale and the shared use of facilities, the FCC felt that the demand of private industry for communication channels which it could design and operate to meet its specific requirements, and the opportunity to introduce competition into the nation's communication system (primarily in the supplying of communication equipment), outweighed the small social loss due to diseconomies of scale and nominal adverse affects upon carrier revenues.

---

21. 27 FCC 359 (1959), 29 FCC 825 (1960).

If it is advisable to allow private microwave systems, as the FCC has concluded that it is, there is no economic reason to disallow interconnecting them with the national carrier network. To disallow interconnection is to emasculate the ruling that allowed private microwave systems to come into existence (except in a few minor cases where the user does not wish to interconnect).

The Absence of Economies of Scale in Local Communication Facilities

Local communication facilities include in-house intercom systems, private PBX's, mobile radio systems, nurse and doctor call systems, etc. (We do not include local exchange telephone service in this discussion.) These services do not exhibit the economies of scale which characterize long haul communication facilities. Nor do the costs of providing local communication services vary substantially with the geography. Potential cream-skimming is not, therefore, of significant concern. Competition in the provision of local and private communication facilities encourages innovation in the services provided, and cost reduction and technological improvement in the equipment supplied. Therefore, in 1969 the carriers, under pressure from their customers and from the FCC to liberalize interconnection restrictions allowed, as a compromise response, interconnection of customer-provided communication systems to *voice-grade* bandwidths, effectively confining the interconnected facilities to local areas.

The Interconnection of Carriers Offering New Communication Services

Irrespective of the claims of the present communication common carriers (e.g., AT&T's advertising reference to itself as the "Anything, Anytime, Anywhere" Network), they cannot serve *all* the communication needs of all potential users. Submarkets (a limited number of users with specialized communication requirements) may be overlooked by the large carriers. In fact, evidence was presented in the previously mentioned Microwave Communications, Inc., case that the demand for low cost communications has not been adequately met.[22] By specializing in the provision of dedicated private channels employing advanced microwave technology, permitting customers to obtain the exact bandwidth required, and providing "short period" leased channels, special service common carriers such as MCI could provide a substantially lower cost service (in spite of the economies of scale of the present carriers) than is available today.

---

22. See, for example, the findings of the Spindletop market research study in *In Re Applications of Microwave Communications, Inc.,* FCC Docket Nos. 16509-16519, Initial Decision of Hearing Examiner Herbert Sharfman, Issued Oct. 17, 1967, pp. 18-22.

If the services of special carriers are to achieve their full potential usefulness to the public, it is necessary for them to be interconnected with the facilities of the telephone companies and perhaps also with those of Western Union. Construction of duplicate "local loop" customer access lines connecting the specialized carrier's long-haul trunk facilities with the customer's premises (using either land-line cable or short-haul radio relay) is undesirable from a public interest point of view. Furthermore, right-of-way constraints would make construction of land-line access facilities into an urban area impractical, and problems of both cost and frequency spectrum scarcity would hamper short-haul low-capacity radio relay links to the customer's place of business. Therefore, new long-haul microwave carriers should be encouraged to make use of existing local distribution facilities of the telephone operating companies, and appropriate measures should be taken by the regulatory agencies to insure that such facilities are not unreasonably denied them. Recognizing the importance of such interconnection, the FCC stated in its approval of the MCI service:

> ". . . Since they [the existing carriers] have indicated that they will not voluntarily provide loop service we shall retain jurisdiction of this proceeding in order to enable MCI to obtain from the Commission a prompt determination on the matter of interconnection. Thus, at such time as MCI has customers and the facts and details of the customers' requirements are known, MCI may come directly to the Commission with a request for an order of interconnection. We have already concluded that a grant of MCI's proposal is in the public interest. We likewise conclude that, absent a significant showing that interconnection is not technically feasible, the issuance of an order requiring the existing carriers to provide loop service is in the public interest."[23]

---

23. *In Re Applications of Microwave Communications, Inc., for construction permits to establish new facilities in the Domestic Public Point-to-Point Microwave Radio Service between Chicago, Illinois, and St. Louis, Missouri,* FCC Docket Nos. 16509-16519, Decision of the Commission, Issued August 13, 1969, p. 16.

As we have discussed earlier in this chapter, technical problems of interconnecting communication systems have been largely resolved; therefore, local loop interconnection, vital to the development of SSCC's, is likely to be forthcoming. However, the growth of specialized carriers may be slowed as the existing carriers force future SSCC interconnection applicants into hearing on a case-by-case basis.

C. CONCLUSION

We have seen that great changes have taken place in the regulatory outlook toward the interconnection of private communication systems with the common carrier network. Several years ago such interconnection was, for the oridinary firm, forbidden. Pressure for change from the data processing community, catalyzed by the FCC's progressive stance in its landmark Carterfone decision, led to significant relaxation of the long-standing interconnection prohibitions.

The implications for the user of the permission to interconnect private systems (and to use foreign attachments) go beyond the specific tariff relaxations discussed in the early parts of this and the prior chapter. At the time of this writing, dialogues between the carriers and other interested parties have begun on several fronts. These include FCC-sponsored discussions on restrictive aspects of the new tariff provisions, and conferences on equipment standards and related areas, between the carriers and industry associations such as the Business Equipment Manufacturers Association (BEMA) and the Electronic Industries Association (EIA). Increased communication and cooperation between the common carriers and industry in general is likely to lead to further relaxation of tariff restrictions and greater carrier responsiveness to the nation's communication requirements.

CHAPTER VII

LINE SHARING/RESALE

Line sharing, as the term implies, involves the shared use of a communications channel or group of channels by more than one user.[1] The obvious purpose of sharing is to provide each communications user with just as much capacity (measured in terms of channel time or bandwidth) as he needs, with no wasted excess, thereby squeezing the maximum communications capacity out of the same physical plant; thus per-unit costs fall. By reducing communications costs we also encourage the entry and expansion of entirely new services such as remote-access computer time-sharing or message-switching systems. Cheaper communications will permit computer time-sharing systems to extend their geographic market coverage,[2] and to overlap and compete with one another to a greater extent than is economically practical today.

Historically, a user of leased communications lines has been forced to buy more communications capacity than he needs in two respects. First, leased lines

---

1. All of the following discussion relates to the use of leased "dedicated" communications lines, although much of the logic could be applied to message toll ("dial-up") service as well.

2. The effect of communications costs upon the "geographical coverage" of a firm offering time-sharing services is illustrated by the comments of Mr. William Emmons of the Keydata Corporation in testimony before the House Small Business Committee.

> Mr. Potvin: [Committee Counsel] Does this charge for the transmission of the data from the establishment to your computer constitute a barrier or impediment to greater usage of your facilities by small firms, in your opinion?
>
> Mr. Emmons: Very definitely.
>
> Mr. Potvin: Could you elaborate on that point?
>
> Mr. Emmons: Well, it is the whole economic problem. That $2.34 a mile is fine for anyone within, generally speaking, a 50-mile radius of Boston. As we go further out, the cost of the communication line added into the cost of our service reaches a point where it is just not salable, if I can use that word — they cannot economically justify the cost of our service much farther out.

(*Hearings on the Activities of Regulatory and Enforcement Agencies Relating to Small Business,* Before the House Select Committee on Small Businesses, 89th Congress, Second Session, 1966.)

have been available only on a "full period" basis, that is, 24 hours per day, seven days per week, for a minimum period of one month; and spare capacity cannot be resold.[3] Many users have not been able to efficiently utilize this capacity; they either have requirements limited to one part of the day, or their fluctuating traffic usage pattern leaves the line idle for long periods of time between communications. Multi-user sharing of lines *on an alternate-use basis* was sought to reduce this wasted excess, and will be referred to as *"time sharing."*

Second, the channel bandwidths available today do not satisfactorily match the bandwidth requirements for all forms of data communications. Being unable to obtain a channel of the proper bandwidth, a user has been forced to take the next larger size and use only a portion of its full bandwidth.[4] The excess, paid for by the user and removed from the carrier's "pool" of available channels, has been wasted. A form of line sharing known as *"channelizing"* was sought to permit the larger channel to be split into subchannels of varying sizes as required, for independent use of several cooperating organizations, each of which would thereby reduce its communications costs. In addition to making use of otherwise wasted channel bandwidth, "channelizing" could in some cases enable the sharing users to take advantage of "quantity discounts" on the price of broadband lines.

Until February 1969, neither the "time sharing" nor the "channelizing" approaches to multi-user line sharing were permitted. One exception to this rule related to the sharing of broadband TELPAK lines by certain limited classes of users: railroads, pipe line companies, other regulated carriers or public utilities, and government agencies.[5] In addition, the tariffs contain "authorized user" clauses which permit a telephone company customer, by prior approval, to transmit information relating to his own business over a leased line connecting

---

3. AT&T Tariff FCC No. 260, para. 3.2 (except special channels for radio and TV broadcast use); also Western Union Tariff FCC No. 237, para. 2.1.

4. The information carrying capacity of a given communication channel depends primarily upon the bandwidth of the channel and the type of modem interface units at the ends of the line. Available channels include telegraph-grade lines operating at up to 180 bits per second (bps), voice-grade lines at 1200-9600 bps, and broadband lines at 50,000 and 250,000 bps. This will be discussed further in Chapter VIII on the adequacy of common carrier communications facilities.

5. AT&T Tariff FCC No. 260, para. 3.2.5 (B) (3) (a).

him with one or more "authorized users." In the event there are several such authorized users connected to the same lines with the customer, the line can be effectively shared among them for communications with the customer himself (but *not* among the various authorized users!). Examples of this in practice are the competing computer-based inquiry services for securities brokerage firms, provided on a nationwide basis by Bunker-Ramo Corporation, Scantlin Electronics, Inc., and Ultronic Systems, Inc. Approximately 150,000 miles of communications lines are leased from the common carriers to service more than 25,000 terminals on these systems, with the trunk lines from the central computers being shared on a time basis among the many subscribers to a particular inquiry service.[6] This type of line sharing provides the economical communications necessary to make such inquiry services possible, and may also be beneficial to the common carriers by increasing the demand for communications facilities.

Another exception to the long-standing line sharing prohibition, authorized by the FCC in 1937, has been the national communications network which Aeronautical Radio, Inc. (ARINC) leases from the carriers and uses to provide communications service to the airlines industry. As noted in Chapter IV on message switching, ARINC is a nonprofit corporation owned jointly by the airlines. ARINC's shared leased line network, which transmits weather, navigation, reservations, and equipment status information for the airlines, was opposed by AT&T, but sanctioned by the FCC, on the grounds that the information transmitted "...aids the lessee in carrying out purposes of its own distinct from those of a communication carrier for hire."[7] Here, the shared use of lines (among the various airlines) avoids duplicate networks and reduces costs, while at the same time facilitating the extensive inter-airline communications required by the nature of the airlines business.

In February 1969, AT&T revised its interstate private line tariff to permit shared use of both telegraph-grade and voice-grade private (leased) channels. Such channels were excluded where they were sold as Foreign Exchange (FX) service, were derived from TELPAK basebands, or were used as a part of a Common Control Switching Arrangement (CCSA) or a Switched Circuit Automatic

---

6. "Stock Tapes Go From Ticker to Quicker," *The New York Times,* Sec. 3, p. 1, February 11, 1968.

7. *In the Matter of Aeronautical Radio, Inc.* v. *American Telephone & Telegraph Company,* Docket No. 3349, Report of the Commission, 4 FCC 155, at 164 (1937).

Network; other private line services such as Series 8000 broadband channels were also excluded. The several subscribers to a shared service are termed "joint users," each of whom indicates, at the beginning of a monthly billing cycle, what fraction of the total channel charge he will pay. The total channel charge for a shared private line is the normal tariff charge plus 10% of the intercity channel rate. Each joint user is billed directly by the telephone company for his share of the total charge. Each joint user must have a terminal station (voice, data, telemetry, etc.), on the shared line and the communications to or from a joint user must relate directly to his business.[8]

---

8. The AT&T Private Line Tariff FCC No. 260, (revised February 1, 1969) states:

> 2.2.9 Joint Use — When the private line service is arranged for joint use, the service may be used for the transmission of communications to or from the joint user and relating directly to the joint user's business.

> * * *

> 3.1.5 Joint use arrangements are offered on those private line services furnished for 24 hours per day, seven days per week between points in the United States which utilize Series 1000, Series 2000, Series 3000 or Series 4000 channels and equipment [telegraph-grade and voice-grade channels] except

> (1) those services...used...in connection with foreign exchange service.

> (2) those services which are furnished in connection with a Common Control Switching Arrangement or a Switched Circuit Automatic Network.

> Joint use is not offered on those services which utilize...Series 5000, Series 6000, **Series 7000,** Series 8000, Series 9000, or Series 10,000 channels [wideband channels]....

> A joint user must have a station and a service terminal or local channel on the private line and the station must be located on the premises of the joint user....

> * * *

> Where a customer requires that a service be arranged for joint use, the charges for the service shall be determined as provided in this tariff plus 10% of the interexchange channel service charge.

> All charges for the service, including the charges for the joint use arrangement and for service terminals, channel terminals, local channels and station equipment furnished for the joint users as part of the private line service, will be computed as though the service were to be billed to the customer. The customer and each joint user will be billed for the components of the service which are furnished exclusively to each of them for his individual use. The charges for components of the service which are jointly used will be allocated for billing purposes in accordance with percentages of use specified by the customer. The specified percentages shall remain in effect for a minimum of one month and such percentages on file on the first day of the customer's billing cycle will be used in computing that month's billing. Without affecting the customer's ultimate responsibility for payment of all charges for the service, each joint user shall be responsible for the payment of the charges billed to him in accordance with this subparagraph.

In March 1969, AT&T proposed the addition of an experimental wideband service to its private line offerings, which could be shared by more than one firm.[9] (At the time of this writing, the proposal is still pending.) The experimental offering, providing for a three year market testing period, would be confined to termination points in the industrial Chicago-to-New York corridor (the following states are included: Illinois, Indiana, Michigan, New Jersey, New York, Ohio, and Pennsylvania).

The experimental service, known as Series 11,000, would offer two types of channels: First, 240 kHz channels, usable either as wideband data channels (approximately 250 kilobits per second) or as 60 individual voice-grade channels; and second, 48 kHz channels, also usable either as wideband data channels (50 kilobits per second) or as 12 individual voice channels.

These proposed new services are distinguished from the telephone company's TELPAK private line services because (1) they would be provided over discrete coaxial and microwave facilities (a TELPAK channel between two points may be furnished over *several* parallel circuit paths), (2) several users would be allowed to share the base capacity of these channels, and (3) a combination of wideband data and voice subchannels would be available.

The rate structure for the Series 11,000 channels would be similar to the costs and financial commitments which a firm would experience in arranging for private microwave facilities. The minimum service period would be 12 months.

The Series 11,000 service would allow the channelizing type of sharing, wherein several users obtain 24-hour access to their own dedicated subchannels. However, the time sharing of Series 11,000 channels or subchannels (i.e., the alternate use of these channels) would not be permitted.

In the remainder of this chapter we discuss, in detail, the motivations for both the "time-sharing" and "channelizing" forms of line sharing, and the operational circumstances in which each may be used. The chapter concludes with the forecast that both forms of line sharing will see increasingly wider usage in future years.

---

9. AT&T Private Line Tariff FCC No. 260 (proposed revisions submitted to the FCC, March 28, 1969), Sec. 3.2.11.

A. "TIME-SHARED" USE OF COMMUNICATION LINES

First, we must define our terms. As stated previously, time-shared use of communication lines refers to multi-user sharing of a leased "dedicated" line or group of lines on an alternate use basis. An example of this is a leased line used by firm A during the day (perhaps for voice) and by firm B at night (perhaps for data transmission). By the term "time-shared" we do *not* refer to the slicing of time into very small increments and using time-division multiplexing techniques (interleaving several low speed bit streams to form a single high speed bit stream); this form of line sharing results in the creation of several logical subchannels for "simultaneous" usage, and will be treated in the discussion of "channelizing" in a subsequent section.

1. Reduced-Cost Motivation

At first glance, it would appear obvious that it is economically desirable to keep a given communications line as busy as possible at all times. The line is a fixed physical part of the total communications plant, and if not kept busy by its lessee, or shared with other communications users, it will lie idle with consequent economic loss to the community at large. In other words, for a given volume of communications the less total plant required to handle them (by more efficient utilization of that plant), the lower the total communications cost.[10]

Until February 1969, leased private lines were available only on a "full-period" around-the-clock basis. As a consequence of the tariff revision at that time, permitting the joint use of low-speed and voice-grade leased private lines, an organization can now make an arrangement to alternate in the use of a leased voice or sub-voice grade line with one or more other organizations, thereby effectively obtaining "short-period" leased channels, i.e., channels used and/or paid for on the basis of actual or estimated hourly use during the day, week, or

---

10. Line utilization is not an isolated parameter which one wants to maximize in every case. In fact, on a message switching or remote processing network one generally does not want to maximize line utilization but rather one wants to obtain a given "response time" for the least possible cost. A given response time implies a specific line utilization and the actual goal of the user is to attain the desired line utilization. But when one seeks a line utilization of say 50%, it should be possible to "adapt" the available services so that this line utilization results, and not a lower utilization which would mean overcapacity and unnecessary costs nor a higher utilization which would mean undercapacity and a failure to meet the desired response time requirements.

month. At the time of this writing there has been only limited experience with such "time-sharing" of communication channels among different organizations but, for a number of reasons discussed below, we expect this form of sharing to become widespread in the next several years.

Many users have requirements unsuited to a "constant" communications capacity; they either have requirements limited to one part of the day, or their traffic pattern leaves the line idle for periods of time between communications. Typically, one's pattern of usage fluctuates widely over time. A user generally designs his leased line network to handle his peak traffic load and, therefore, will necessarily have excess capacity during the off-peak traffic periods. Several examples occur frequently in practice. The communication network of a stock brokerage firm may handle very heavy traffic during the hours that the stock market is open and have excess capacity afterwards. A nationwide firm's communication traffic loads may shift geographically during the day corresponding to the business hours in each time zone across the country. Also, a great many firms using leased lines for voice communications during the business day have considerable idle time on them during the night—time which some of these firms are able to use for data transmission, but which in many cases is wasted. By sharing, both large and small users can sell excess capacity and reduce communication costs.

Before the advent of computers and data communications, there was relatively little that could be done to increase line utilization. One would have difficulty scheduling use of idle time on a leased line between random voice communication users; and there was little demand for evening time. Data communications changes this. The high data transfer rates possible today make even a few minutes on an idle line valuable, and available computer technology makes it possible for data users to alternate, on either a scheduled or a contention basis, in the use of the line. Also, many data communications applications are ideally suited for evening utilization of lines which are used for voice during the day, and large companies use their leased lines in this fashion. For example, a retail branch outlet might transmit each evening the day's sales information, accumulated on punched paper tape, to central headquarters. However, not all organizations have sufficient internal communications demand to keep their lines busy by interleaving traffic and by alternate day/night use. By sharing communication lines small firms may be able to obtain the same economies as larger firms, which have been effectively sharing a corporate communication network among their several divisions.

The International Communications Association (ICA), or regional associations of communications managers, might serve as information clearinghouses for circuit sharing. Such associations could collect data on the communication paths of member firms, and attempt to identify firms with parallel routes and complementary traffic requirements. These firms might then make private arrangements to share channels.

2. Message Switching Services

As discussed in Chapter IV, there is considerable support for the proposal that unregulated private firms should be permitted to offer computer-based message switching services. The economic viability of many such services would require the use of shared communication channels: shared multi-drop or multiplexed lines connecting several different subscribers to a central switching computer, and/or shared trunk lines interconnecting switching computers in a large distributed network. Prior to February 1969, only the carriers could offer a message switching service which involved the "time-sharing" of a communications line among the subscribers to the service, giving the carriers a substantial competitive advantage over potential non-carrier message switching services. For example, Western Union's INFO-COM, a message switching service, employs multi-drop communication channels for certain segments of its service.[11] Here, several organizations are "time-sharing" the multi-drop line connecting their respective terminals to the switching computer. RCA Global Communications, Inc., also provides a computer-based message switching service, AIRCON; however, because of RCA Globcom's status as a non-carrier within the domestic United States (it is an international carrier), the ban against sharing has (prior to February 1969) forced each of the AIRCON subscribers to provide his own non-shared communication link to the switch. Consequently, Western Union's INFO-COM service was substantially less expensive to the user. Now, since the carrier tariffs permit several organizations to "time-share" a communication channel, the disparity between the charges associated with a carrier-provided and a non-carrier-provided message switching service should be reduced.

---

11. Western Union's INFO-COM service has been designed to provide three classes of service, to high, medium, and low volume users respectively. Class 2 service, for medium volume users, employs polled, multi-drop communication channels which are effectively "time-shared" among the Class 2 subscribers.

There does remain one distinction, however, between carrier and non-carrier provided message switching services — AT&T's private line tariff requires that all "joint users" have terminal stations connected to the shared communication channel. If a non-carrier service organization wanted to provide a distributed message switching system, the trunk lines interconnecting the switching nodes could not be classified as jointly-used lines because the joint users would not have terminals directly attached. The service organization could not lease the trunk channels and permit the subscribers to the message switching service to use these trunk lines. This would constitute *resale,* which involves (1) using a private line for a purpose for which compensation is received, and (2) using it in the collection, delivery, and transmission of communications for others; both of which are explicitly prohibited in the private line tariffs.[12]

3. The ARPA Computer Network: An Example

An example of a service which may be precluded or seriously inhibited by the prohibition against non-carrier provision of a distributed message switching network is a network for linking together geographically dispersed time-sharing computer centers. (Note that the term "time-sharing" is used here to apply to a central computer.) A network for this purpose is being designed by the Advanced Research Projects Agency of the Department of Defense. ARPA is the primary driving force behind much of the advanced computer research and development in the United States, and is sponsoring computer time-sharing projects at universities such as M.I.T. (Project MAC), Stanford, Carnegie-Mellon, the University of California at Berkeley, and at several firms in the computer industry such as the System Development Corporation and Bolt Beranek and Newman. Partly as a result of ARPA's support, a rapidly growing industry of time-sharing computer centers, which make the power of the modern computer widely and conveniently available, has emerged.

ARPA is developing a communication network to interconnect the heterogeneous time-sharing computer facilities of each of its contractors (see Table 1) as

---

12. The AT&T private line tariff, FCC No. 260, provides that:

> Private line service shall not be used for any purpose for which a payment or other compensation shall be received by either the customer or any authorized or joint user, or in the collection, transmission, or delivery of any communications for others...This provision does not prohibit an arrangement between the customer and the authorized or joint user to share the cost of the private line service.

TABLE 1
TIME-SHARING COMPUTER CENTERS TO BE INTERCONNECTED BY THE ARPA NETWORK

Organization	Site Location	Computers
Carnegie-Mellon University	Pittsburgh, Pennsylvania	UNIVAC 1108, IBM 360/67, G-21
Dartmouth College	Hanover, New Hampshire	GE 635
Harvard University	Cambridge, Massachusetts	SDS 940, IBM 360/50, DEC PDP-1
Massachusetts Institute of Technology	Cambridge, Massachusetts	IBM 7094, DEC PDP-6/10, GE 645
Stanford University	Stanford, California	DEC PDP-6/10
University of California at Berkeley	Berkeley, California	SDS-940, SCC 6700
University of California at Los Angeles	Los Angeles, California	SDS Sigma-7
University of California at Santa Barbara	Santa Barbara, California	IBM 360/50
University of Illinois	Urbana, Illinois	Burroughs B-6500/ ILLIAC IV
University of Michigan	Ann Arbor, Michigan	IBM 360/67
University of Utah	Salt Lake City, Utah	UNIVAC 1108
Washington University	St. Louis, Missouri	Special Equipment
Advanced Research Projects Agency	Washington, D. C.	DEC 338
Bell Telephone Laboratories	Murray Hill, New Jersey	GE 645
Bolt Beranek & Newman, Inc.	Van Nuys, California	SDS 940, DEC PDP-10
M.I.T. Lincoln Laboratory	Cambridge, Massachusetts	TX-2, IBM 360/67
RAND Corporation	Santa Monica, California	DEC PDP-6, IBM 1800
Stanford Research Institute	Palo Alto, California	SDS 940 (2)
System Development Corporation	Santa Monica, California	IBM 360/50-65

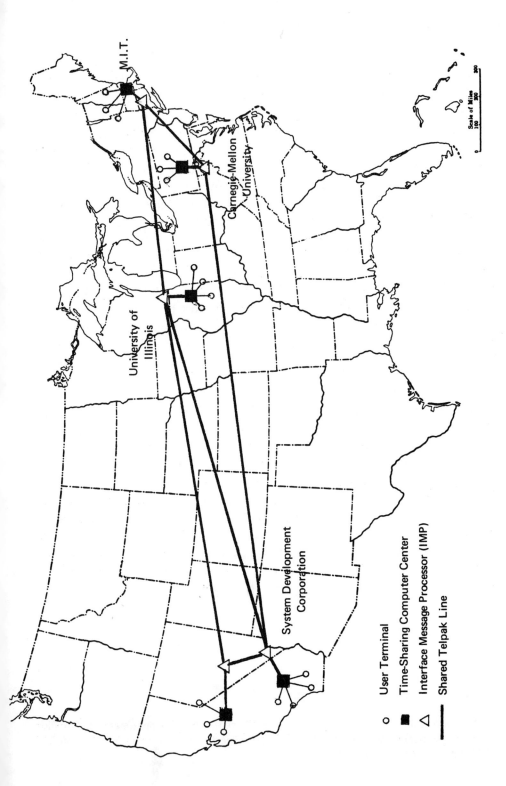

FIGURE 6   THE ARPA NETWORK (GREATLY SIMPLIFIED)

illustrated in a simplified fashion in Figure 6. Each time-sharing computer (called a Host) will be connected to a nearby Interface Message Processor (IMP), which is a dedicated message switching computer. The IMPs, in turn, will be interconnected via a network a full-duplex (simultaneous two-way transmission) broadband lines, and are used to provide high-speed store-and-forward switching for messages exchanged between the Host computers. The inter-IMP network will be dynamically configured (lines added or removed on various routes between IMPs) through the use of ESS circuit-switching exchanges (not shown on diagram) to meet the shifting load requirements on the network.

Based upon TELPAK D tariff rates, the implementation and use of this network, while adding substantially to the utility of each of the participating time-sharing computer centers, adds only approximately 15% to the operating cost of each.[13] For the estimated traffic load, the cost to send a 6000 character message over the ARPA network will be approximately one cent. In contrast, a user can only send 3-4 characters over the Telex network for one cent, for typical transmission distances. Additionally, the average message transit delay in the ARPA network will be less than one second.

Bolt Beranek and Newman, Inc. (BBN), one of the ARPA contractors who will ultimately be a participating node in the network, has received the ARPA contract to design the full nineteen-node network and to install an initial four-node test network. This includes design and construction of the IMPs (Honeywell DDP-516 computers with specialized communications interface hardware), development and testing of the IMP software, and installation of four IMPs at Stanford Research Institute, the University of California at Los Angeles, the University of California at Santa Barbara, and the University of Utah. At the time of this writing, BBN has developed the IMPs and is installing them at the initial host sites; the four-node test network will be operational in early 1970, for further testing and development before expansion to the remaining fifteen host installations.

---

13. Annual machine rental cost for the time-sharing centers totals approximately $12 million; total communication line and IMP costs are expected to be slightly greater than $1.5 million per year.

The reasons for establishing a network of time-sharing computers are several: First, specialized application programs and translators (compilers, assemblers, etc.), located at the computer center with the hardware most suitable for the particular program, will be accessible to all subscribers. Second, the network will permit interpersonal communication among the users of the various computer centers linked to the network. Third, a data base can be located at a single, rather than multiple locations, and can be shared by subscribers to all computer centers. Fourth, the several computers may be used for "load sharing" as is done among the electric power utilities. When the computational burden on any one computer center became too heavy, tasks can be transmitted to other computer centers under-utilized at that moment.

These capabilities will permit geographically separated team members of a research project to work together on a common problem, rather than working separately and duplicating each other's work. Researchers will be able to develop "communities of intellectual interest" rather than working according to geographically determined groupings.

> "Once it is practical to utilize programs at remote locations, programmers will consider investigating what exists elsewhere. The savings possible from non-duplication of effort are enormous. A network could foster the 'community' use of computers. Cooperative programming would be stimulated, and in particular fields or disciplines it would be possible to achieve a 'critical mass' of talent by allowing geographically separated people to work effectively in interaction with a system."[14]

Although the technical design and cost areas seem to be well in hand, ARPA's plans for this network to interlink time-sharing computer centers may run afoul of the existing common carrier tariff restrictions regarding "third party" use of communication lines. Initially, the communication lines will be leased from the carriers by the Department of Defense, for the joint use of the participating computer centers; these centers being DOD contractors, and the network being operated by DOD to serve them, "third party" use of lines does not occur—the network is considered to be an "in-house" private activity. After the develop-

---

14. Dr. Lawrence G. Roberts, ARPA, "Multiple Computer Networks and Intercomputer Communication," (Unpublished Paper, 1968), p. 2.

mental phase, however, ARPA plans to turn the operation of the network over to a contractor who would operate the IMP computers and lease the lines from the carriers; the participating centers would, in turn, "buy" the service from the contractor. (The contractor could presumably be Bolt Beranek and Newman, Inc., who designed the IMPs and the prototype network, or it could be some other firm in the computer or communications industry.) This, according to the carriers, would amount to the provision of a communication service for hire by a non-carrier; this is illegal under their interpretation of the Communication Act, and certainly in violation of their tariff prohibitions regarding resale of communication lines. This difficulty would not result, of course, if the organization operating the network were a common carrier.

It is not clear that operation and systems management of the ARPA network by a common carrier would optimize other factors, however. Competition in contractor selection would be effectively eliminated, with probable adverse effects upon both cost and system performance. Although the carriers certainly have or could acquire the expertise to operate and continue the evolutionary development of this network, the problems which must be solved are fundamentally those of computer systems design and operation rather than problems of the traditional communications engineering nature. Accordingly, it is plausible to assume that certain firms in the computing industry may be better qualified and/or motivated than the common carriers to perform this task well. Furthermore, such organizations, being more familiar with and responsive to the needs of the computing industry, might be more likely to carry this interactive computer network concept into other areas of business and scientific application, thus fulfilling a major objective of ARPA research project support: to develop and encourage the application of advanced technological concepts, in industry as well as in the military. Therefore, ARPA favors the removal of the prohibitions against communications line resale, to reduce communications cost and increase flexibility in the supply of end-product communication services such as the ARPA network.

4. Time-Shared Use of Lines: Conclusions

The above arguments suggest that time-sharing of communication channels is a useful capability and should be extended to include broadband lines. This type of flexible use of the nation's communications plant is likely to result in reduced overall communications cost. Neither the telephone companies, nor Western Union, have adequately met this need, possibly because "sharing" would reduce their revenues. For example, Western Union attempted to persuade the FCC to suspend AT&T's tariff provisions permitting the sharing of channels of voice-grade or narrower bandwidths. Western Union said:

> "To the extent that existing Western Union non-government private wire customers could establish joint use of intercity circuits under a matching Western Union tariff provision—filed solely to meet the competition of an effective AT&T joint user provision—*in lieu of multiple individual circuits,* losses of approximately $15,000,000 of private wire service circuit revenues can result." (Emphasis added.)[15]

Western Union losses can, in this instance, be translated into direct savings to Western Union customers who are able to jointly use a single communication line where previously "multiple individual circuits" were necessary. Since Western Union's private wire service revenues for non-government customers are approximately $35,000,000 per year, the telegraph company is, in effect, admitting that there is substantial opportunity for its customers to reduce the number of physical communication channels required to provide a given communications capacity. Western Union private line customers could thus reduce their costs by approximately 43% in the aggregate ($15 million savings out of $35 million previous costs), less whatever costs were incurred for sharing equipment.

B. CHANNELIZING

Channelizing, as the term implies, is the process of creating a number of independent communications channels from some larger block of bandwidth capacity, often referred to as the "baseband." Thus, for example, a 1000 kHz baseband may be subdivided into 240 voice-grade channels of approximately 4 kHz each; a 48 kHz baseband may be subdivided into 12 such voice-grade channels; or a single voice-grade channel may be subdivided into perhaps 20 to 100 telegraph-grade channels. Channelizing may be thought of conceptually as the inverse of multiplexing, for in the former we subdivide a larger baseband into smaller channels or "subchannels," whereas in the latter we transmit several independent subchannels over a single larger channel; from whichever direction the process is approached, however, the result is the same. Therefore, the terms are often used interchangeably, which is only slightly inaccurate.

---

15. Western Union Petition to the FCC re Revisions of AT&T Tariff FCC No. 260 (Private Line Services), January 20, 1969, p. 24.

In order to transmit two or more different signals over a single communications circuit, they must be separated so that they do not interfere with each other. This can be done by separating them either in *frequency* or in *time*. The former procedure is known as frequency-division multiplexing (FDM), and is the older technique, whereas the latter is known as time-division multiplexing (TDM). These techniques were described in the discussion of Western Union's SICOM service in Chapter IV, and will be discussed further in Chapter VIII on the adequacy of common carrier communications services.

With respect to TDM, we must distinguish this from the concept of "time-sharing" of a communication line, as discussed in the preceding section. If a single communication channel is used by several users, each of whom transmits for a period of time and then relinquishes the channel to the next user, this is "time-sharing." If, however, digital information is being transmitted by the users, and the repetitive time slice assigned to each user (by an electronic "commutator" device, or TDM unit) is so small as to approach the duration of an individual digital pulse, or bit, then the line is being "channelized" rather than "time-shared." That is, when the individual bit or character streams from several lower-speed lines are interleaved to form a single higher-speed bit or character stream which is then transmitted over a high-speed line, this high-speed line is effectively subdivided into several lower-speed subchannels. Time-sharing of a line *by bits or characters* (known as time-division multiplexing) is therefore identical, in the end result, to the frequency-division multiplexing done by the telephone companies, and the two are discussed together in this section.

On the other hand, time-sharing of a communication line which occurs by means of alternating the line among the *messages* of several users (as would be the case in the trunk lines of a distributed message switching system) should *not* be considered as constituting channelizing, and this discussion of channelizing does not apply.

The discussion here will focus upon the advantages to the communications user of engaging in the channelizing form of line sharing, and the pro's and con's from a public policy viewpoint of permitting such activity. Since, in general, a private line lessee can channelize his lines for his own use without restriction,[16]

---

16. TELPAK channels, the baseband of which must be channelized using telephone company equipment (customer channelizing in TELPAK is limited to voice-grade subchannels and below), constitute an exception to this rule.

provided that the subchannels thus created are not made available to other users, we will not treat this case. For the purpose of this discussion, therefore, channelizing may be thought of as "capacity sharing" — parallel to the concept of "time-sharing," in that time and capacity (amount of information transmitted per second) are the two dimensions by which a channel can be shared.

Capacity sharing occurs when several different organizations share a leased private line baseband by channelizing it into several dedicated subchannels. For the purpose of this discussion, it is generally immaterial whether such channelizing is done by a group of users acting in concert, or by an entrepreneur who preforms the service and sells the subchannels to the users. Prior to February 1969, capacity sharing was prohibited by the common carrier tariffs. At that time, AT&T modified its interstate private line tariff (AT&T Tariff FCC No. 260), to permit capacity sharing by means of customer-provided channelizing equipment, as well as the time-sharing discussed in the previous section, on private voice-grade leased lines. Also, capacity sharing of certain wideband channels was proposed on an experimental basis in a limited geographical area. However, *resale* of subchannels by a channelizing entrepreneur, as suggested above, continued to be specifically forbidden — for, in the view of the carriers, line *sharing* and line *resale* do constitute entirely different activities.

From a public policy viewpoint, there are three primary reasons why capacity sharing/resale is desirable; why AT&T's tariff relaxations in February 1969 are important to the public; and why these relaxations should perhaps be extended to include broadband channels as well as voice-grade and smaller channels. First, the ability to channelize and dispose of excess line capacity by making it available to other users gives the individual private line lessee additional flexibility in his use of common carrier facilities. Second, elimination of the carriers' monopoly in channelizing (and consequent virtual monopoly in equipment, since AT&T's 83% share of the national telephone system uses Western Electric Company equipment almost exclusively) will greatly increase the forces of competition in the supply of channelizing equipment, which should have several benefits for the user. Third, user channelizing and sharing/resale of communications line capacity will tend to reduce the carriers' ability to practice price discrimination, and it will tend to cause channels derived and sold by the common carriers to be priced according to their true cost. Additionally, from the user's viewpoint, the ability to channelize using the most cost-effective equipment available commercially means substantial potential savings in communications costs, based upon today's common carrier tariff rates for the various sizes of leased private channels. We will discuss these factors below.

1. Bandwidth Optimization

As described in the following chapter dealing with the adequacy of presently available communication services, there is a need for communication channels of bandwidths currently unavailable from the carriers. Some of these demands are of a sufficiently widespread nature that the FCC should recognize them and require the carriers to provide such service. Other demands are for channel bandwidths to accommodate *specialized* data communications and remote processing applications and will not be sufficiently widespread to warrant general common carrier offerings. Most of the presently available computer input/output devices operate at transmission speeds in excess of the maximum capacity of a voice-grade telephone channel, as shown in Table 3, Chapter VIII, pages 157-159. Many of these devices operate at speeds intermediate between those of a voice-grade line and the next larger size available from the carriers, a 50,000 bits per second (50 kilobit) broadband channel, whereas others operate at still higher speeds. (A user may also obtain a 19.2 kilobit channel by leasing a 50 kilobit line with a 19.2 kilobit subchannel.)

Today, the user who needs a non-standard bandwidth greater than voice-grade must buy the next larger capacity channel, which may be an order of magnitude in excess of his requirements.[17] Since he cannot "resell" the excess channel capacity to others, his cost of communicating is greatly increased. Permission for capacity sharing of broadband channels would allow him to sell this unused channel capacity, and as with the time-sharing of communications lines described above, would be in the overall social good, for a greater volume of communications would now be possible with the same physical plant.[18]

---

17. A mismatch between a user's bandwidth requirements and the bandwidths of available channels occurs frequently because there are large gaps in the spectrum of carrier offerings. Low speed channels have a nominal bandwidth of 300 Hz, voice-grade channels jump to 3000 Hz, and broadband channels are available at 48,000 Hz, 240,000 Hz, and 1,000,000 Hz.

18. This statement is not completely correct, for additional hardware at the ends of the line would be required to accomplish the channelizing; but for lines of moderate length (greater than approximately 100-300 miles) the channel cost savings would far outweigh the cost of such additional equipment.

2. Effects Upon Equipment Supply and Usage

The second motivation for permitting capacity sharing is to encourage competition and innovation in the supply of channelizing equipment, and thereby to pressure the carriers to use the most advanced and efficient channelizing equipment available.

Presently the major telephone carriers "purchase" channel derivation equipment from their manufacturing subsidiaries, Western Electric for the Bell System, and Automatic Electric and Lenkurt for the General Telephone System. The purchasing is often not done on an "arms length" basis, and outside manufacturers are either foreclosed from substantial markets or sell equipment on an OEM basis in a monopsony (single buyer) market. When users share communications capacity through the use of their own channel derivation equipment, the carriers' hold upon the channelizing equipment market is weakened. The FCC has encouraged similar market development in its "*Above 890*" decision, in which one of the stated reasons for permitting private microwave systems was the fact that they would open up new markets for microwave equipment manufacturers, thereby encouraging competition and innovation:

> "With the opening of a new market for microwave equipment, it seems clear that the resultant competitive situation among manufacturers *will provide the incentive for developing better equipment for meeting the needs of private users and a concomitant improvement in the communications art.* Our experience in the mobile communications field amply demonstrates this point." (Emphasis added.)[19]

Secondly, permitting communications users to channelize lines and share the subchannels with other users, or resell the subchannels (the effect is the same), will produce considerable pressure upon the carriers to use the most efficient and economical channelizing equipment available. The user who shares or resells communications channel capacity becomes, in effect, a competitor to the carrier who performs the same function and makes a profit on it; thus, the carrier would prefer to do *all* channelizing and sell channels only on a "retail" basis. If the user

---

19. 29 FCC 825, at 854. (1960).

(or some sort of "communications broker" who might offer this service) is permitted to engage in channelizing for sharing or resale, the carrier must employ equally efficient channelizing equipment in order to compete.[20]

Also, there would be an incentive for the carrier to take into account the rate of technological change when designing and building his equipment, and when depreciating it. Lower cost equipment can be built by designing for more reasonable lifetimes, which should be no longer than the expected time until the equipment becomes technologically obsolete. Under the present regulatory incentive system, it is in the carrier's interest to build expensive equipment which requires little maintenance, because the equipment cost enters into the carrier's rate base and therefore adds to his profit, whereas the maintenance cost, largely labor, is an operating expense on which the carrier earns no profit. Although the costs of building extremely reliable equipment with a long lifetime may be substantially higher than the costs of building less reliable equipment which requires more frequent preventive maintenance to achieve the same level of service, the carrier will tend to build long-lived reliable equipment, because of his rate base reward structure. Therefore, it is desirable to provide the carriers with competition in the supply of channelizing equipment, for this will pressure them to design equipment with the lowest annualized cost rather than with the largest book value.[21]

---

20. The carriers argue that this form of price competition between the carrier and non-carriers, caused by permission of line sharing/resale, would be a form of "cream-skimming" — harmful to the carrier and to those of its customers unable to share. This is, however, untrue. There are two components of cost for a channel derived from some larger "baseband" channel. First, there is the cost per mile of the baseband; in addition, there is the cost, at both ends of the line, for the channelizing equipment which derives the desired subchannels from the baseband. The cost of the baseband provided by a common carrier between any given locations is fixed, and must be incurred by all who would provide channelizing service and offer subchannels for sale (including the carrier itself). If channelizing and resale by non-carriers were permitted there would *only* be competition on the cost of the channelizing equipment.

21. Annualized cost is the sum of the annual depreciation expense for the equipment plus the operating expense associated with its use during the year.

As an example of the importance of technological change in the design of equipment, consider the channelizing equipment employed by the common carriers. With the exception of limited recent use of T-carrier systems, it is almost entirely the old, frequency-multiplexing type, which before the advent of digital technology was the only channelizing method available. Today, however, time-division multiplexing offers an increasingly attractive alternative — providing higher-quality channels for data and digitally-encoded voice signals, at lower cost.[22] A major reason for common carrier reluctance to introduce use of this equipment appears to be the large investment in frequency multiplexing equipment, which was (and is) designed and built with physical lifetimes and depreciation rates considerably longer than the "technological" lifetime of the equipment.

If channelizing of broadband lines and resale of subchannels by non-common carriers were permitted, and the telephone companies could not compete because of their large investment in frequency multiplexing technology, then either of two alternatives are possible. The telephone companies could abandon the data communications part of the channelizing market, which is highly unlikely (since it is the most rapidly growing area of communications) as well as undesirable. Alternatively, they would be forced to offer their channelizing services by means of the same new technology, which would require scrapping some of their older channelizing equipment. This in turn would result in the write-off of the remaining book value of the equipment less whatever salvage value remains. In the short run there would be a loss which could, with the FCC's approval, be passed on to the public rather than to the carriers' stockholders. After that point the carriers and their potential competitors would be on an equal footing with respect to their ability to make economical use of the newest technology. In the long run the public would benefit, in that competitive channelizing would be an effective incentive to prevent the carriers from over-investing in channelizing equipment (to build up the rate base).

Over time, we would find that the entrepreneurs who were initially able to offer the channelizing service at rates below those of the carriers would perhaps be driven from the market as the same low-cost techniques were installed by the

---

22. This subject will be discussed at greater length in Chapter VIII, Section D, "A Digital Communications Network."

carriers. To the extent that there are economies of scale in this operation, the carriers should be able in the long run to perform time-division channelizing of data channels at the lowest cost.[23]

3. Discriminatory Pricing

The third motivation for permitting capacity sharing is to ensure that channels derived and sold by the common carriers will be priced according to their true cost. Permitting several organizations to share a broadband communications line by channelizing it into dedicated subchannels would help to prevent the carriers from employing a discriminatory pricing structure for broadband and derived channels. If the prices charged for the derived channels are not in line with the prices charged for the basebands, then market arbitrage will force the carriers to adjust the price ratios so that the charges for the derived channels reflect the actual cost of deriving these subchannels.

Price discrimination is defined in economic theory as the selling by a producer of a single product or service to different buyers at two or more different prices, for reasons not associated with differences in cost.[24] Successful price discrimination requires that the seller have control over the supply of the product to a particular buyer (which makes monopoly power in some form a necessary condition) and that the seller can prevent the resale of the commodity from one buyer to another. Price discrimination allows the monopolist to increase his profit above the level which could be attained by charging a single monopoly price, and it permits him to charge lower prices in a particular market segment to defeat actual or potential competition in that market segment. As a consequence,

---

23. The provision of time-multiplexed channelizing by private firms would be one step that would encourage the carriers to develop and offer a network based on a digital philosophy with its attendant improved performance and lower costs. (Paul Baran, "The Future Computer Utility," *The Public Interest,* Summer 1967, p. 83.)

24. In the common carrier communications industry, this is known as "value of service pricing," and is a deeply-entrenched marketing principle for AT&T and other carriers; however, it has recently come under close scrutiny and attack in several FCC proceedings, such as the AT&T Rate Investigation (Docket No. 16258) and the TELPAK Sharing case (Docket No. 17457). See footnote 11, p. 55, for further explanation of this concept.

> "The results of the application of monopolistic price discrimination are an artificial stimulation of consumer demand in the low-priced low profit submarket and an artificial restriction of consumer demand in the high-priced, high profit submarket. It tends to misallocate the flow of resources away from the most profitable alternative and toward the least profitable alternative, and frequently encourages the supply of the favored submarkets at prices below their economic costs."[25]

The artificially low rates to favored users may prevent the entry of more efficient competitors and may result in the monopolist's providing services which could be more cheaply provided by someone else. Dr. William H. Melody, FCC economist, notes with respect to broadband communication lines: "Firms may be purchasing private line services from Bell at TELPAK rates which are below the cost of providing this service by their own private microwave systems. TELPAK was instituted by AT&T to counter competition from private microwave. But the cost to Bell of supplying these private lines may be greater than the cost of supply by private microwave."[26] Thus socially undesirable mis-allocation of resources results, as explained in the Averch-Johnson thesis (discussed in Chapter III, and repeated here for the convenience of the reader):

> "The purpose here is (a) to develop a theory of the monopoly firm seeking to maximize profit but subject to such constraint on its rate of return, and (b) to apply the model to one particular regulated industry — the domestic telephone and telegraph industry. We conclude in the theoretical analysis that a "regulatory bias" operates in the following manner: ... The firm has an incentive to *expand into other regulated markets, even if it operates at a (long-run) loss in these markets;* therefore, it may drive out other firms, or discourage their entry into these markets, even though the competing firms may be lower-cost producers." (Emphasis added.)[27]

---

25. Testimony of Dr. William H. Melody (Economic Studies Division, FCC Common Carrier Bureau), FCC Docket No. 17457 (TELPAK Sharing; Common Carrier Bureau Exhibit 1), April 15, 1968, p. 7.

26. *Ibid.,* p. 13.

27. Averch, Harvey, and Leland Johnson, "Behavior of the Firm Under Regulatory Constraint," *American Economic Review,* Vol. 52, No. 5, December 1962, p. 1052.

With reference to communications markets, and leased private lines in particular, the ability of the carrier to indulge in this obviously undesirable cross-subsidization activity depends upon its ability to prevent arbitrage — or "capacity sharing," when the larger-capacity communication lines are underpriced relative to the smaller-capacity lines. Dr. Melody concludes,

> "In essence, a requirement of unlimited TELPAK sharing allows the arbitrage function to be performed by permitting all users access to the low priced submarket. In order to prevent the monopolistic price discrimination made possible by the existing TELPAK sharing provisions [i.e., sharing permitted only by regulated right-of-way companies and a few others] from reoccuring in a new form in the future, *the Commission should encourage the performance of the arbitrage function in all communications markets. This could be effectively performed by permitting communications users to act as wholesalers of service and resell some of their purchases.* Their entry into the market would establish competition in the establishment of efficient rate structures. As long as monopolistic price discrimination did not exist, there would be no arbitrage. *When discriminatory rates were established, the arbitrage would quickly force its elimination.*" (Emphasis added.)[28]

From the user's viewpoint, low-speed telegraph-grade lines leased by the common carriers are over-priced relative to voice-grade lines (approximately 1/10 to 1/20 the capacity, for up to 3/4 the price) just as voice-grade lines are overpriced relative to broadband TELPAK lines. Arbitrage would function similarly in both instances: arbitragers would buy the relatively underpriced larger lines, channelize them, and sell the subchannels at prices below the rates currently charged by the carriers. The general public would benefit in both cases.

The carriers, themselves, admit that the public would benefit by sharing private line facilities. As noted earlier, Western Union attempted to persuade the FCC to suspend AT&T's tariff modifications permitting the sharing of channels of voice-grade or narrower bandwidth. Western Union, in effect, admitted that there is substantial opportunity for its customers to reduce the number of physical

---

28. Melody, *op. cit.,* p. 23.

communication channels required to provide a given communications capacity. Western Union private line customers could thus reduce their costs by more than 40% in the aggregate less whatever costs were incurred for sharing equipment (e.g., multiplexers).[29]

4. Carrier Opposition to Customer Channelizing

One of the reasons for the telephone company's traditional opposition to channelizing in general may be found in AT&T's internal *Broadband Report* of September 1960:

> "Permitting the customer to channelize one of our basebands means that he must physically connect fairly complex bays of carrier equipment to our facilities, even including the radio and coaxial paths used for both network television and the transcontinental message toll system. If we can tolerate such risks to the reliability of our backbone plant, it will be extremely difficult to explain why we cannot take the risk of interconnecting our PBX trunks to a PAX, our teletype circuits to other teletype systems, or a customer's voice circuit (including those furnished by private microwave) to our general switching network."[30]

This statement is in regard to the channelizing of TELPAK lines, but the same sort of thinking has also applied to channelizing of any telephone company lines by its customers. The telephone company has long feared anything which would set a precedent encouraging interconnection; however, presumably the AT&T tariff changes introduced in January 1969, to permit interconnection of customer-provided systems to the dial telephone network and to voice-grade and low-speed leased lines, will eliminate part of the basis for such a fear. Although a lessee can today channelize a line for his own use, there is substantially less motivation for him to do so as compared with the motivation of a number of users who might share a baseband by channelizing it.

---

29. See page 135.

30. AT&T, *Broadband Rate Planning Group Report,* September 1960 (company confidential; made public record as MCI Ex. 30, FCC Docket Nos. 16509-16519, 1967), hereinafter referred to as *Broadband Report,* Section 5, pp. 5-6.

There are various technical arguments posed by the carriers against customer channelizing in general. For example, in the *Broadband Report,* AT&T states:

> "Our entire long distance network and a growing fraction of our exchange network is based on derived circuits. These circuits are obtained from carrier and radio systems in which many channels are amplified or carried by a single facility. Use of these derived channels is basic to our whole telephone economy. Like all good things derived circuits have a bad side. The bad side is that the common equipment, which is what makes these systems economical, inevitably gets the different channels mixed up to some small extent. The mixing process is called 'generation of intermodulation products.' Briefly it means that every tone or frequency mixes with every other one to produce new and spurious tones that did not exist before.

* * *

> "... This is why it is more accurate to think of our communications pipes as so many 'troughs' rather than isolated pipes.

* * *

> "... It [channelizing] can be effective provided the pipe is channelized in such a way as to distribute the signal energy through the spectrum in a fashion that will throw the intermodulation products into frequencies which do not interfere with other users of the same broadband facilities. This is why we strongly believe that whenever we provide a pipe, we can give better overall service to the customer and incur less overall maintenance difficulties if we also derive whatever channels are derived from the pipe."[31]

AT&T goes on to say that is difficult to tell users how to derive channels so as not to interfere with other users because (1) if deriving voice circuits the user will end up doing it the same way as the telephone company at twice the cost, and (2) if data circuits are derived, the required procedures might be contrary to the customer's interests; e.g., the customer may have to use more bandwidth than simple transmission theory would indicate. Use of less bandwidth per derived channel would not necessarily degrade the user's own service, but it would impair the service of others.

---

31. *Ibid.,* Section 5, p. 3.

This argument is purely academic. Since the common carriers permit users to channelize and, since 1969, to share leased voice-grade lines, and since user channelizing of broadband private microwave channels is commonplace, there should exist no technical difficulties in permitting broadband capacity sharing.

Also, the argument that a user who wishes to derive his own voice channels, or any other type of channel, may not be able to do so as cheaply as the telephone company is no reason for denying him the opportunity if he so desires. If in fact the telephone company can do it more efficiently, and *if* these economies are passed on to the user, then presumably, telephone company channel derivation will always be used.

C. THE ECONOMIC IMPACT OF LINE SHARING

The carriers argue that permitting unrestricted line sharing (both time sharing and capacity sharing) would distort the entire communications pricing philosophy. If users with complementary requirements were permitted to share lines, fewer lines would be leased, and carrier revenues on leased lines would fall — although the cost of the carrier's physical plant would remain constant. Therefore, other communications customers would be forced to pay higher prices to make up the lost revenue, in effect subsidizing those customers whose requirements permitted them to share lines. Also, if line sharing were permitted, the carriers argue, certain users of the message toll (dial) network will switch over to share lines on the leased network and volume on the dial network would decrease, causing price increases there as well. The household consumer would, as a result be subsidizing the businessman, for only the businessman could take advantage of leased line sharing. Are these objections really valid?

1. Line Sharing: An Example

An examination of this question must consider both the *short run* and the *long run* results of opening the door to "time sharing" and "capacity sharing" of lines. To begin, let us suppose that unrestricted line sharing/resale were permitted. What would happen? Users could get together either directly or through entrepreneurs who would serve as "placement centers" or "line brokers" trying to find users with complementary needs for communication channels. Firms could lease broadband lines and channelize them into whatever "packages" users wanted. The use of a line could be alternated among users during the day. The specialized needs of "minority" groups could be met. Line utilization on leased channels would increase. Competition in the equipment market would encourage the use of the most advanced equipment for channel derivation, and would pressure the telephone companies to rapidly introduce new channelizing technologies.

Short-Run Effects

For further analysis let us make a few assumptions. First, assume that initially the physical plant is fixed, the demand for communications capacity is insensitive to price (price inelastic) and the cost of the network is fixed. If the utilization per line increased, then clearly the number of lines used would decrease. The telephone company earns its revenue from the lease of the lines, so the total revenue received by the telephone company would fall. Yet the total costs of maintaining the physical plant would remain constant (the operating cost is only negligibly reduced if fewer than the total available lines are actually in use). Therefore, in order to recover its costs the telephone company would have to raise the tariff rates per line. The net effect, *in the short run,* would be that the prices charged per line would rise across the board; i.e., all users would be paying for the fact that some users were sharing lines. The non-sharers would be paying a larger amount for the same service they received before sharing was permitted. Sharers would be paying less. The total revenue to the telephone company would have to be the same to cover the same operating costs (roughly) and to earn the same percentage profit on the rate base (the size of which is unchanged in the short run) as before.

That is what would happen to the *pricing structure* in the short run. Now what would happen to the utilization of the network in the short run? Since we are assuming that the users' demand is insensitive to price, the non-sharers would use as many lines as they had before, and these lines would be utilized (kept busy) to the same extent. But the sharers would require fewer lines and the utilization on the lines used by the sharers would necessarily increase. The net effect would be to reduce the total number of lines actually used. In other words, as sharing began to occur, there would be evident excess capacity in the nation's communication network. Note that the common carriers have not been hurt so far. If anyone has been hurt, it is the non-sharer who has to pay more for the same service.

Now, suppose that the pricing structure was altered to avoid impacting the non-sharer. This could easily be done by charging a higher price for a line which was going to be shared than for a line which was not going to be shared.[32]

---

32. AT&T has, in fact, done this in its 1969 tariff revisions permitting sharing of voice-grade and smaller leased lines. A 10% surcharge is applied to a line which is to be shared.

Properly-set price levels might yield significant savings to users taking advantage of line sharing, while minimizing the "injury" to the non-sharer. Note here that large corporations already effectively share lines with themselves, their subsidiaries, and their assorted divisions in a vertically integrated structure. Small firms would benefit most from these proposed changes in the regulations regarding line sharing.

Long-Run Effects

Now let us look at the *long run* effects. The size of the physical plant, in the long run, is not constant; in fact, it is currently increasing at more than 13% per year.[33] If sharing were permitted, then the number of lines required to serve the nation would be fewer or would increase at a slower rate. This would be contrary to the interests of the telephone company stockholders, for the telephone company's rate base and earnings are directly proportional to the size of its physical plant. But from the public interest point of view any reduction (or even stabilization) in the size of the physical telephone plant (i.e., rate base) without any reduction in the service provided is desirable: The per unit cost of communications would fall. Even if the demand for communication capacity were price inelastic (little or no increase in demand at lower line charges), as earlier assumed for purposes of analysis, this would be beneficial. But almost certainly the demand for communications capacity is elastic (see the discussion below), and if the costs per unit of communication capacity were lower, the demand for communication capacity would increase. Demand may even be sufficiently elastic that the short-term "excess plant capacity" assumed above would be immediately absorbed.

Summary

This example shows that unrestricted line sharing would result in more total communications traffic per unit of physical plant. In the short run the net effect would be merely a shifting of traffic from the dial telephone network to leased lines. But in the slightly longer run, which at a rate of growth of demand of 10-15 percent per year will not be long, an increase in communications volume per unit of plant will occur. Expansion of the plant may be temporarily slowed somewhat — by an amount dependent upon the price elasticity of demand for communication capacity — but the public will certainly benefit from lower costs.

---

33. At the end of 1968, AT&T's plant investment stood at $34.8 billion, with annual expenditures of more than $4.5 billion for new plant.

2. Price Elasticity of Demand

Since allowable profits are determined by the size of the common carrier's rate base, the carrier's objective is to increase the size of the capital investment in plant and equipment, which means taking those steps which will lead to increased capital equipment requirements in the long run. With this reward structure, there is little incentive to increase utilization of existing physical plant, unless it can be seen that the per-unit reductions in communications cost which this allows will in turn lead to such an increase in volume that a larger overall plant will be required. Whether or not this will happen depends upon several factors, the first of which is price elasticity of demand. Price elasticity of demand is defined as the percent change in quantity demanded divided by the percent change in price; thus an elasticity factor of greater than one means that total revenue (price times quantity) will increase for a decrease in price. If demand for communications is elastic (as we will see that it generally is), sharing and its resulting effective price reduction could increase the carrier's "obtainable" revenue from its customers (it is possible, but unlikely, that sharing would reduce the carrier's "obtainable" revenue by reducing the number of lines leased); however, its "allowable" revenue is determined by the size of its rate base, so that unless sharing increases the size of the required physical plant as well, the carrier does not profit by allowing it. Whether sharing will lead to increases in the size of the capital investment in plant and equipment depends upon both the price elasticity of demand and the degree to which the existing plant is being utilized, i.e., the amount of "slack" which increased demand must absorb before new plant would be required. However, the consumer benefits by sharing whether the rate base increases or not.

Historically, AT&T has been very reluctant to make the assumption that decreases in the cost of communicating (through tariff rate reductions, sharing, etc.), would lead to increased requirements for physical plant (perhaps because of knowledge that the existing level of plant utilization is very low?) and instead takes a very conservative view of demand. The validity of this outlook may be questioned, in the light of the results of several recent telephone rate reductions. During the past decade several reductions have been ordered, over AT&T protests, by the FCC and, less frequently, by various state public utility commissions. The market reactions have in some cases caught AT&T off guard, such as the impressive increase in volume (which even overloaded certain switching facilities) when evening long distance rates were cut a few years ago. FCC Commissioner

Nicholas Johnson has commented, "... the evidence has been that each rate reduction has been followed by an increase in telephone usage *and company revenues* — benefitting shareholders and subscribers alike." (Emphasis added.)[34]

Demand is especially elastic in the data communications field where the cost of communications channels is increasingly becoming the limiting factor in the installation of computer systems which involve data communications. Currently the communication channels alone comprise from 4% to 24% of total system cost in the typical "communications-served data processing" system.[35] And the costs of computer equipment are falling sharply each year, more rapidly than are communications costs.

In the four years from 1962 to 1966, the cost efficiency of newly-introduced computers increased by a factor of 2.6 times (i.e., costs for performing a given operation fell to 38.5% of their 1962 value) for commercial computations. For scientific computations, which are better able to take full advantage of recent advances in computer technology, the relative increase in efficiency was 4.7 times, that is, costs fell to 21.3% of their 1962 value.[36]

As the cost of communication channels becomes an increasingly significant fraction of the cost of large computer systems, the decision by management as to whether or not to install a large system will increasingly turn on the cost of the communication channels. In economic terms this merely says that the elasticity of demand for data communication channels is increasing. For example, if communications costs represent 20% of total systems cost, then a 30% reduction in communications charges results in a 6% reduction in total systems cost. But if communications costs represent 50% of total systems cost, then the same 30% reduction in communications charges results in a 15% reduction in total systems cost.

---

34. *Telecommunications Reports,* Vol. 34, No. 16 (April 8, 1968), p. 27.

35. IBM *Response,* Vol. II, pp. 24-30.

36. Calculated from Kenneth E. Knight, "Evolving Computer Performance 1963-67," *Datamation,* Vol. 14, No. 1 (January 1968), pp. 31-35.

D. LINE SHARING/RESALE: SUMMARY

We have described two forms of multi-user sharing of communication lines: First, "time sharing" of lines, in which several subscribers alternate in the use of a given line; and second, "capacity sharing," in which several users lease a baseband line and channelize (subdivide) it into several logically independent subchannels, each subchannel dedicated to a different subscriber. Both forms of line sharing effectively amount to "reselling" of communication channels, which is allowed by the common carriers only in certain limited circumstances; both forms of sharing are, however, socially beneficial because they reduce communication costs. "Time sharing" of lines would facilitate future non-carrier provision of computer-based message switching services either as a stand-alone offering or in combination with a remote-access data processing or information service. "Capacity sharing" of lines would allow non-carriers to lease broadband channels and derive subchannels with bandwidths (information carrying capacity) that "match" the unsatisfied requirements of the data processing industry. Capacity sharing would also allow competition in the provision of channelizing equipment, thereby opening new markets to industry, encouraging technological innovation, and providing an incentive for the carriers to employ the most cost-effective equipment and techniques. In addition, the arbitrage function provided by capacity sharing would reduce the carriers' ability to artificially segment communications markets and practice cross-subsidization pricing (price discrimination) to maximize monopoly revenues.

We have attempted to show that opposition to either form of line sharing on grounds of technical dangers, system integrity, etc., has negligible validity. The real arguments pro and con are economic. We have discussed the short run and long run effects of permitting unrestricted multi-user line sharing and have concluded that in the short run it would lead to a degree of excess capacity in the national network and would require a restructuring of the pricing system. In the longer run, unrestricted line sharing would permit more efficient utilization of the national network and would reduce communications costs in general.

CHAPTER VIII

ADEQUACY OF COMMUNICATION SERVICE

Historically, the national communications network and its underlying philosophy have been geared to the needs of the traditional voice and telegraph customers. However, accompanying the recent development and rapid growth of the computer industry a need has arisen for the efficient transmission of information in digital form. The transmission of digital information requires communications facilities and services quite different from those required for voice and conventional forms of record transmission. Although the carriers have made some significant contributions in adapting the existing communication network to the requirements of the data processing industry, many shortcomings still remain.[1] These, combined with the anticipation of the continued rapid growth of data communications, have even led several respected authorities to propose that a separate digital communications network be considered.[2]

Continued growth of data communications is anticipated in forecasts by a number of organizations. Booz, Allen & Hamilton has estimated that while the rate of increase in the number of computer systems has been about 25% since 1963, the rate of increase in the number of on-line computer systems since 1963 has been about 75% per year.[3] Additionally, Booz, Allen estimates that the

---

1. The carriers are fully aware of the misfit between the capabilities of the existing communications network and the requirements of the computer industry. In 1957, Bell Telephone Laboratories studied the use of the telephone network for data transmission and found:

    "The telephone network was developed for speech transmission, and its characteristics were designed to fit that objective. Hence, it is recognized that the use of it for a distinctly different purpose, such as data transmission, may impose compromises both in the medium and in the special services contemplated."

    P. Mertz and D. Mitchell, "Transmission Aspects of Data Transmission Service Using Private Line Voice Telephone Channels," *Bell System Technical Journal*, November 1957.

2. For example, Paul Baran has said, "Is it time now to start thinking about a new and possibly non-existent public utility, a common user digital data communication plant designed specifically for the transmission of digital data among a large set of subscribers?" (*On Distributed Communication Networks*, RAND Corporation Paper P-2626, September 1962, p. 40.)

3. Booz, Allen & Hamilton, *op. cit.*, p. 178.

volume of data communications is growing at a rate of 50% annually.[4] In contrast, the volume of voice communications is currently growing at a rate of only 10% annually. The carriers themselves recognize the growing significance of data communications. Frederick Kappel, Past Chairman of the Board of AT&T, estimated that by 1975 the volume of data communications would *exceed* the volume of voice communications.[5]

The communications common carriers offer a variety of switched and leased communications channels, of several different frequency bandwidths, and with several different pricing structures. Table 2 on the following page outlines the current communications offerings of interest to the user of data communications. The governing common carrier tariffs are indicated in parentheses following each service title, and the data transfer rates available to users of each service are indicated in the right hand column.

Subsequent discussion of the adequacy of the common carrier offerings is organized into four major sections. The first section deals with the question of whether or not the channel bandwidths provided by the carriers "match" the channel bandwidths required by the data processing industry. The second section is concerned with the reliability, or error rates, of the available channels. The third section examines the incompatibility of the pricing structure for certain services with the usage patterns typically found in data transmission. The fourth, and last, section discusses the concept of a "digital" communication network.

A. CHANNEL BANDWIDTHS

The present communication network was designed to carry voice and record (non-voice textual) information. Therefore, the available channels have bandwidths suited to these needs. Record information was transmitted primarily by means of teletypewriter equipment, which required channels with frequency bandwidths less than 300 Hz. To provide adequate voice transmission quality, voice channels were designed with a bandwidth of 3000 Hz. For efficiency in long haul transmission, voice and sub-voice channels were gathered by means of

---

4. *Ibid.*, p. 179.

5. F. R. Kappel, before the Annual Meeting of the North Carolina Citizens Association, Raleigh, North Carolina, March 22, 1961.

TABLE 2

CURRENTLY AVAILABLE COMMON CARRIER COMMUNICATIONS OFFERINGS USEFUL FOR DATA TRANSMISSION

	Data transfer rate in bits per second
National switched networks	
Telegraph grade	
TWX (AT&T Tariff F.C.C. No. 133)	45-150
Telex (W.U. Tariff F.C.C. No. 240)	50
Voice grade	
Message toll telephone (AT&T Tariff F.C.C. No. 263)	1,200-2,000*
WATS (AT&T Tariff F.C.C. No. 259)	1,200-2,000*
Broadband exchange (W.U. Tariff F.C.C. No. 246)	1,200-2,400
Dataphone 50 (AT&T)	50,000
National leased network (AT&T Tariff F.C.C. No. 260, W.U. Tariff F.C.C. No. 237)	
Telegraph grade	45-180
Voice grade	1,200-9,600*
Broadband	
12 voice channels (Series 8000)	50,000
60 voice channels (TELPAK C)	250,000
240 voice channels (TELPAK D)	500,000

---

*One of the factors limiting the data transfer rate on a given communications channel is the performance characteristics of the modem interface device at the endpoints of the line. In the near future, commercially available modems will allow data transfer rates of 3,600 bps on switched, voice-grade lines. Modems operating above 4,800 bps on voice-grade lines are infrequently used today because of their high cost and sensitivity to time-varying channel characteristics.

frequency-division multiplexing into groups, the amount of consolidation increasing, in general, with distance. Twelve voice grade channels were consolidated into a "carrier group," 60 voice grade channels (five groups) into a "super group," and 240 voice grade channels (four super groups) into a "master group." Transmission equipment was designed for use with each of these "packages" of channels. When the need for data communications channels arose, the existing channels and "packages" of channels were simply adapted to enable the transmission of digital information.[6] The bandwidths, and therefore the maximum data transfer rates of the channels, were unchanged. Naturally, the bandwidths of the available channels do not "match" the bandwidths required by the large variety of data transmission devices, ranging from very slow teletypewriters up to high speed magnetic tape drives.

According to theory (the Shannon-Hartley law) the capacity, or maximum rate of information transmission at an arbitrarily small error rate, of a bandlimited communication channel is determined by its bandwidth and by the signal-to-noise ratio present on the channel.[7] Using efficient signal encoding procedures and other techniques to obtain high signal-to-noise performance, data transmission rates of more than ten times channel bandwidth, that is, 30,000 bits per second over a 3000 Hz voice-grade telephone channel, are theoretically possible. In practice, however, such speeds are not attained, due to the limitations of the state of the art, and the complexity and consequent cost of encoding equipment necessary to approach theoretical transmission limits. For example, on a leased voice grade line the maximum data transfer rate actually possible today is 2400 bits per second, using modems available from the carriers, or 9600 bits per second using modems available from several independent manufacturers.

Table 3 on the following pages indicates the actual data transfer rates possible on available channels (using carrier-provided interface devices), and compares these with the data transfer rates of typical data processing devices.

Based upon an examination of the chart, we can draw several conclusions about the adequacy of the bandwidths, or data transfer rates, of the available channels. We are not concerned here with the price structure, which will be discussed separately in a subsequent section.

---

6. See discussion of modems in Chapter VI on foreign attachments.

7. See, e.g., James Martin, *Telecommunications and the Computer* (Englewood Cliffs, N.J.: Prentice-Hall, Inc., 1969), Chapter 11; also John M. Wozencraft and Irwin M. Jacobs, *Principles of Communications Engineering* (New York: John Wiley & Sons, Inc., 1965), Chapters 5 and 6.

TABLE 3

COMPARISON OF DATA PROCESSING EQUIPMENT OPERATING SPEEDS WITH AVAILABLE TRANSMISSION LINE SPEEDS[8]

Data processing equipment[1]	Operating speed (bits/sec.)	Available transmission line	Present transmission line speed (bits/sec.)
Card reader			
300 CPM[2]	3,200	Voice: switched	2,000
		leased	2,400
600 CPM	6,400	Voice: switched	2,000
		leased	2,400
		Broadband: leased only	50,000
1000 CPM	10,600	Voice: switched	2,000
		leased	2,400
		Broadband: leased only	50,000
Card punch			
300 CPM	3,200	Voice: switched	2,000
		leased	2,400
500 CPM	5,300	Voice: switched	2,000
		leased	2,400
Paper tape reader	75	Telegraph or Teletypewriter: switched and leased	110/180
	2,800	Voice: switched	2,000
		leased	2,400
	4,000	Voice: switched	2,000
		leased	2,400
		Broadband: leased only	50,000
	8,000	Voice: switched	2,000
		leased	2,400
		Broadband: leased only	50,000

---

8. From Booz, Allen & Hamilton, Inc., *Study of Interdependence of Computers and Communications Services,* Exhibit XXXII, following p. 168.

TABLE 3 (Continued)

Data processing equipment[1]	Operating speed (bits/sec.)	Available transmission line	Present transmission line speed (bits/second)
Paper tape punch	75	Telegraph or Teletypewriter: switched and leased	110/180
	800	Voice: switched	2,000
		leased	2,400
Printer			
300 LPM[3]	6,000 to		
600 LPM	10,600	Voice: switched	2,000
		leased	2,400
		Broadband: leased only	50,000
1000 LPM	19,400	Voice: switched	2,000
		leased	2,400
		Broadband: leased only	50,000
Teletypewriter	45-150	Teletypewriter: switched and leased	45-150
Cathode ray tube	8,000	Voice: switched	2,000
		leased	2,400
		Broadband: leased only	50,000
Magnetic tape transport	150-3,000	Teletypewriter: switched and leased	110/180
		Voice: switched	2,000
		leased	2,400
	120,000	Broadband: leased only	50,000
		Broadband (Telpak C) leased only	250,000

TABLE 3 (Continued)

Data processing equipment[1]	Operating speed (bits/sec.)	Available transmission line	Present transmission line speed (bits/sec.)
Magnetic tape transport (continued)	240,000	Broadband (Telpak C) leased only	250,000
	480,000	Broadband (Telpak D) leased only[4]	500,000
	720,000	Broadband (Telpak D) leased only[4]	500,000
	960,000	Broadband (Telpak D) leased only[4]	500,000
	1,440,000	Broadband (Telpak D) leased only[4]	500,000
	2,720,000	Broadband (Telpak D) leased only[4]	500,000
Disk units	1,248,000	Broadband (Telpak D) leased only[4]	500,000
	2,496,000	Broadband (Telpak D) leased only[4]	500,000
Drum units	1,000,000	Broadband (Telpak D) leased only[4]	500,000
	8,000,000	Broadband (Telpak D) leased only[4]	500,000
Central processors	2,000,000	Broadband (Telpak D) leased only[4]	500,000
	6,400,000	Broadband (Telpak D) leased only[4]	500,000
	16,000,000	Broadband (Telpak D) leased only[4]	500,000

1. Includes most commonly used data processing equipment.
2. CPM: cards per minute
3. LPM lines per minute
4. There is no standard modem tariffed for use with Telpak D service at this time. Modems for Telpak D require special order from the common carrier.

1. Low-Speed Channels

The bandwidth requirements of the data processing industry for low speed channels, with data transfer rates under 180 bits per second, are adequately satisfied by available carrier offerings; however, as will be seen, these requirements could be met using less costly technology. For data transmission purposes, these channels are useful solely for teletypewriters and low speed paper tape devices, and are available both in leased and dial-up forms. These channels were designed long before the advent of computers, for "record" transmission between teletypewriter devices. Modern keyboard terminals with character-by-character "typewriter" output are, fundamentally, no different than the original teletypewriter units and, therefore, the channels designed for record transmission satisfy the requirements of today's keyboard terminals.

However, as noted in other sections of this paper, the frequency-division multiplexing techniques used by the carriers today for channel derivation are less cost-effective for digital data transmission than the newer time-division multiplexing approach, and in this respect the carriers' currently-available low-speed channel offerings are deficient. The authors are aware of only one type of carrier channel offering utilizing this new technology: the Western Union SICOM and INFO-COM message switching services, which use TDM trunk lines to connect remote concentrators (Western Union Dalcode units) to SICOM and INFO-COM switching computers. Although Western Union could offer TDM subchannels independently, they have not done so. This monopolistic tie-in between switching services and channel facilities appears to be contrary to U.S. antitrust law; otherwise, these services are commendable beginnings, and similar use of TDM technology by AT&T and other telephone companies should be encouraged.

The channels available for use with keyboard terminal devices have one deficiency, however. Visual display terminal units are becoming increasingly popular for use with remote access computer systems. These terminals have keyboard input and visual display output (using a cathode ray tube). The rate of data *entry* is limited by the typing speed of the operator, which is well within the capacity of available low speed lines.[9] However, graphic output devices can

---

9. Low-speed leased lines are available at speeds ranging from 45 to 180 bits per second, or approximately 60 to 200 words per minute; switched service is available at 60 to 100 words per minute (TWX) and 66 words per minute (Telex). Few typists can exceed 60-70 words per minute.

sustain much higher data transfer rates from computer to terminal — up to 8000 bits per second, depending upon the design of the display terminal being used. Neither a low speed line, nor even a voice grade line is satisfactory for output to most such visual displays.[10] An *asymmetric line* with a low speed in one direction and a high speed in the other is required, but not yet available.

2. Medium-Speed Channels

From Table 3 we see that very few data processing peripheral or input/output devices operate at the data transfer rates of voice grade lines. The paper tape readers and punches that do operate at voice-grade line speeds may, in fact, be operating at other than their most efficient data rates. They may have been designed to operate at these speeds because the available lines operate at these speeds. It is therefore unfortunate for the data user that voice grade channels constitute the bulk of the communications channels in the network, especially since a large portion of the information transmitted in the future will be data, not voice.[11] However, one should not conclude that the data processing community cannot make effective use of the voice grade communication lines available today. Data users often fill requirements for multiple low-speed leased lines by using multiplexing equipment to derive these channels from a single voice-grade line; user-derived channels are often cheaper than comparable channels leased directly from the carriers.[12] For example, the frequency-division equipment used by the carriers can derive up to 25 telegraph-grade subchannels from a voice-grade channel. Time-division equipment, frequently used today in commercial time-sharing systems, can derive up to 175 telegraph-grade subchannels from the same voice-grade channel.

---

10. Certain limited-capability graphic display devices (e.g., those using bistable storage CRT's) and graphic display terminals which include expensive local display regeneration and control units, perform adequately in some applications using voice-grade telephone channels. In most cases these channels are less than optimal for the purpose, and are used only because channels of two to three times voice-grade bandwidth are not available, and the broadband 50-kilobit channels are too large and therefore too costly.

11. See, e.g., remarks by F. R. Kappel, former chairman of AT&T, cited in footnote 5, *supra*.

12. This is evidence that the channel derivation technology employed by the carriers lags behind that of private industry. Specifically, the carriers use frequency division channel derivation techniques while the private firms use time division multiplexing which, as discussed previously, is more suitable and economical for digital communications.

Many of the commonly used data processing peripheral and terminal devices operate at data rates in the range of 3,000-20,000 bits per second. These devices include:

Card readers	3,200-10,600 bits per second
Card punches	3,200- 5,300 bits per second
Paper tape readers	2,800- 8,000 bits per second
Printers	6,000-19,400 bits per second

However, the common carriers do not provide channels, suitable for data transmission, in the range between 2,400-50,000 bits per second.[13] The carriers do provide so-called "program circuits" for broadcast stations which are equivalent to several combined voice grade circuits; however, the electrical characteristics of these channels make them unsuitable for data transmission.[14] Yet, according to one of the largest non-Bell suppliers of data transmission equipment, it is well within the present state of the art to combine voice channels to provide proper conditioning for data transmission.[15]

So long as the absence of suitable channels in the range of 2,400-50,000 bits per second persists, users of remote devices operating in this range will be forced to lease a higher speed line than actually required and therefore pay for idle capacity, or operate their peripheral and terminal devices below their most efficient speeds. Current tariff provisions add insult to injury, by prohibiting the user from channelizing a broadband line to obtain the desired intermediate speed channels.[16]

---

13. BEMA *Response,* p. 143. As noted earlier, a more sophisticated (and more expensive) independently-supplied high-speed modem can be used to extend the capacity of a voice-grade channel to 9600 bps, but relatively few such units are in use today.

14. Rixon Electronics, Inc., *Response,* p. 6.

15. *Ibid.*

16. For example, AT&T Tariff FCC No. 260, regarding Series 8000 wideband channels (50,000 bits per second or 12 voice channels), provides:

"The customer may not create additional channels from channels furnished under this series except that the customer may create additional channels from channels of voice-grade to the extent permitted under the provision of 2.2.6." (Section 3.2.8 (B) (1) (f)).

3. High-Speed Channels

Data communication channels are currently not generally available above 250,000 bits per second. Yet, there is an increasing need for the long distance transfer of data between high-speed computer storage devices such as magnetic tape units, magnetic disks and drums, and processor core memory. The technology required to economically provide channels for such high-speed data rates exists now, with the T-carrier systems developed at Bell Telephone Laboratories. The problem lies with the very slow rate at which T-carrier transmission systems are being introduced. This will be discussed further in a subsequent section.

B. CHANNEL RELIABILITY

The overall reliability of a communications channel consists of both its freedom from total failure (outage) and freedom from the loss of information (introduction of errors) during transmission. Total failure or "outage" is extremely rare, to the credit of the common carriers, and will not be considered here. However, microwave fading often causes momentary outages.

Loss of information during transmission is due to noise on the channel and to distortion of the transmitted data signal caused by the electrical characteristics of the channel. There are two forms of noise: steady background noise (thermal noise, crosstalk, etc.) and impulse noise. Background noise is objectionable during voice communications but is tolerable during data transmission because discrete digital bit streams can be detected in the presence of steady noise levels. However, impulse noise, which consists of transient signals peaks (caused by lightning, maintenance procedures, or by switching circuitry), may possibly cause the loss of digital information.[17]

Errors in data transmission may also be caused by the electrical characteristics of the channel, which introduce several forms of distortion of the data signals: attenuation-frequency distortion, phase-frequency or delay distortion, frequency offset, and bias and characteristic distortion. These difficulties are most pronounced in high-speed data transmission (including high transmission rates on voice-grade channels), and can often be reduced by line equalization and other electrical compensation techniques.

---

17. A. A. Alexander, R. M. Gryb, D. W. Nast, "Capabilities of the Telephone Network for Data Transmission," *Bell System Technical Journal,* May 1960, p. 25.

Typical error rates on voice-grade telephone lines are one bit wrong in 500,000 transmitted at 600 bits per second, one bit wrong in 200,000 at 1200 bits per second, and one wrong in 100,000 at 2000 bits per second; with today's modems the error rate at higher speeds may increase rapidly. Low-speed telegraph lines have typical error rates of one bit wrong in 50,000 to 100,000 bits transmitted.[18] Only by incorporating redundant information in the digital data stream can transmission errors be detected and corrected. Frequently, especially for information whose accuracy is critical, transmitted data is encoded with redundant information in the form of "parity" bits to detect transmission errors.[19] A message containing a detected error can be retransmitted. Data can also be coded with additional redundancy so that detected errors can automatically be corrected without retransmission.

The errors during transmission, caused by impulse noise, are tolerable but costly where accuracy is required because of the necessity of transmitting non-information-carrying bits for error control. Higher quality lines would reduce the costs by minimizing retransmission and the need for redundancy. One technique for obtaining higher quality communication lines is to employ a digital communications network rather than the present analog network. This will be discussed in detail in a subsequent section.

C. PRICE STRUCTURE

The carriers offer various communications services, each with a different pricing structure. Communication lines can be *leased* for full time, 24 hour use for a minimum period of a month,[20] with charges based upon bandwidth and distance. Switched lines are available for any time duration simply by dialing to establish a connection. For dial-up voice-grade lines, charges are based upon time

---

18. James Martin, *Telecommunications and the Computer*, (Englewood Cliffs, N.J.: Prentice-Hall, Inc., 1969), pp. 363-64. The interested reader is referred to the thorough discussions of noise and distortion in data transmission (Chapter 12) and of error rates (Chapter 21) in this book.

19. A "parity" bit is a non-information-carrying bit appended to a group of bits so that there will always be, by convention, either an odd number or an even number of bits in that group in an "on" state.

20. There are some minor exceptions to the mandatory full-time lease arrangement, e.g., press and video channels.

and distance with a minimum charge period of three minutes. For dial-up low-speed lines (TWX and Telex teletypewriter exchange service), charges are also based upon time and distance. Telex has no minimum charge period and TWX has a minimum of one minute. In February 1968, AT&T announced a switched broadband service, charges for which are based upon time and distance.[21]

Within this general framework, exact prices for communication lines vary according to whether they cross state boundaries. If they do, they come under *interstate* tariffs regulated by the FCC; otherwise *intrastate* tariffs, filed before the various state regulatory commissions, apply. In contrast to the interstate rates, which are uniform across the country, intrastate rates vary widely among the states (although they are invariably much higher than corresponding interstate rates).

There are many weaknesses inherent in the present communications pricing structure, particularly for data communications. First, the periods for which leased lines are available are not sufficiently flexible. As discussed in Chapter VII on Line Sharing, all users do not require the use of a leased line for a full 24-hour day but rather, for example, only during business hours or in the evening. Second, the available services do not allow for charging *according to the amount of information sent* — rather they charge for line holding time (SICOM and INFO-COM are exceptions).[22] Third, the tariffs do not charge according to the error characteristics of the lines. And fourth, the minimum charge times for switched services are inappropriate when the service is used for data communications.

1. "Short-Period" Leased Lines

There are two ways in which leased lines could be made available for periods of time less than the full 24 hours per day: First, as recommended in the previous chapter on line sharing, users could be granted greater opportunity to share lines. This would allow the maximum flexibility, although an evident requirement is the existence of "line brokers." A second alternative would be for the carriers to file

---

21. This service was an experimental offering, only available between New York City, Los Angeles, Chicago, and Washington, D.C.

22. *Line holding time* is the period during which a switched circuit is *dedicated* to the caller; i.e., the duration of the call.

new tariffs for leased services available for time periods less than the full 24 hour day; the minimum *contract* duration could remain one month. For example, such a tariff could provide economical lines for data transmission, available only during the off-peak evening hours when the carrier has idle capacity.

Either of the above alternatives would alleviate a carrier problem that is becoming increasingly critical: the long line holding times which often occur when the switched network is used for data communications. The switched carrier network is designed with sufficient inter-exchange trunk circuits and other commonly-used facilities to provide acceptable service to subscribers using the network for voice telephone calls, which have average line holding times of three to five minutes (during which shared common facilities are tied up and unavailable to other subscribers). Recently many remotely accessed computer time-sharing systems have begun to employ the switched telephone network, with typical line holding times running up to several hours. The growth of such traffic may "overload" central exchange and trunking facilities not designed to accommodate it. Since many long-holding-time applications occur at predictable times rather than at random, the user does not want to pay for the extra common carrier plant required in the switched network to handle random traffic, and would therefore use less expensive "short period" leased lines, were such available.

2. Charging by the Bit

As noted above, in present common carrier communication services other than SICOM and INFO-COM, the user is not charged according to the amount of information that is actually sent. Instead, he is charged for the time that the channel is actually dedicated to him, irrespective of whether he is sending or receiving any traffic. There is a minimum charge to cover the cost of establishing the connection, such as the well-known three-minute minimum charge period on the long distance switched voice network. This pricing policy is a direct result of the circuit switching network philosophy wherein a dedicated "copper wire" connection is provided between two parties. For voice communications there is at present no feasible alternative to the circuit switching communications approach, and the consequent pricing policy of charging according to line holding time. Nor is this pricing philosophy objectionable, since during a voice telephone call the parties are conversing for the duration of connection.

However, in many *data* communication applications the line is idle for large portions of time during the call; e.g., users of interactive time-sharing systems spend much of their "on-line" time thinking. It would be inconvenient for them to hang-up and re-dial because of the long "connect time" (15-25 seconds) and the time-consuming process of re-establishing their legitimate access to the time-sharing system. (Logging-in at M.I.T.'s Project MAC, for example, requires the use of a password and consumes several seconds of costly CPU time.) The problem could be avoided if the parties were provided with a non-dedicated communication path, charges for which were based upon usage rather than total line holding time.[23]

As explained in Chapter IV, store-and-forward message switching is an alternative method for permitting record (data) communications (1) between terminal stations, (2) between a terminal station and a computer, and (3) between computers. The first situation is illustrated by the typical corporate computer-based message-switching system. Here, generally, several terminals are connected to a leased, multi-drop, low speed or voice-grade communication line terminating in a central store-and-forward computer. Terminals with traffic to be transmitted either contend for the shared line, or respond to polling commands from the message-switching computer according to a pre-arranged line discipline (e.g., each terminal is called every five minutes to determine if it has any traffic to send). Messages are sent to the store-and-forward computer to await the availability of the outgoing communication line to which the destination terminal is connected. (See Figure 2, p. 53.)

The second situation is similar to the first except that here either the sending or receiving "terminal" is a computer (which may be programmed for engineering computations, data collection, inquiry processing, etc.).

---

23. Professor Richard G. Mills of Project MAC at M.I.T. has observed that:

"There is a clear need for a different tariff structure, which implies a need for a different basic communication system design, to achieve closer compatibility between the communication system capabilities and the communication requirements of this application. It would be desirable to "time share" the facilities of the communication system in the same way that we now time share the computer."

***

"The transmission requirements for multiple-access systems probably cannot be met by straightforward extensions of the services now available through the present common-carrier systems. A message store-and-forward system appears to satisfactorily meet the new transmission requirements if delays are kept short."

("Communications Implications of the Project MAC Multiple-Access Computer System," *IEEE International Convention Record,* Part I, 1965, pp. 237-238).

The third situation is illustrated by the ARPA Network described in Chapter VII on line sharing. Here, a message switching network serves a number of computer centers, each of which has an interactive time-sharing computer with a number of users linked to it. A message from a participating computer center is sent to a local "Interface Message Processor" (IMP), which may store it temporarily, and subsequently forward it to another IMP which, in turn transmits the message to the destination computer.

The basis for pricing in these situations would be somewhat different from that employed under a circuit switching network philosophy. Here, the user should pay, first, for the line time which he actually uses; second, for a share of the idle line time; and third, for that portion of the processing time and storage capacity in the message switching computer(s) which he uses. Both the processing time and storage capacity required are approximately proportional to the number of alphanumeric characters handled by the store-and-forward switch. The line usage time per message is also proportional to the number of characters in the message. Therefore, under a message-switching network philosophy the user may be charged "by the bit," that is, his charges are roughly proportional to the amount of information transmitted. Distance is a factor, but is often a small one and for simplicity can be ignored.

Priorities could be established and a premium charged for messages requiring rapid transit times (e.g., a few seconds or minutes). Likewise, discounts could be allowed for deferred-delivery messages (e.g., overnight).[24] A message switching network for use with interactive computer systems would require the capacity to provide very short message transit times or "response times." For example, in the distributed message switching network proposed by Paul Baran, the mean response time is one-half second.[25]

---

24. Delay-intolerant traffic imposes a greater burden upon the network, as it requires line operation at lower utilization levels (the ratio of the channel time actually used for data transfer, to the channel time available). The closer the line utilization comes to 100%, the longer the average message transit time, and the lower the cost per message. In effect, high priority messages would be charged a larger proportion of the cost of idle time than their traffic volumes warranted.

25. Paul Baran, "On Distributed Communications: Introduction to Distributed Communication Networks," RAND Memorandum, RM-3420-PR.

There are two ways in which message switching facilities could be provided: First, they could be provided on a large scale by the common carriers, possibly in the distant future supplanting the present circuit switching network. Second, they could be provided on a small scale, custom-designed for specific industries or applications, by entrepreneurial firms and/or by carriers. We believe that *both* are desirable. As digital technology (both transmission and switching) is phased into the existing carrier network (discussed in the following section of this chapter), message switching may become an integral part of the carrier's offering. However, as long as message switching systems are "superimposed" upon the present analog circuit-switched network, the authors believe that non-carriers should be permitted to offer these services. This was discussed previously, in Chapter IV on message-switching, which concluded that these services should be unregulated and permitted to be offered by a common carrier only through a separate subsidiary organization. The arguments given were essentially that competition among firms offering message switching services would be practical (because of limited economies of scale) and desirable because computer-based message switching is a service in which substantial innovation is possible and which is "sensitive" to changing technology. In addition, the diversity of specialized message switching services which will be required can best be provided by a variety of service firms, each expert in its respective client industry.

3. Pricing According to Channel Error Characteristics

At present, the charges for communication channels do not reflect the error characteristics of the channels.[26] Since there is now only one "grade" of line available at a given bandwidth, no other charging scheme is currently possible. If, however, carriers offer higher-grade lines, as many respondents to the FCC Inquiry have urged that they should, charges would vary with the error characteristics of the service. If message switching services are offered, by either carriers or non-carriers, charges could vary with (1) the probability of incorrectly received characters, and (2) the probability of an aborted message.

---

26. The carriers do offer "conditioned" lines at additional cost which improve channel capacity (e.g., permit a 2000 bps line to operate at 2400 bps) and reduce the probability of errors.

170   Adequacy of Communications Services

4. Shorter Minimum Charge Times

The minimum charge time of three minutes on a switched, voice-grade circuit and one minute on a switched telegraph-grade (TWX) circuit are both inappropriate for many data communications applications. Very short calls are often employed in private computer-based inquiry systems and data collection systems. Also, central computers can be equipped with dialing units capable of automatically placing 120 calls per hour on one telephone access line.[27] (The limitation is due to the 15-25 second "connect time" required by present central exchanges. When ESS exchanges are widespread, "connect time" will be reduced to 5-15 seconds and more calls per hour could be placed.) Presently, because of the minimum charge period on dial-up calls, data communication system designers are, in some cases, forced to employ leased line networks although the lines are used very sporadically and are idle for much of the day.

Of course, any change in the minimum charge times, particularly on the voice network, would have a substantial effect upon carrier revenue. However, switched lines are being used quite differently today than they were in the past, and the trend is expected to continue. Therefore, the existing price structure is becoming increasingly unsuitable and should be modernized based upon the present and anticipated distribution of the duration of switched calls and the "fixed cost" of establishing the connection.

D. A DIGITAL COMMUNICATIONS NETWORK

The existing common carrier communication network is based upon analog transmission techniques and frequency-division multiplexing, which historically have been most efficient for the transmission of voice signals. However, a different network philosophy, based upon digital carrier systems, time-division multiplexing, and store-and-forward techniques is more efficient for the transmission of data signals.

1. Signal Transmission and Multiplexing

All signals, digital or analog, lose power as they pass through a transmission system, and must be repeated at regular intervals to restore their strength. An *analog* network employs linear amplifiers at intervals along the transmission path,

---

27. BEMA *Response*, p. 150.

to keep the analog signal above a pre-determined level. However, all spurious noise is cumulatively amplified along with the desired signal. A *digital* network, on the other hand, which carries a stream of binary pulses from end to end, uses non-linear, regenerative repeaters (which cannot be used on analog waveforms). These repeaters detect the presence of a pulse and regenerate the pulse shape and amplitude. They substantially improve error performance since they do not pass noise which is below a given level. Digital carrier systems are efficient means of transmission of both digital and digitally-encoded analog information: data in binary form, and pulse-code-modulated (PCM) voice, television, and facsimile signals.[28]

The offsetting disadvantage of digital transmission is the amount of bandwidth it requires to reproduce a voice or video signal. Digital transmission systems have to date been able to overcome this disadvantage only where the transmission medium is not "bandwidth limited" (i.e., on cable as opposed to microwave channels). However, digital microwave equipment designed for data and PCM voice transmission has been recently developed and will begin to find applications along with the cable-oriented digital transmission systems.

The bandwidth requirements of digital systems will eventually be reduced by redundancy-removal equipment that identifies the unchanging elements in a series of voice or video signals (e.g., the timbre of a telephone caller's voice or the background of a television scene) and transmits only those signals that change over time. Today this equipment is either very expensive or else degrades the quality of the signal.

Use of large scale integrated (LSI) circuitry promises to reduce the costs of digital hardware and improve reliability and performance characteristics relative to that of analog hardware. These developments will eventually lead toward greatly increased use of digital transmission in the national communication network.

In addition to possessing superior transmission error characteristics, which is of great importance in data communications, and improving the quality of voice transmission, digital carrier systems permit a substantially greater amount of

---

28. The interested reader is referred to overviews of PCM technology in: Pietro Bronzini, "Some Notes on Pulse Code Modulation," *Telecommunication Journal,* Vol. 32, No. 11 (November 1965) pp. 464-470; W. Jack Hill, "Multiplexing: The Science of Mixing Voice," *Electronics World,* February 1968, pp. 43-46; and Jorgen H. Bistrup, "PCM—Advantages and Disadvantages," *Telecommunications,* Vol. 3, No. 6 (June 1969) pp. 9-14.

information to be carried over a given physical circuit. The Bell System has developed and is installing digital transmission facilities, called T-carrier systems, whose capacity for data transmission is orders of magnitude greater than existing analog carrier systems.[29] GT&E, Stromberg-Carlson, and other manufacturers are also producing such equipment for use by the independent telephone companies.

To obtain economies in long-haul transmission, many low-capacity channels are combined (multiplexed) into a single high-capacity channel. In an analog network multiplexing is achieved through frequency-division multiplexing (FDM) techniques.[30] In a digital network time-division multiplexing (TDM) is employed.[31] TDM can be used with streams of pulses (binary data or digitally-encoded voice) but not with voice signals in analog form.

When the subdividing of long-haul channels is accomplished by time-division multiplexing, modems are not required for data transmission, thereby eliminating a costly item in data communication systems. On the other hand, when TDM is used with voice signals, equipment to digitally encode and re-create speech is required at the ends of the communication line; however, the economies of digital transmission often outweigh this cost.

2. Store-and-Forward Switching

Given today's state of the art, transmission of both data and voice communications is more cost-effective in many cases (depending upon traffic volumes and distances), using the new digital rather than the older analog multiplexing and

---

29. Bell has installed 200,000 circuit miles of T-1 carrier, which has a capacity of 1.544 megabits per second. T-2 carrier will transmit 6.3 megabits per second over two telephone wire pairs and will be introduced in 1970. T-4 carrier will transmit 600 megabits per second over a coaxial cable and will be introduced in the early 1970's.

30. In frequency-division multiplexing several subchannels are each assigned different, but adjacent (stacked), frequency bands which are then raised to a higher frequency and transmitted over a single carrier channel.

31. In time-division multiplexing several low-speed bit streams are interleaved to form a single high-speed bit stream which is synchronized so that the low-speed bit streams can be separated out at the receiving end of the line. AT&T's T-carrier systems, for example, employ time-division multiplexing.

transmission systems; and rapid advances are making digital technology continually more attractive. Consideration must be given to the possibility of all-digital common carrier communication networks in the future. In addition to the use of time-division multiplexing and digital transmission systems, a completely digital network philosophy would employ store-and-forward switching of "blocks" of message traffic to "connect" any two points, rather than the present circuit switching method of establishing connection.

Paul Baran, of the RAND Corporation, completed in 1964 a several-year study for the U.S. Air Force, whose goal was to design an "ideal" communication system.[32] An ideal system is, according to Mr. Baran,

> "...one that permits any person or machine to reliably and instantaneously communicate with any combination of other people or machines, anywhere, anytime, and at zero cost.
>
> It should effectively allow the illusion that those in communication with one another are all within the same sound-proofed room—and that the door is locked."[33]

Mr. Baran further noted that,

> "Present-day networks are designed to do one particular set of tasks well. In the future, we shall make even greater demands upon our networks and shall consider new ways of building communication networks *taking advantage of the newly emerging computer-based technology.*" (Emphasis added.)[34]

The product of Mr. Baran's communication network design effort was a completely digital, distributed message switching (store-and-forward) network for both data *and voice* communications. A completely digital network, in addition to its higher cost-effectiveness compared to the present network, will "meet

---

32. Paul Baran *et al.*, "On Distributed Communications," RAND Memoranda, RM-3420-PR, RM-3103-PR, RM-3578-PR, RM-3638-PR, RM-3097-PR, RM-3762 through RM-3767-PR.

33. Paul Baran, "On Distributed Communications," RAND Memorandum RM-3767-PR, p. 1.

34. *Ibid.,* p. 2.

future requirements of military security, physical survivability, digital data flexibility, and ease of adding new services."[35] Perhaps AT&T should be encouraged to devote a portion of its communication facilities to the establishment of a separate and completely digital network. New, specialized common carriers, to be discussed in the following chapter, offer another means of rapidly and efficiently achieving a separate digital network.[36] Initially a separate network might be used for only data communications, because the present volume of data communications is smaller than that of voice (but growing much more rapidly), and because the "payoff" of such a digital system is higher for data communications. Ultimately, however (perhaps sooner than we might think, given AT&T's development of Picturephone, which requires very high-speed channels most efficiently provided by a digital system), the present carriers can be expected to use digital transmission technology for *all* types of information — not just data.

The carrier-supplied digital network proposed here for the long term future does not preclude non-carriers, in the near future, from offering computer-based message switching services. The scale and efficiency of non-carrier offerings would be limited, and in many cases specialized toward the needs of particular classes of users.

Additional analysis should be undertaken to consider the feasibility of digital networks and to recommend measures that will encourage the application of digital transmission and switching techniques in the national communication system. Heretofore the Federal Communications Commission has only perfunctorily concerned itself with trends in communications technology, making little or no attempt to influence the development and implementation of new communication concepts by the carriers. The Commission's selection in 1968 of Stanford Research Institute to provide assistance in evaluating material received in its Computer Inquiry, and to conduct a separate study of problems associated with the Land Mobile Radio field, was accompanied by FCC statements of intent to continue such studies in the future. In that regard, the application of a burgeoning digital technology to the communications industry is a promising area for further inquiry.

---

35. *Ibid.*, p. 8.

36. Indeed, the pending University Computing Company proposal to establish a new nationwide data-oriented common carrier network embodies many of the digital system concepts which we have discussed in this section.

CHAPTER IX

SPECIAL SERVICE COMMON CARRIERS

A development of specific interest to the computer industry is the emergence and potential growth of a new class of communications common carriers in the United States, which we will refer to as "Special Service Common Carriers" (SSCC). Acting as a complementary carrier to the existing network of telephone companies and the Western Union Telegraph Company, this new type of common carrier would lease inter-city trunk communications channels to business, educational, and government organizations. Common user exchange service (dial telephone, Telex, etc.) would not generally be offered; nor, in most cases, would local subscriber loops be provided to connect users with their inter-city leased lines.[1]

The growth of a new class of carrier with a special range of services would simply augment the spectrum of carriers already in existence. In addition to existing telephone and telegraph carriers, the following types of carriers are presently in operation:

1. Community antenna television (CATV) carriers with one-way point-to-point microwave and cable systems which provide remote communities or cities in poor signal areas with strong TV signals and more channels.

2. Mobile radiotelephone services to mobile units in fixed geographic areas which operate on land mobile radio frequencies and include such users as taxis, trucks, corporate fleets, etc.

3. Marine radiotelephone services to commercial and private craft operating in the marine environment for purposes of two-way communications, safety, and search and rescue requirements.

---

1. After this chapter was completed University Computing Company announced plans to file an FCC application for permission to construct an ambitious, nationwide, common carrier network specifically designed for data transmission. UCC thus proposes to enter the special service common carrier market; however, certain of its service characteristics would differ from those of the representative SSCC firms discussed on the following pages. A description of the proposed UCC data network appears in the final section of this chapter.

4. Radio paging services which provide one-way signals to individuals who are then required to telephone a predetermined number to receive their messages.

These specialized services are provided by small corporations which entered the submarket in the absence of such a service by existing carriers. These small carriers frequently exist in competition with similar services started at later times by telephone or telegraph carriers. Special service common carriers (SSCC) would simply add one more distinct class of carriers providing a range of services not presently offered by common carriers.

This class of common carrier would offer to business and data users intercity trunk communications channels which could be leased in the bandwidth and (within certain limits) the time desired, to which they could connect their own business and computer equipment and utilize in whatever manner they chose.

In order to offer these services, special service common carriers must obtain the approval of the FCC. According to the FCC's *Rules and Regulations:*

"Applications [for common-carrier radio license] will be granted only in cases where it is shown that....

(a) the applicant is legally, financially, technically, and otherwise qualified to render the proposed service,

(b) there are frequencies available to enable the applicant to render a satisfactory service, and

(c) the public interest, convenience or necessity would be served by a grant thereof."[2]

The three basic differences between this class of special carrier and existing common carriers are:

1. They would not provide customer services on an end-to-end basis as are presently provided by telephone companies and

---

2. *FCC Rules and Regulations,* Part 21 — Domestic Public Radio Service, Subpart I — Point to Point Microwave Radio Service, Sec. 21.700-Eligibility.

Western Union, but would offer inter-city trunk circuit capacity customized to specific subscriber requirements. They would provide trunk capacity to all businesses, large and small, with a reliability and cost commensurate with the wide range of needs of these communications users.

2. There would be virtually no tariff restrictions or prohibitions against the interconnection of privately owned communications systems, or links, with the SSCC's trunk circuit facilities, or against the shared use of such SSCC facilities.

3. A single mode or type of carrier facility (i.e., microwave) would be utilized in the provision of the inter-city communications channels which such carriers would offer.

The first distinguishing feature is the provision of the benefits of private microwave systems to all users on a common carrier basis, permitting a full range of flexibility and utilization of carrier capacity in virtually any quantity.

The second distinction is more in the nature of an institutional shift away from the historical pattern of telephone companies' restrictions limiting the interconnection of private systems with channels leased from the carriers, the multi-user shared use of such leased channels, or the use of customer-owned terminal equipment (i.e., foreign attachments) on common carrier channels. Although there may have been some historical rationale for the existence of such prohibitions, today's state of the art has developed to the point where provisions limiting such interconnection and shared use of facilities by business and data users do not reflect an enlightened view of either the needs or the capabilities of this particular submarket of communications, as we have discussed in Chapters V, VI, and VII.

The third distinguishing feature of these specialized common carriers is the exclusive utilization of microwave technology in the proposed carrier systems. Telephone companies have grown over the years through several successive stages of transmission technology and their systems are at the present time composed of a mix of old and new carrier technologies, with a preponderant weighting toward the older, higher cost, less efficient types of carrier systems. The result of this mix of old and new carrier systems is an average unit cost per voice frequency channel mile which is considerably higher than the average unit cost of similar capacity provided by the exclusive use of microwave as a transmission medium. The newer microwave carrier systems also provide channels of more uniform and generally superior quality for data transmission purposes than do the older carrier systems designed only for voice transmission.

These combined distinctions define the limits and characteristics of the submarkets which this new class of carriers will serve, by offering a range and versatility of use other carriers do not provide. Although the market is largely unknown, since it is at present not served on a common carrier basis, the characteristics which caused the growth of private microwave construction seem to define the needs, if not the size, of the potential market for the point-to-point channels which SSCC-type carriers propose to offer. Private microwave systems have been constructed only by corporations of a size capable of both financing the construction and having the intra-corporate requirement for the quantity of communications capacity thus constructed. These users were seeking lower costs, complete flexibility of bandwidths, freedom of choice of terminal facilities, and special transmission characteristics unavailable from the common carriers. But even these large corporate concerns can gain only limited areas of freedom from reliance on existing common carriers, since geographical considerations alone would often prevent a single corporate user from building all the intra-corporate communications capacity it needs.

The natural evolution of the private microwave systems concept would be the extension of these same benefits on a wide scale, through the provision on a common carrier basis of such services to all business voice and data communications users who require them, but for whom individual construction of private microwave facilities is impractical due to size and financial constraints. Microwave Communications, Inc. (MCI) of Joliet, Illinois, saw this business opportunity and applied to FCC in 1963 for a "Certificate of Convenience and Necessity" to establish, on a route between St. Louis and Chicago, the first "Special Service Common Carrier." Six years later, the FCC approved the MCI application, in a precedent-setting decision[3] which is excerpted in Section F of this chapter. As will be discussed in Section G, other SSCC license applications have been filed with the FCC, and more are surely in the offing.

---

3. *In Re Applications of Microwave Communications, Inc.,* FCC Docket Nos. 16509-19519, Decision of the Commission, Issued August 13, 1969.

Certainly this submarket of communications must be served, and although the precise configuration of the network which may be developed cannot be predicted today, the needs of its potential users are currently constrained by lack of the services proposed by the SSCC applicants. Since similar benefits of freedom, flexibility, and low cost are not likely to come from existing common carriers, potential subscribers may press for a separate leased-line voice and data communications network open to all users.

A. THE USE OF MICROWAVE FOR LONG-HAUL COMMUNICATIONS

The major technological development which led to the inception of the special service common carrier concept was the development and use of microwave radio as a point-to-point telecommunications medium. Microwave transmission technology was developed during World War II for anti-aircraft radar purposes, with Bell Telephone Laboratories and the M.I.T. Instrumentation Laboratory (which later formed the M.I.T. Lincoln Laboratory) largely responsible for the invention and application of the klystron, the super high frequency transmitting tube which made microwave radio possible.

Microwave was soon adapted for point-to-point communications purposes, offering line-of-sight radio channels of much greater capacity, transmission quality, and reliability than previously available. Microwave became a viable alternative to wire and cable transmission systems for overland communication between points approximately 20 or more miles apart, providing substantially lower-cost channels of equal or better quality, with very low incremental cost for increases in system capacity. These systems can be constructed by placing relay towers every 20 to 30 miles along the transmission route; installation costs are low compared with the cost of laying cable, a minimum of maintenance is required, and only a short time is needed for actual erection prior to operation.

Microwave, as a common carrier transmission system, was adopted shortly after World War II by Western Union, which constructed an experimental route between New York City and Philadelphia. Although Western Union's innovation in the utilization of microwave was not immediately effective, due to an adverse FCC ruling which refused Western Union permission to carry television programs over this system, it became apparent to the Bell System and the independent telephone companies that microwave, as a medium of low-cost, high-capacity, long-haul transmission, had important potential for the national telephone network.

With the gradual acceptance of microwave by the telephone companies, a network of microwave carrier systems[4] was constructed in the United States. Telephone companies today employ microwave as only one type of carrier system in their national network, which consists of a mix of equipment ranging from low-to-medium capacity open-wire carrier systems to coaxial cable systems of extremely large capacity. Microwave became simply another medium for providing long-haul inter-exchange trunk and private line circuitry over specific routes and types of terrain where its advantages are most pronounced.

Early microwave systems were less costly than the older types of carrier systems then in operation, and as microwave technology has progressed ("third generation" transistorized equipment is now available) the costs per voice-frequency channel mile have continued to decline. The intrinsic characteristics of microwave permit a very low unit cost, due in part to the fact that the system consists of a series of widely-spaced relay stations and in part to the fact that the system is capable of expanding its channel capacity up to many thousand voice-frequency channels by the simple addition of modular units of transmission and multiplexing equipment on an existing tower network.

With the development of a carrier technology with cost characteristics such as those possessed by microwave came the interest of large-volume communications users in providing their own communications systems separate from the existing common carriers networks. The reasons for this desire were twofold:

1. A separate system owned by a private enterprise could be used in a much more flexible and potentially more efficient manner than could similar facilities, with their attendant restrictions, leased from a common carrier.

2. A large-volume user could reduce his communications costs for point-to-point facilities considerably below the level of prices charged by the common carriers for leased private line channels.

---

4. A "carrier system" may be defined as a means of obtaining a number of channels over a single electrical path, using appropriate multiplexing equipment. Any transmission medium, such as wire, cable, or radio—microwave, troposcatter, or high frequency radio—may be employed in a carrier system.

Private use of microwave appeared first in 1948 with the application by a railroad to construct a private system on an experimental basis to test against its leased private line service from common carriers. The following year, a pipeline company received a similar experimental license, and broad interest in this medium began to grow among numerous public utilities and transportation companies.

New uses for microwave communications followed in 1952 with the application by a community antenna television (CATV) company to experiment with microwave to relay television programs picked up in Tennessee and relayed to Kennett and Poplar Bluff, Missouri. Three years later the FCC found in favor of the CATV applicant, granting a private system license. Such systems grew slowly because of the difficulty of obtaining a license; however, by the late 1950's, licenses could be had as common carriers, private carriers or specialized common carriers.[5]

In 1959, in what has become known as the *"Above 890"* decision,[6] the FCC concluded that private microwave applicants should be granted access to a dedicated portion of the radio spectrum, as were mobile radio systems for industrial and transportation private users, without impact upon the common carriers. Sharing of these systems presented regulatory problems, so this privilege was restricted to state and municipal governments, public utilities and right-of-way companies.[7]

Each of these new uses of microwave was opposed vigorously by the Bell System and the independent telephone companies and often also by Western Union; but in each case, the carriers eventually lost. Yet the markets developed by the new entrants proved so vigorous that the established common carriers subsequently attempted to institute similar services.

AT&T's response to the construction of private microwave systems was to institute the TELPAK tariff, providing drastically reduced rates for bulk lease of channels. By reducing its rates, AT&T hoped to lessen the cost discrepancy for

---

5. Beelar, Donald C., "Cables in the Sky and the Struggle for Their Control," *Federal Communications Bar Journal*, Vol. 2, No. 1, 1967, p. 35.

6. *In the Matter of Allocation of Frequencies in the Bands Above 890 MC,* 27 FCC 359 (1959), 29 FCC 825 (1960).

7. Beelar, *Op. Cit.,* p. 37.

large users between private line service and the lower cost of constructing and operating private systems. However, after permitting TELPAK to exist for several years as an experimental offering, the FCC concluded that its low rates were not based upon actual lower cost of providing the bulk service, and were therefore discriminatory; the smaller TELPAK channel offerings (TELPAK A and B, providing 12 and 24 voice channels, respectively) were therefore withdrawn, and the prices of the remaining TELPAK offerings (TELPAK C and D, at 60 and 240 voice channels, respectively) have been increased substantially. The cost discrepancy between private systems and available common carrier services has thus been partly restored.

The major prerequisites for the construction of a private microwave system are heavy requirements for point-to-point carrier capacity over a given route plus the financial capability to invest large amounts in the system's construction in addition to its operating expenses. The size of the capital investment alone would preclude most corporations from using private microwave systems as a means of satisfying their communications needs. If, however, numerous corporations can use the same system, it then becomes financially feasible, permitting each user to enjoy the benefits of low unit cost and expanded flexibility. If several small businesses can lease one common carrier channel and use it at alternate times (share it) during the day or week, then they could also share the use of a private system at even lower cost. As we have noted a few pages earlier, the development of a new type of common carrier, the SSCC, which would offer the benefits of a shared private system, without the burdens upon the user of heavy capital investment and the responsibility for system operation and maintenance, is a natural outgrowth of this trend.

The physical and cost characteristics of microwave transmission technology have made feasible the operation of small (relative to the massive telephone and telegraph company facilities) common carrier networks of this nature. Previous long haul transmission technology would have made such a venture impractical; today only institutional barriers must be overcome if such systems are to become a part of our national telecommunications complex. As will be seen in the remainder of this chapter, the major barrier of this nature is the prevailing regulatory attitude based upon theories about the benefits of a single monopoly carrier.

B. CHARACTERISTICS OF THE PROPOSED SERVICES

The basic service concept of what may be a number of special service common carriers is to provide flexible, low-cost voice and data communication links between urban centers for lease by business and industrial users.

Conceptually, SSCC services, provided on a common carrier basis, would have the following characteristics, most of which are similar to those available through the ownership and operation of a private microwave system:[8]

1. Inter-city radio frequency baseband capacity, or individual smaller channels for either voice or data communication, would be leased to any organization wishing to connect to the system.

2. Such capacity would be offered in increments of 2 kHz for channel capacity and in increments of 48 kHz up to 1000 kHz for radio frequency channels. This capability is not provided by the telephone companies or Western Union.

3. Customers would have complete control over the use of their leased channels, and would be permitted to (a) use any type of terminal equipment, and (b) share the channels with other customers. The SSCC would not manufacture or supply customer terminal equipment, confining itself to the provision of transmission facilities.

4. Customers could connect directly to the SSCC microwave terminals via leased lines of other common carriers or through construction of their own point-to-point radio systems.

5. Customers would be given assistance in installing, connecting and maintaining access links to the system and assuring that their "local loop" systems were adequate to provide the service required.

---

8. This list is taken from the proposal of Microwave Communications, Inc. (MCI), in its recently approved application before the FCC to become the first SSCC.

These services represent a significant departure from any service offered by the existing common carriers in that virtually all of the restrictions placed on use of their facilities are eliminated. No attempt to provide end-to-end service is contemplated. The provision of long-haul point-to-point capacity is the only service that is offered, and since there is no apparent intent to own, lease or manufacture terminal equipment, the basic service can receive the full attention of these carriers.

1. Rate Level

Aside from the distinctions in service concept between the existing common carriers and the proposed SSCC's, the rates charged for lease of SSCC facilities would be markedly lower than present common carrier levels. For example, let us consider the tariff proposed by Microwave Communications, Inc. (MCI) in its recently approved license application before the FCC. Although it is somewhat difficult to draw exact comparisons between MCI services and somewhat comparable private line capacity leased from telephone carriers, the rates proposed by MCI represent prospective reductions of 54 percent or more over existing rates of common carriers. An example given by MCI during its hearing was as follows:

> "AT&T's charge for end-to-end service from St. Louis to Chicago is $988.00 per month for two voice channels....These channels are available only on a 24-hour per day, 7-day per week basis...Alternate [voice/data] use of these channels requires an additional monthly charge (AT&T Tariff No. 260, pp. 140-141). Simultaneous operation is restricted (AT&T Tariff No. 260, p. 18). Conditioning of the AT&T channels requires an installation charge plus an additional monthly charge (AT&T Tariff No. 260, p. 62). Duplex operation costs 10 percent more (AT&T Tariff No. 260, p. 55)."[9]

This long and involved quote illustrates the complexities of tariffs plus the difficulty of giving a positive response to the simple question of "How much will service cost?" Obviously one has to know what the specific service is, the distance covered, the equipment leased, the tariff under which it is provided, the part of

---

9. Microwave Communications, Inc., "Proposed Findings of Fact and Conclusions of Law of Microwave Communications, Inc.," FCC Docket No. 16509–16519, July 28, 1967, p. 33.

the country in which it is located, etc., etc. Attempting, however, to draw a comparison, MCI states:

> "MCI's end-to-end charge from St. Louis to Chicago would be $462.37 per month for two voice grade channels plus $33.00 per month for local loops if provided by the local telephone company. MCI would allow up to five subscribers to share this channel, which would reduce this rate to $97.40. MCI's proposed tariff provides for half time use which would further reduce this monthly charge to $74.30. MCI subscribers may install their own switching for any alternate use with no increase in cost and have full simultaneous use of MCI channels. Conditioning of MCI channels is not required. MCI microwave channels are of one maximum grade quality, and the only variable is the bandwidth. Duplex operation is standard on MCI channels.[10]

Using only the figures for AT&T of $988.00 per month and for MCI of $462.37 per month for full-time non-shared channels, a reduction of about 54% in total charges results from comparing the two services provided in similar ways but under different lease conditions.

It is not the intent in this treatment of the subject to attempt to determine the magnitude of the rate level differentials between existing telephone company private line services and the proposals of MCI and other SSCC license applicants. That such differentials exist is clear, but it is not clear that these differentials are even necessary to develop the submarket which the special service common carrier applicants propose to serve. Other distinctions of flexibility, freedom of terminal equipment selection and use, and part-time or shared use may be enough to develop a substantial market. Such considerations have certainly been pertinent for those corporations which have constructed their own private microwave systems.

Private microwave systems have, however, provided distinct cost benefits in most cases to those corporations which constructed and used them. If a single corporate user can obtain cost advantages by using a private system at something

---

10. *Ibid.*

less than full capacity, joint use of a system through the common carrier principle should permit even greater reductions in average unit cost (due to the decline in unit cost of microwave as additional units of capacity are added) and hence comparable decreases in the rate level.

C. COMPETITIVE BENEFITS OF SPECIAL SERVICE COMMON CARRIERS

The introduction of competition into the private line service market was one of the basic benefits stressed by the FCC Common Carrier Bureau, the Hearing Examiner, and the full Commission in agreeing that the MCI application be granted. Also, although delineating the telephone exchange markets (local exchange and long distance) as having monopoly characteristics which at the present stage of development would not benefit from competition, the President's Task Force on Communications Policy stated specifically that introduction of competition into other communications markets would have salutary effects.[11]

Independent manufacturers of communications equipment are foreclosed from about 85% of their total potential market because of the exclusive monopoly of Western Electric as manufacturer, supplier, and installer of Bell System equipment. The entry of additional carriers into the rapidly growing business and data communications market could potentially open significant portions of this growth market to the introduction of communications equipment manufactured by others than Western Electric.

The motivation behind the introduction is not solely to open wider markets to small businesses, but to spur the rate of innovation in the manufacture of such equipment. The potential of entry into a sizable market may have the twin effects of improving the quality of existing types of equipment, and speeding the rate of introduction of innovations, as yet unknown, into special application areas. The research and development requisite to the creation and application of new technology will only become feasible for a larger number of firms when there is assurance of potential entry and sale of equipment in quantities adequate to justify the R&D expense.

As the number and variety of business voice and data communications equipment suppliers increases, the competitive impact on the market presses

---

11. The President's Task Force on Communications Policy, *Final Report* (Washington, D.C.: U.S. Government Printing Office, 1969), Chapter 6, pp. 10-20.

down the level of equipment prices. Since the markets for new uses and applications of communications equipment which reduce costs have generally proven relatively elastic, the expectation is that future growth in the market will absorb the new entry without drastic impact on the established suppliers.

The judgments which are willing to permit competition in private line markets are in part based upon the fact that these submarkets account for only a small percentage of total revenue of the Bell System and other existing carriers, and thus the revenue impact of competition should be minimal. Of more significance is the resulting impact of competitive benefits upon the users and the economy by opening these submarkets to a wider range of communications options for a larger number of users.

D. ARGUMENTS AGAINST THE SSCC CONCEPT

The theory upon which U.S. regulatory actions have been based in the past is that the establishment (and maintenance) of a regulated monopoly provides more efficient utilization of communications resources, at lowest cost, without the duplication of carrier facilities associated with unregulated competition. It is upon this theory that most arguments against the establishment of a group of special service common carriers are based.

The historical pattern of communications development in the United States has evolved from an early system (1895 to 1920) of a large number of telephone companies of localized origin which were sometimes in competition and sometimes held a geographical monopoly, and several telegraph companies, competing over some routes and monopolizing others. From this has emerged a national telephone system dominated by the Bell System, which achieved its present status through a systematic program of acquisition, merger, or interconnection. The telegraph companies eventually merged into a single national network operated by Western Union.

Regulatory policy of the Federal Communications Commission has operated since 1934 within a market structure which had long since become a series of localized geographical monopolies. The Bell System and Bell Associated Companies presently control about 50% of the geographical area and about 84% of total U.S. telephones. Independent telephone companies operate in localized markets in the other half of the geographical area of the U.S. with about 16% of total telephones.

The case against allowing additional carriers to enter the communications field follows a pattern set early in the development of the present structure of geographical telephone monopolies, and contains the following four arguments.

First, the entrance of a new common carrier requires that it must have a network of carrier facilities over which it can supply services to its customers. The requirement for interstate service would include the construction of long-distance carrier facilities and some means to obtain access to the customer's premises over a local loop system. Construction of facilities between cities where a carrier system already exists results in duplication of facilities, and carries with it several economic consequences, some of which are helpful and others which are considered harmful.

Second, construction of duplicate and parallel carrier systems prevents one carrier from constructing the entire point-to-point capacity at lowest unit cost and achieving maximum economies of scale. Maximizing such economies depends upon building point-to-point capacity using a single carrier system which has the full range of transmission characteristics requisite to serving all of the various and diverse needs of voice and data transmission.

The continued use of "economies of scale" as a buffer argument against the introduction of duplicate facilities needs to be reexamined in the light of the common carriers' unit cost levels today — which exist in the *absence* of duplication of facilities by other carriers. MCI rates, which are based on full cost, are so significantly lower than rates of existing carriers that the "single carrier" concept is open to question. Present unit costs may be high because the assumed economies of scale are not being realized, or because the present structure (vertically-integrated monopoly) of the common carrier industry may not encourage optimal operating policies and practices.

Third, the existence of two systems means that each must be supported by earned revenues. If competition by both carriers is in the same market then one carrier may resort to "price cutting" below actual full costs to gain a larger share of the market and greater revenues. The price cutting results in a rate war and service begins to suffer as each competing carrier attempts to cut costs for further price cuts. The result, if uncontrolled and allowed to continue, could be harmful to communications users. The carrier which proves strongest financially would drive the other carrier to ruin.

This argument assumes, of course, that no FCC regulation exists and that such pricing actions by either carrier would be permitted without some basis in cost. The other assumption is that both carriers are in the identical market without distinction, which may not be the case for special service common carriers.

Fourth, "cream-skimming" of lucrative markets is an additional change which the carriers warn against, an action which has an effect upon both the cost and rate levels of the existing carriers. By entering only high density routes where unit cost of service is low, it is claimed that new carriers will drain off the excess revenues in these markets, which help to support the present carriers' "average" rate levels on higher unit cost routes. This attacks the nationwide rate averaging principle (uniform nationwide rates charged for interstate leased channels) which the Bell System and other carriers have adopted to avoid using more costly and cumbersome point-to-point pricing schemes.

"Cream-skimming" can be viewed in an alternative way as an argument against the introduction of cost reducing innovations on point-to-point routes, possibly served at present by an inefficient carrier system. Assuming an existing carrier is serving high density routes in an efficient manner so that each submarket is adequately provided with the capacity and transmission characteristics it requires, it would seem difficult for another carrier to enter and penetrate such a market. If, however, new carriers enter special service business and data communications markets and develop services other carriers are unable or unwilling to provide, the argument of cream-skimming hardly applies.

The basic question which must be raised is whether or not we need to reexamine the fundamental concepts upon which regulatory actions in the United States are based. The history of successful entrants into the arena of communications markets in the United States has been stark indeed. Many aspiring, and eventually perspiring, candidates have attempted to penetrate segments of the submarkets for terminal equipment, special attachments to existing equipment, and other areas of real communications need, only to be blocked by the carriers with the tacit approval of the FCC. The Bell System has consistently pleaded that it can supply all services more efficiently and at a lower cost, under the theory that a regulated monopoly is most efficient.

MCI has made a strong case, in at least one communications submarket, that this theory is potentially very wrong, and if allowed to hold through denial of

SSCC license applications, could result in an important opportunity loss for business communications.

In approving the MCI application, the FCC Common Carrier Bureau, the Hearing Examiner, and the full Commission found the standard arguments of "cream-skimming," duplication of facilities, et al, to be outweighed by the potential benefit to communications users and they therefore granted the requested common carrier license.

The changing viewpoints about communications policy, predicated upon knowledge of market characteristics and requirements rather than an unflinching faith that telephone monopolies have the answer for every communications submarket, may prevail at the FCC. A thorough knowledge about the cost characteristics, relevant demand, and special applications of technology to solve communications problems could do much to expand the conceptual framework upon which communications policy is based. Should this lead to opening other submarkets to more rapid adaptation of innovations by new carriers, it is conceivable that future communications growth will greatly exceed present expectations.

E. COMPETITIVE RESPONSE OF THE EXISTING CARRIERS TO SSCC'S

The economic rationale for entry of specialized carriers into specific submarkets is that they will offer service(s) distinct from those already in existence, based on legitimate operational and cost characteristics. However, to permit SSCC's to come into existence for this purpose and then to allow existing carriers to modify without constraint their practices and prices on established facilities to combat the SSCC's would defeat the purpose for establishing them.

The President's Task Force on Communications Policy has recognized this potentially difficult issue, and recommended:

"The sound response of policy, we believe, is to provide flexible opportunities for entry, matched by a policy of allowing the established carriers sufficient pricing flexibility to respond economically to the challenge of the new services. However, in placing the new entrant and the established carrier on equal terms, the new entrant should be protected against the threat of non-compensatory or "predatory" pricing on the part of a carrier who has a monopoly market. The danger of non-compensatory pricing is real. Under a system of regulation dominated by criteria of fair return on the entire rate base of the carrier, the possibility always exists that the carrier would use its superior position in sheltered markets to cover losses in the competitive sector. What is needed is a minimum price standard calculated with reference to the "long-run incremental costs" for a particular service (including the cost of capital and the profits allowed for the incremental capital associated with the service), rather than for the system as a whole. With such a pricing standard, users of non-competitive services would not subsidize the users with competitive alternatives. The competitive services of the existing carriers would still pay the added costs they impose on the system."[12]

This recommendation is the standard response to regulation of all competitive market situations, well founded in economic theory, but practically impossible for the FCC to perform in the present market situation. There are virtually no historical cost data, cost allocation methods, or time series upon which the FCC can base an independent decision on whether a "cost" figure for a service, submitted by the Bell System or another carrier, properly reflects the long-run incremental (marginal) costs of the service.

With the exception of the "Seven-Way Cost Study," which was the fully allocated cost of all facilities used by the Bell System for seven individual services for a 10-day period in March 1964, all other cost studies submitted by the Bell System have been derived by arbitrary "costing" of specific portions of plant alleged by Bell to be used in the service. Without having an historical time series of cost components by category on a full cost basis, there is no way the evidence of shifted costs (i.e., overhead costs, allocated fixed costs, or allocated variable costs) can be measured by the FCC independent of whatever claims are made by the carrier. Due to this lack of information, the Bell System and other carriers can arbitrarily allocate any portion of old or new investment to the rate base of a specific service, allocate any portion of existing or increased operating expenses to the same service, and then claim that these are full incremental costs, marginal costs, or any other cost, dependent on the terminology used in whatever pricing policy is adopted by the FCC. Without some sort of cost trend data, there is no way these claims can be independently verified.

---

12. The President's Task Force on Communications Policy, *Final Report* (Washington, D.C.: U.S. Government Printing Office, 1969), Chapter 6, pp. 17-18.

It is difficult for the Bell System to assign or dedicate specific portions of its plant to a special service due to its common usage by all services. Although the Bell System plant investment program assigns a percentage of new plant to "Special Services," the assignment is arbitrary and may bear no relation to actual use, due to the vast commonality of the system and alternative uses which are made of it. This feature of the plant means the incidence of all costs less than a fully allocated cost for a special service must fall on other classes of communications users who are serviced by the same carrier and equipment.

Under these conditions a more informed policy would seem to dictate that the permissible range of any pricing response which should be allowed the existing carriers must be limited in the first instance to actual cost reductions, and in the second instance to a cost which includes a pro-rata proportion of overhead costs on the basis of plant use, just as must be borne by all other services which use the common plant. To do otherwise establishes the basis for rate discrimination among other classes of users, and refutes completely the economic rationale for establishing the special service common carriers.

Telephone and telegraph carriers will, of course, continue to offer their present private line services and should be encouraged to expand them on a selective basis with appropriate tariff modifications as dictated by the needs of the markets which they service. They should not, however, be allowed to introduce price reductions unsupported by cost reductions. Only limited changes in prices by these carriers should be permitted and only then with sufficient, demonstrable evidence that the costs or operating characteristics, on which leased line service charges are based, have changed.

F. EXCERPTS FROM THE FCC'S MCI DECISION

As this book went to press the FCC issued its decision in the previously mentioned MCI case. The decision was a close one (the vote was four to three, with the FCC Chairman issuing a strong dissenting opinion), and is both precedent-setting and controversial. To provide the reader with additional understanding of the policy problems addressed in this chapter and in the MCI docket, and certain to be addressed in subsequent FCC proceedings, we include below selected excerpts from the FCC's decision. We have omitted those passages which deal solely with MCI's financial and technical qualifications, rather than with the generally applicable policy considerations:

FCC Decision

Adopted: August 13, 1969   Released: August 14, 1969
Commissioner Bartley for the Commission: Chairman Hyde dissenting and issuing a statement; Commissioner Robert E. Lee dissenting and issuing a statement in which Commissioner Wadsworth joins; Commissioner Johnson concurring and issuing a separate statement.

    This proceeding involves applications filed by Microwave Communications, Inc. (MCI) for construction permits for new facilities in the Domestic Public Point-to-Point Radio Service at Chicago, Illinois, St. Louis, Missouri, and nine (9) intermediate points. MCI proposes to offer its subscribers a limited common carrier microwave radio service, designed to meet the interoffice and interplant communications needs of small businesses. Its subscribers would be able to lease microwave channels in varying bandwidths in increments of 2 kc for either the entire length of its system or any segment thereof. For broad band users, channels may be leased in increments of 48, 250, and 1,000 kc. MCI, however, does not plan to provide its subscribers with a complete microwave service. The proposed service would be limited to transmissions between MCI's microwave sites, making it incumbent upon each subscriber to supply his own communications link between MCI's sites and his place of business (loop service).

    MCI contends that it will offer its subscribers substantially lower rates than those charged for similar services by the established carriers and that subscribers with less than full-time communication needs will be able to achieve additional savings through the channel sharing and half-time use provisions of its proposed tariff. Up to five subscribers will be permitted to share each channel on a "party line" basis with a pro-rata reduction in rates. MCI will lease channels for half-time use between 6:00 a.m. and 6:00 p.m. with a 25% reduction in rates; and between 6:00 p.m. and 6:00 a.m. it proposes to combine the channels leased for half-time use into broad band channels of 48, 250, or 1,000 kc for high speed data transmission. MCI further asserts that its proposed tariff contains fewer restrictions than those of the existing common carriers, so that greater flexibility of use will be possible, particularly with respect to channel bandwidth, splitting channels for voice and data transmissions, and in the attachment of customer equipment.

    MCI's applications are opposed by Western Union Telegraph Company (Western Union), General Telephone Company of Illinois (General), and the Associated Bell System Companies, American Telephone and Telegraph Company, Illinois Bell Telephone Company, and Southwestern Bell Telephone Company (Bell), which presently provide microwave services to the geographical area which MCI proposes to serve....We designated the MCI applications for hearing on issues to determine *inter alia:* (a) whether the established common carriers offer services meeting the needs which MCI proposes to meet in the area which MCI proposes to serve; (b) whether the grant of MCI's applications would result in wasteful duplication of facilities; (c) whether MCI is financially qualified to construct and operate its proposed facilities; (d) whether there is a need for MCI's proposal; and (e) whether operation of MCI's proposed system would result in interference to existing common carrier services.

* * *

Need for MCI's Proposal

...While we ... believe that there is a basis in the record for a finding that the cost to MCI's customers for comparable through service would be less than that charged by the existing carriers, particularly since customers may use some individually owned equipment, we do not deem this factor to be a critical consideration here. The significant fact remains that the existing carriers do not offer a 2 kc voice channel and MCI will; so that a subscriber may achieve a substantial saving in his communications costs by utilizing MCI's services. Further savings may be effected by the sharing and part-time provision of MCI's proposal. ... No comparable offerings are made by the existing common carriers.* Thus, a grant of the applications under consideration will make microwave service available to potential users who have no need for and cannot afford the full-time, non-sharing, and more sophisticated services to which they are limited under present tariffs.

Bell asserts that it had previously marketed 2 kc voice channels but was forced to abandon the offering because of customer dissatisfaction. Utilizing a 2 kc voice channel, MCI cannot be expected to match the quality of voice communications offered by Bell with its 4 kc channels; but the record establishes that MCI's 2 kc channel will provide an acceptable voice communications service. As a result, MCI will attract users with less stringent requirements as to quality but who are in need of an economic communications service. Subscribers for whom quality of service rather than cost is the primary consideration will continue to utilize existing services rather than those of MCI.

---

*[Subsequent to the closing of the record in the MCI docket AT&T revised its private line tariff to permit shared use (termed "joint use" by AT&T) of leased lines of voice-grade or lower bandwidth.]

... Furthermore, MCI imposes fewer restrictions on the nature of the subscribers' terminal equipment and on the use of its channels. The absence of restrictions gives each MCI subscriber the same flexibility to vary its stations' capability and use as if it were its own private system. Thus, each subscriber may adapt the system to its particular needs and equipment, lease shelter and tower space from MCI, and use the MCI trunk system for the carriage of voice, facsimile, and high speed or lower speed data transmissions, or a combination thereof in a manner which best suits its business requirements. No comparable degree of flexibility is offered by the existing carriers. Thus, while no new technology is involved in MCI's proposal, it does present a concept of common carrier microwave offerings which differs from those of the established carriers. Therefore in determining the question of need, the controlling consideration is not whether existing communications services are being utilized by potential subscribers of MCI, but whether the proposed operation would better meet the particular needs of potential subscribers. *We believe that MCI's offering would enable such subscribers to obtain a type of service not presently available and would tend to increase the efficiency of operation of the subscribers' businesses.* [Emphasis added.]

* * *

MCI is offering a service intended primarily for interplant and interoffice communications with unique and specialized characteristics. In these circumstances we cannot perceive how a grant of the authorizations requested would pose any serious threat to the established carriers' price averaging policies. Lower rates for the service offered is not the sole basis for our determination that MCI has demonstrated a need for the proposed facilities, but the flexibility available to subscribers, and the sharing and the part-time features of the proposal have been considered to be significant factors as well....It may be, as the telephone companies and Western Union argue, that some business will be diverted from the existing carriers upon the grant of MCI's applications, but that fact provides no sufficient basis for depriving a segment of the public of the benefits of a new and different service.

Moreover, if we were to follow the carriers' reasoning and specify as a prerequisite to the establishment of a new common carrier service that it be so widespread as to permit cost averaging, we would in effect restrict the entry of new licensees into the common carrier field to a few large companies which are capable of serving the entire nation. Such an approach is both unrealistic and inconsistent with the public interest. Innovations in the types and character of communications services offered or economies in operation which could not at once be instituted on a nationwide basis would be precluded from ever being introduced. In the circumstances of this case, we find the "cream skimming" argument to be without merit.

* * *

The Feasibility of Loop Service

...In general, MCI's potential subscribers have no interest in providing their own communications link between their facilities and MCI's transmitter sites. Therefore, MCI's ability to market its service will be dependent on the ability of its subscribers to secure loop service from the other common carriers serving the service area.

We are not unmindful of the fact that the carriers maintain that loop service is not technically feasible and that there is no provision for such service in their tariffs. However, insufficient evidence is contained in this record to support a conclusion that the proposed interconnection is not feasible, and we are not disposed to deny MCI's applications on the basis of the unsubstantiated allegations which have been advanced herein by the telephone companies and Western Union. What seems a more likely obstacle to interconnection is as the Hearing Examiner indicated, the "carriers' intransigence, manifested in this case...." In these circumstances, the carriers are not in a position to argue that consideration of the interconnection question is premature. Since they have indicated that they will not voluntarily provide loop service** we shall retain jurisdiction of this proceeding in order to enable MCI to obtain from the Commission a prompt determination on the matter of interconnection. Thus, at such time as MCI has customers and the facts and details of the customers' requirements are known, MCI may come directly to the Commission with a request for an order of interconnection. *We have already concluded that a grant of MCI's proposal is in the public interest. We likewise conclude that, absent a significant showing that interconnection is not technically feasible, the issuance of an order requiring the existing carriers to provide loop service is in the public interest.* [Emphasis added]

Summary

This is a very close case and one which presents exceptionally difficult questions....We wish to make clear...that the findings and conclusions reached herein apply only to the frequencies specified, and for the areas described, in the applications now pending before

---

**The carriers did not request the addition of issues respecting loop service and they have refused to discuss the matter with MCI, asserting that consideration of such a request must await the outcome of this proceeding.

us....However, it would be inconsistent with the public interest to deny MCI's applications and thus deprive the applicant of an opportunity to demonstrate that its proposed microwave facilities will bring to its subscribers the substantial benefits which it predicts and which we have found to be supported by the evidence in this proceeding. We conclude, on the basis of the record as a whole, that the public interest will be served by a grant of MCI's applications.

Dissenting Statement of FCC Chairman Rosel H. Hyde

The decision of the majority is diametrically opposed to sound economics and regulatory principles. It likewise is designed to cost the average American ratepayer money to the immediate benefit of a few with special interests.

Of course, this agency must be alert to new developments which would promote the public interest in more efficient or better service. Clearly, we should be responsive to competitive innovation which meets the same test. Thus, I strongly favor introducing new competition to the present carrier system, such as through domestic satellite authorizations, which will benefit the public interest.

But the law is equally clear that the public interest is *the* test — that this agency should not authorize new service simply because it constitutes "competition." Thus, in the leading case in the common carrier communications field, *FCC v. RCA Communications, Inc.,* 346 U.S. 86 (1952), the Supreme Court made clear that there could be no talismanic reliance by the Commission upon "competition" *per se,* but rather that the Commission, as the expert agency, was required to search the record and facts of each case, and to authorize new competitive service where it could warrant that competition would serve some beneficial purpose.

The record in this case simply does not permit the agency so to warrant.

* * *

Why does the majority condone this obviously grossly inefficient use of the spectrum? Because, they say, of the low cost and "flexibility" of the facilities proposed by applicants. It should be noted that the so-called "flexibility" cited by the majority is here, as it was in the TELPAK case, a mere euphemism connoting lower rates. But how is it that applicant is able to propose lower rates than the existing common carriers for private line service? For no other reason than that it is proposing a typical "cream skimming" operation. Thus, it has selected a major route, Chicago to St. Louis, with heavy traffic density characteristics and the concomitant lower unit costs. The existing common carriers, on the other hand have been encouraged by the Commission, primarily for social reasons, to base their rates both for message toll and private line services on nationwide average costs. Thus the small user in the hinterlands is afforded the same rates as the large users in the major cities. The evidence in this record tends to show, and there is no basis in our experience to believe otherwise, that AT&T and Western Union could offer lower rates for private line service between Chicago and St. Louis than those proposed by MCI, were they to base such rates on their costs for that route alone.

The chances are, and the record so indicates, that AT&T and Western Union will be constrained to lower their private line rates to meet the competition of MCI. This, however, means that other users, either the small private line users who are not so fortunate as to live in large cities, or message toll users, or both, will have to make up the difference if total revenue requirements are to be met. This is in addition to the higher costs which are bound to fall on all communications users by virtue of the inefficient use of the frequency at stake.

The effect of the majority decision is to destroy the principle of nationwide average rate making. Perhaps, as some economists have urged, this is a desirable result. But it certainly should not be accomplished through the vehicle of a grant of a radio authorization which represents a wasteful use of our scarce spectrum space. This entire case is a classic example of an inefficient, uneconomic misallocation of resources.

...I am compelled to say, based upon this record, that the majority has acted not upon considered judgment of the evidence but rather, as in the *RCA* case, simply upon a reflex judgment that competition, particularly to AT&T, is *per se* good. The action, because it is unsupported by the facts of record, is inconsistent with the statutory standard.

Concurring Statement of Commissioner Nicholas Johnson

...What do the proponents of regulation fear about this "radical" experiment with competition? "Cream skimming." ...The suggestion is that if competition were ever to be extended to such a sensitive area as pricing, then the "correct" prices might not be charged. The services for which costs are low might be priced lower than those for which costs are high. The currently exorbitantly high profits from the low cost services would be "skimmed" by a company that would fail to provide the higher cost services as well.

* * *

...To make the [cream skimming] argument is to suggest that any deviation from present pricing policies is somehow immoral, illegal and unprecedented, and will result in higher prices to all consumers. It is none of the above. The telephone rate system is riddled with subsidies. In the first place, the telephone company itself deals in "cream skimming." The really high-cost-low-revenue subscribers — those who live in rural America — would never have had telephone service had they waited for Bell to ring. They had to get government assistance through the Rural Electrification Administration, their own cooperative telephone services, and non-Bell microwave carriers. In the second place, Bell has notoriously "undercharged" for those services in which it is engaged in competition — making up the difference by soaking those subscribers over whom it has a monopoly. That is what the TELPAK, Private Line, Telegraph Investigation, and Phase 1B (Dkt. No. 16258) cases have been concerned with. Finally, Congress has just given its approval to the principle of "free or reduced cost" service to educational radio and television interconnection — over the protests of the telephone company.

We can never know that a new product or service offered in the marketplace will become economically viable. Henry Ford couldn't know for sure. We cannot know for sure about MCI. Those who do business with MCI are getting a special service; it may be, for some, less satisfactory than the services offered by Bell; it will be, for all who use it, much cheaper. Presumably the consumers who will now have a choice will exercise it in the manner best suited to their interests. We have long authorized private microwave systems for individual companies servicing their own needs; MCI will simply be doing for a small group of users what each could have done for itself, had it the volume of business to justify the investment. Competition need not be so unusual. We are even experimenting with increased competition in the sources of equipment that can be "plugged into" the telephone system — equipment known to the Bell Telephone-Western Electric complex as "foreign attachments."

No one has ever suggested that government regulation is a panacea for men's ills. It is a last resort; a patchwork remedy for the failings and special cases of the marketplace....But I am not satisfied with the job the FCC has been doing. And I am still looking, at this juncture, for ways to add a little salt and pepper of competition to the rather tasteless stew of regulatory protection that this Commission and Bell have cooked up.

G. LONG-RANGE POTENTIAL OF SPECIAL SERVICE COMMON CARRIERS

The 1969 FCC decision granting MCI's application for a common carrier franchise between Chicago and St. Louis, excerpted in the previous section, was a precedent-setting step. The decision may be modified by subsequent legal action arising from appeal by the common carriers, but it stands as an indication of the winds of change which are beginning to sweep through the communications industry. Regardless of the specific near-term outcome of the MCI controversy, the need for such services and therefore the impetus for their establishment remains.

The establishment of special service common carriers would appear to have a greater long-range impact upon the computer and business communications user community and equipment suppliers than upon the common carrier industry. In some ways, the proposed SSCC services complement the present range of end-to-end services offered by the existing carriers. Thus, the revenues of this new class of carrier may come largely from the generation of new business in specific submarkets of the communications industry and from the tailoring of services to other new rapidly growing markets which account for only a small percentage of the existing carriers' revenues and interests. On the basis of this premise, those favoring the SSCC concept have predicted that little diversion of existing common carrier revenues will occur. In their opposition to MCI's license application, however, the existing carriers have prophesied disruptive effects (which we have discussed earlier) upon their revenues and rate structures. Little *hard* data is available at present to conclusively support either point of view. As MCI goes into operation, however, the extent of revenue diversion, if a significant amount, should be apparent.

We have seen that the lease of channels for data transmission would be one of the primary activities of the SSCC. Demand for such channels is growing at a tremendous rate today, and is highly price-elastic relative to other classes of communications demand. One indicator of this rate of growth can be found in the recent annual growth rates for common-carrier-provided data sets in the United States. Considering only leased private lines (the market segment which the SSCC would serve), during the period 1964-69 the number of data sets (modems) leased by the Bell System for low-speed (telegraph-grade) lines increased at a rate of more than 85% per year. For voice-grade channels, the rate of growth was more than 55% per year, and for broadband channels (1966-69 only) it was 60% per year.[13]

---

13. Calculated from data supplied by AT&T, presented in Stanford Research Institute, *Report to the FCC in Docket No. 16979*, Vol. 4, "Patterns of Technology in Data Processing and Data Communications," Table 4, p. 63.

An additional characteristic of data communications demand today is its high degree of price-elasticity, relative to other classes of communications demand. We have discussed this factor in Chapter VII on line sharing/resale, page 151. With the lower prices and increased flexibility of use which the SSCC would offer, the rate of growth of data communications demand can only be increased. Adding such a stimulation of demand to the already quite impressive monopoly-market growth rates presented above, it is reasonable to expect the SSCC to thrive upon a portion of the *new* leased-line communications markets.

The FCC's approval of MCI as the first special service common carrier is likely to attract a large number of independent firms to enter this market. It may eventually result in the creation of a separate nationwide communication network. A firm affiliated with MCI has already filed an application for authorization to offer similar services between New York City and Chicago. Other MCI affiliates are preparing applications for services between St. Louis and New Orleans, and between other major metropolitan and industrial areas. As noted earlier, while FCC action on MCI's application for the St. Louis-Chicago route was still pending, Interdata Communications, Inc. (ICI), a new firm, filed an application to provide similar service between New York City and Washington, D.C. Also, as discussed in the following section, University Computing Company, a rapidly growing diversified computer firm, is planning to enter this market.

It is likely that a separate communications network — designed with data transmission capability as its paramount objective, although voice transmission will be an integral part of the system — will emerge as this new class of carrier develops. Some coordination and cooperation between the various carriers comprising this network will be necessary, to provide through service to a single customer over the routes of several different carriers, and perhaps to provide joint billing to such customers, as well as to coordinate plans for plant expansion, and so forth. The larger the number of new specialized carriers, naturally the more difficult this coordination would become. However, the problems do not appear unsurmountable. Ultimately, the resulting integrated or semi-integrated network would be conceptually similar to the present network of some 2100 independent telephone companies, and would probably function much as that system functions.

The need for each of the individual companies to have sophisticated coordination agreements with the other SSCC's will be much less than those presently required by the interconnected telephone companies, since the basic SSCC service will be the leasing of trunk capacity for point-to-point use. Rates may not be completely uniform per circuit mile of leased capacity, but terminal equipment charges, special service charges, and intricate end-to-end service tariffs are likely to be eliminated.

SSCC system development will logically be concentrated initially between large urban centers where the demand exists for interbusiness voice and data links. The existing common carriers will probably continue the cry of "cream-skimming" on each license application but may, because of the threats of competition, become more responsive to market requirements. For example, shortly after the MCI oral testimony before the FCC in 1968, in which MCI argued that permitting subscribers to share channels was one of the distinctive features of its service, AT&T announced that it would allow customers to share voice and sub-voice grade leased channels. (Other factors, such as the filings in the FCC's inquiry into the interdependence of computers and communications, also were contributing influences in this AT&T policy shift.)

Microwave transmission capability which is potentially useful for SSCC operations already exists in many parts of the country. Existing CATV microwave systems are all potential paths for two-way transmission of data and voice, using existing relay sites and towers. Two-way transmission over such facilities would require only the addition of dish or horn antenna and transmission equipment in the opposite direction to the present one-way signal capability. The geographical location of some of these CATV systems (generally rural areas) may not be optimum for development of a national grid stressing direct links between major urban centers; however, interconnecting links and alternate routes will be required as the system grows, and CATV microwave systems may be used for these purposes. Existing CATV local distribution facilities in urban areas (broadband coaxial cable networks) may also prove quite useful as "local loop" channels to link individual SSCC subscribers with the microwave terminal points of the SSCC long-haul channels.

The major part of the new long-haul system will have to be designed and constructed as new plant on new routes. Because there would be little or no switching equipment in the system, the construction would be limited primarily to relay towers and their support facilities. This will undoubtedly be an important ingredient in the shortness of time required to complete a nationwide system capable of offering the lease of carrier capacity to a series of multiple points. In any event, a national network of special service common carrier facilities could possibly emerge in skeletal form within three to four years from the first license grant.

Whether voluntarily or involuntarily, the realities of the submarket will require that connecting carriers coordinate and cooperate. The sophistication of business voice and data communications users will dictate an equal degree of sophistication from interconnecting carriers in all matters, ranging from ease of getting the customer tied into the system quickly and efficiently, to providing capacity and flexibility under all conditions, to the billing and administrative details. Thus, the technical and administrative matters of importance to customers are but one of the areas where would-be new microwave common carriers must prove themselves capable and efficient, for they must each perform well in marketing, pricing, interconnection, system design, and certainly not last or least, relations with the Federal Communications Commission and possibly numerous state commissions. The ability of the participating carriers in the special service common carrier network to react positively to each of these challenges, plus their ability to enter into cooperative arrrangements, will have much to do with both the shape of the physical plant and the structure of the industry which may evolve.

H. UNIVERSITY COMPUTING COMPANY'S PROPOSED COMMON CARRIER DATA TRANSMISSION NETWORK

As this book went to press, University Computing Company (UCC), a Dallas-based computer service bureau and peripherals manufacturing firm, was in the late stages of planning a new nationwide common carrier network designed for data transmission and using state-of-the-art digital transmission and computer-based switching equipment. As explained to the authors by UCC management,[14] a wholly-owned subsidiary to be called Data Transmission Company (formerly known as Microwave Transmission Corporation) would offer both leased and switched data transmission services with features not currently available on a common carrier basis, and with prices less than one-half those of the present common carriers.

---

14. Discussions with Mr. Sy Joffe, Executive Vice President, University Computing Company, and President, Data Transmission Company; and Mr. Edward A. Berg, Vice President of Operations, Data Transmission Company.

UCC has, for several years, shown interest in entering the communications common carrier field. In May 1968, the company announced a tender offer for 750,000 shares of Western Union common stock; Western Union successfully opposed this move. In August 1968 UCC formed a new subsidiary, Microwave Transmission Corporation, and agreed to acquire TransAmerican Microwave, Inc., of Los Angeles, a large microwave system which serves CATV operators between Los Angeles and San Francisco. Since UCC had no previous operations in CATV or related fields, there was widespread speculation that its entry into CATV microwave transmission was a prelude to system expansion to include point-to-point common carrier service. From UCC's point of view, common carrier data transmission services would be a natural adjunct to its remote access computer services.

UCC's proposed data network is sweeping in scope, and will undoubtedly generate considerable public discussion over the next several years. This proposed common carrier service offering, which requires FCC approval, will certainly be strongly opposed by the telephone companies and by Western Union, who risk substantial impact upon their most rapidly growing market segment — data communications. An equally strong reaction in support of the UCC proposal is likely to come from the remote-access data processing community, who feel that today's common carrier communications services are too costly, that the facilities employed in the provision of these services are of inadequate quality for efficient data communications usage, and that the present carriers are unresponsive and ill-equipped for quickly and properly serving the specialized needs of data transmission users.

The proposed data network would consist of microwave transmission facilities and computer-based switching equipment interconnecting the major urban concentrations of data processing activity across the country. The system as currently configured would cost over three-quarters of a billion dollars to build. Routing for the proposed network is designed to minimize system cost in light of the predicted volume of traffic among the cities which will be served. The initial microwave route takes the form of the letter "W," and connects San Francisco, Los Angeles, Dallas, Kansas City, Minneapolis, Chicago, Atlanta, Washington, D.C., Boston and other major cities. A total of 52 cities would be served by the initial skeletal network.

Transmission Technology

Transmission and multiplexing equipment most suited for digital data signals rather than analog voice signals are important components of the UCC data network. As discussed in the previous chapter, an all-digital transmission system (such as in the UCC network) offers lower costs and significantly improved performance (lower error rate, etc.), for data transmission. Excluding switching equipment, the two components of such a system are time-division multiplexing (TDM) equipment and digital transmission facilities (such as coaxial cable, which can carry high-speed digital pulses, or digital microwave equipment). TDM equipment can also be employed in conjunction with conventional analog channels for data transmission[15] (the composite multiplexed signal is converted to analog form by a high-speed modem before transmission over the long-haul channel), but greater advantages can be obtained through use of TDM with digital transmission media.

Digitized voice signals (e.g., using PCM) generally require more bandwidth for transmission than do such signals in analog form; therefore, telephone company development work for the new T-carrier digital systems (which are designed primarily for transmission of voice, rather than data) has concentrated upon the use of wire and coaxial cable transmission media, rather than microwave facilities, since microwave channels are bandwidth-limited. Digital *microwave* systems for voice communications are thus in their infancy. For a system dedicated to data transmission, however, digital microwave and time-division multiplexing are an attractive combination. Using digital rather than conventional analog microwave equipment, for example, reliable communications can be maintained with a much lower signal-to-noise ratio (the digital signal is regenerated at each repeater, so that noise and distortion are non-cumulative and are therefore lesser problems); therefore, lower transmitter power levels can be used (thus reducing interference to adjacent microwave stations), and system degradation and outages due to transmission path fading occur less frequently.

Since UCC has designed its proposed network expressly for data transmission, and was not compelled to use existing communications facilities, the

---

15. Indeed, most present applications of TDM, such as its use by commercial time-sharing computer services and by the Western Union SICOM and INFO-COM communication services, employ voice-grade analog channels as the basic transmission media. The new T-carrier systems discussed in the previous chapter, however, transmit signals in digital form over wire and cable transmission facilities.

company has chosen to employ TDM exclusively (rather than the frequency-division multiplexing technology upon which today's telephone network is based), together with state-of-the-art digital microwave transmission facilities. (Coaxial cable may also be used for portions of the network.) The microwave system for the backbone UCC network would have a capacity of 4000 to 16,000 data channels, each channel with a data rate of 4800 bits per second.

The UCC system is designed for an end-to-end error rate of slightly less than one bit in error per 10^7 bits transmitted, and the tariff offering will *guarantee* that the error rate will not exceed approximately one bit in error per 10^6 bits transmitted. The exact guaranteed error rate will not be determined until the system is further along in design and/or construction.

Voice transmission over the proposed network is possible, using vocoders (a device which digitizes an analog voice signal), and would be available to all subscribers, but is not the primary objective of the system. Most likely voice transmission would be used only in conjunction with various data transmission services.

Local distribution of signals from the UCC terminal in a city to the individual subscribers located nearby would not rely upon interconnection with and use of telephone company facilities. In fact, UCC disclaims any desire to use such existing telephone lines, and instead plans to install and operate its own local loop transmission facilities, and provide them on a leased basis to UCC subscribers ($15/month is the planned fee for a local subscriber access connection). Such local distribution would be by means of one-hop microwave systems operating in the 18 GHz common carrier radio band, and using omni-directional or narrow-beam antennas to transmit directly from the UCC site to the subscribers' nearby office or plant locations. Only through operational control of the local access facilities, UCC claims, can it guarantee the high-quality end-to-end error performance discussed above. It would appear, however, that realistic planning would indicate use of telephone company facilities as well as local microwave links. Not only would lease of telephone lines be more economical in many cases, but also use of such lines would extend the radius of local area coverage beyond that available through exclusive use of a one-hop microwave distribution system.

Switching

The use of all-digital transmission facilities distinguishes, the UCC proposal from the MCI-type special service common carriers discussed in the earlier part of this chapter (which plan to provide voice, video, and other analog services in addition to data transmission, and thus have chosen to use conventional analog microwave facilities). The other distinction is that UCC plans to establish a number of switching centers along its microwave network, in order to provide dial-up communication services in addition to leased channels, whereas the MCI-type carrier would limit itself to the provision of leased channels. The initial UCC data network plan includes 40 switching centers located in 36 cities. Additionally, five tandem switching offices (higher-level offices for switching the trunks among the 40 first-level switching centers) would be located in Los Angeles, Dallas, Atlanta, Chicago, and New York City. Both circuit-switching and message switching service would be provided at each switching center by a single computer with a circuit-switching reed relay bank attached. End-to-end connect time is designed to be three seconds or less. Up to 4000 subscribers could be served by each switch, with a "grade of service" (probability of encountering a "system busy condition") of $P = 0.01$.

Services to be Provided

The proposed UCC network is oriented exclusively to the provision of data communication services on a common carrier basis. Both leased and switched channels would be offered for transmission at data rates up to and including the following: 150 bits per second (bps), 4800 bps, 9600 bps, and 14,400 bps. In addition, 48,000 bps channels would be offered on a leased basis only. Asymmetric combinations of these transmission speeds (e.g., 150 bps in one direction, and 4800 bps in the other) would also be offered. Multi-user sharing of channels would be permitted without restriction.

UCC has defined its potential market as encompassing all data communications requirements, rather than restricting itself to the intra-corporate communications market (the traditional private line market and the "customized services" market), served by the MCI-type special carriers. Thus, a UCC subscriber may use the switched service to call any other subscriber having compatible terminal equipment; and UCC plans to provide code- and speed-converting capabilities at its switching centers so that incompatible terminals may converse. The firm also plans to lease terminal equipment to its subscribers.

Pricing for the switched communication service would be based upon a modified "postalized" rate structure. That is, pricing for inter-city switched service would be largely *independent of distance,* and would be based primarily upon line holding time and channel bandwidth. Such a rate structure would obviously be a boon to many data users, such as time-sharing computer services eager to serve customers in distant cities, but may raise some questions of possible discrimination against those users who need only short-distance channels. An additional feature of the pricing for switched service is the very short minimum charge time: rather than Bell's 3-minute minimum period (a 1-minute minimum has been tested in a few cities), UCC plans to use a minimum charge time of approximately 6 seconds. Line holding time beyond the minimum period will also be charged for in 6-second increments. Both the postalized rate structure and the short minimum charge time are common carrier service characteristics requested by data users in the FCC's Computer Inquiry, as discussed in the previous chapter.

Leased channels would be priced on the basis of channel bandwidth and airline mileage, similar to the present common carrier philosophy. Per-mile prices are expected to be approximately proportional to rated transmission speed. However, prices for leased channels relative to those for switched channels will encourage use of the switched service; i.e., a greater amount of usage will be required to justify a leased line than is the case with the present common carrier services.

Outlook for the UCC Proposal

UCC's proposal to establish itself (through a wholly-owned subsidiary) as a new nationwide common carrier dedicated to data communications service is a revolutionary idea. The planned UCC network and the communication services which would be offered using it incorporate many of the characteristics desired by the computer industry for data communication services. Some of these characteristics are found also in the special service common carrier concept discussed in the first sections of this chapter, while others are unique to the UCC proposal. The need for such services is clear, and is increasing with the exponential growth in the use of computers and especially of remote-access computing. However, UCC's ambitious goals will not be easy to achieve. Considerable technical, legal, and institutional obstacles must be overcome before the UCC data network could become a reality.

Strong opposition to the UCC proposal from AT&T and Western Union is inevitable. To the extent that the proposed UCC services meet the needs of a large segment of data communications users, they would threaten the expected rapid growth of the existing carriers' data transmission revenues. While the overall effect upon the telephone companies would be very small for years to come, Western Union private line, TWX/Telex, and specialized data services revenues (and, more importantly, the future growth of these revenues) could be more seriously affected. Viewed from a public interest standpoint, possible competitive effects upon Western Union should not bar UCC's entry into the communications common carrier industry, unless these effects would so weaken Western Union's overall viability as to impair its ability to furnish essential public services, such as the Public Message Service (telegram service, which some will argue has outlived its usefulness anyhow). Even if this were the case, it is doubtful whether the public interest would be served by foreclosing more efficient new suppliers from the rapidly growing data communications field. The FCC can be expected to consider this issue carefully.

An additional important issue which must be resolved before UCC's proposal can be approved by the FCC concerns the company's potential dual status as a regulated common carrier and an unregulated supplier of computer equipment and services. If the requested UCC common carrier license is granted, the company will be subject to most of the considerations regarding common carrier provision of data processing services, discussed at length in Chapter III. (These considerations were discussed from the standpoint of common carriers entering the unregulated data processing market, but they apply equally to diversification in the reverse direction, as UCC is attempting to do.) The UCC common carrier subsidiary would sell communications services to the parent company and to other UCC subsidiaries, and would purchase computer time, programming support, and other services from the unregulated parent and subsidiaries. UCC management has stated its unwillingness to foreclose either of these types of intra-company business relationships (although it intends to conduct them on an "arms length" basis), so the FCC must come to grips with the policy problems associated with such a regulated/unregulated business combination.

In summary, the proposed UCC data network represents the most ambitious attempt to date to exploit modern technology in the provision of common carrier data transmission services. The UCC network would offer many of the service characteristics needed by the computer industry, and as such, deserves serious consideration. Perhaps even more than the special service common carriers upon which this chapter has primarily focused, UCC's proposal raises important policy issues which the FCC must address. Even under a favorable turn of events it is unlikely that the UCC data network could be operational before the mid 1970's.

CHAPTER X

PRIVACY AND SECURITY

The problem of privacy in remote access data processing systems has several dimensions. First, use of the system must be restricted to authorized personnel. Second, the individual user's information files (programs and/or data) must be protected from both intentional and unintentional access by other users or by the operators of the system. Third, precautions must be taken to prevent the loss or destruction of information due to system failure, human error, etc.[1] The problems of privacy *per se* are most acute in conversational time-sharing systems, which have a multiplicity of users, most of whom maintain private information files within the system, and some of whom may wish to share their files on a limited basis with certain other users.

A. TECHNIQUES FOR SOLUTION

Briefly stated, there are a number of safeguards which may be built into the design of a remote access data processing system to avoid compromise of security in general and the invasion of privacy in particular. First, access to the system may be controlled by the issuance of a secret password to each authorized user; each user would be required to enter this password from his terminal whenever he

---

1. These problems suggest the necessity for providing at least the following safeguards:
   - Safety from someone masquerading as someone else;
   - Safety from accidents or maliciousness by someone specifically permitted controlled access;
   - Safety from accidents or maliciousness by someone specifically denied access;
   - Safety from accidents self-inflicted;
   - Total privacy, if needed, with access only by one user or a set of users;
   - Safety from hardware or system software failures;
   - Security of system safeguards themselves from tampering by nonauthorized users;
   - Safeguard against overzealous application of other safeguards.

   (From R. C. Daley and P. G. Neumann, "A General-Purpose File System for Secondary Storage," *Proceedings—Fall Joint Computer Conference* [Las Vegas, Nevada, November 30, 1965] p. 214.)

"logs in" to use the time-sharing system. For example, in an inquiry system the user could be required to enter an access code number before he would be permitted to query the system's files.

Additional means of guarding against unauthorized use of the remote access system include physical limitations on access to the terminals (e.g., in a commercial environment, only authorized system users will have terminals in their offices) and verification by the system when a user logs in that he is using a terminal approved for his use.

Privacy of information in the system can be assured in a number of ways. Cryptographic devices located at the terminals and at the central computer may be used to ensure the privacy of information transmitted over the communications links. Data stored at the central computer may also be encrypted if desired.[2]

In many time-sharing systems, users may wish to share their data and/or program files with certain other users, or they may wish to make "public" certain files. This may be accomplished simply by letting each user specify who (if anyone) may have access to each of his files, and what level of access shall be permitted (e.g., permission to read the file but not modify or erase it, permission to execute a program but not to print the program, or permission for unrestricted access).[3] Each request for memory access (including reading from memory, writing into memory, or executing instructions stored in memory) by a user of the system is checked to determine if the memory address specified is within the user's own file storage area or not. If it is, the access request is granted. Otherwise, the access request is granted by the system only if, for the file in question, access permission of the required type (read, write, or execute) has been established for that user. When permission is granted to access another's files, procedures may be triggered to record the "borrower's" identity, the file name, the time of day, the nature of the access, etc. Subsequently, "audit trails" may be prepared, listing which non-owned files were accessed and by whom.

---

2. This includes both storage in the computer's on-line memory (internal core memory and direct-access storage devices such as magnetic disks, drums, and data cells) and storage on off-line media (such as magnetic tapes and punched card files). The data is encrypted by the computer before being placed into storage.

3. Similar precautions may be taken for the portion of memory in which resides the system supervisory control program itself: authorized "system programmers" are permitted access, perhaps in varying degrees based upon their assignments, while others are excluded.

Implementation of these access control procedures may be accomplished by use of both hardware and system software techniques — the combination of which results in much greater data security than is usually found with manual storage of information.[4] For example, there exist computer memory devices which are physically of a "read-only" or "execute-only" nature. Once information is placed there it cannot be modified or (in the "execute-only" devices) even be examined under program control; likewise hardware "trapping" techniques facilitate the foolproof keeping of an "audit trail" record of accesses.[5]

There are many approaches to the problem of safeguarding from loss or destruction the information in a remote access data processing system. Many systems, especially of the real-time variety, employ backup equipment (e.g., reserve central processing units and mass storage devices) and auxiliary sources of electric power. This permits uninterrupted operation and avoids the possible loss of information due to system failure, especially important in real-time applications such as military command and control systems, and airline reservations systems.

It is standard procedure to periodically "dump" the contents of the computer's mass storage memory onto magnetic tape for off-line storage, both to satisfy requirements for permanent records and to permit recovery from system catastrophes which might cause the destruction of information in the on-line files. In addition to these periodic memory "dumps," all transactions between "dumps" which modify the memory files may be recorded on tape in real time as they occur; this transaction history plus the last "dump" may be used to recreate memory files immediately prior to any catastrophic failure. Also, duplicate copies of the backup tapes may be kept, vault storage may be provided for them, etc., depending upon the nature of the application and the importance of absolute information security.

---

4. The combination of both hardware and software security techniques may be made immune to virtually all forms of tampering. This is evidenced by the fact that the MULTICS time-sharing system at Project MAC is being seriously considered for clearance to store classified military information (from a discussion with Professor Malcolm M. Jones, Assistant Director of Project MAC, Massachusetts Institute of Technology).

5. For a discussion of hardware and programming aspects of file security in a multi-user time-sharing system, see Robert M. Graham, "Protection in an Information Processing Utility," *Communications of the ACM,* Vol. 11, No. 5, May 1968, pp. 365-369.

B. LEGAL CONTROLS

A desire to prevent an interference with individual rights is the motivation for governmental attention (by means of legislation or less formal approaches such as regulatory agency action) to problems of privacy and security in their various manifestations. At present the computer service field does not seem to exhibit the need for such attention, not because privacy is not a real concern there, but because adequate safeguards and remedies already exist in the private sector. As we have seen, numerous hardware and software techniques and system operating practices are available to provide any desired degree of security protection to a multi-accessed data processing system. In-house private systems make use of these techniques to insure adequate system and information protection, and users of commercial data processing services are equally concerned about the level of protection afforded to their data by the systems being used. As a result, such systems of necessity also employ appropriate safeguards, and the authors are unaware of any significant complaints or abuses which have arisen from their activities. No requirement for legislative or other governmental action seems to exist at present. Indeed, it could be argued that inappropriate or inflexible governmental action regarding protection of privacy might do more harm than good—by hampering the rapid development and diversification of all types of remote-accessed data processing and information services.

However, it must be admitted that the above conclusion may become less valid in the more distant future, as multi-access systems and the applications of these systems proliferate and begin to affect the consumer rather than exclusively the commercial users found today. Looking forward to the day when the "computer utility" may become a reality, the absence of adequate security safeguards may endanger not only the interests of large-scale computer users but the interests of every individual who wishes to enjoy the advantages of an advanced technology. It may therefore become necessary at some point in the future to guarantee that available techniques and procedures for safeguarding users' privacy be employed in a consistent, competent, and uniform manner by data processing service organizations. This potential requirement has at least two dimensions: First, what means may be employed to insure that technical safeguards are incorporated into multi-access computer systems? And second, what means may be employed to insure that the companies offering these services and the individuals operating the systems meet minimum standards of competence and integrity?

It is important to note that while the question of privacy with respect to "data stored in computers and transmitted over communication facilities" was addressed in the FCC Inquiry,[6] the jurisdiction of the FCC in these matters is rather narrowly restricted. The Communications Act of 1934, the FCC's enabling legislation, provides that "no person not being authorized by the sender shall intercept any communication [by wire or radio] and divulge or publish the existence, contents, substance, purport, effect, or meaning of such intercepted communication to any person..."[7] This relates to the transmission rather than to the storage and processing of information, and thus covers only a small part of the overall privacy problem discussed here. At present the great majority of all computer installations, both private systems and for-hire services, do not even use communication lines to provide remote access capabilities to their users; but they have a concern for privacy nonetheless. Communications capability only magnifies certain aspects of the problem.

The result is that if and when governmental attention should be directed to the subject of privacy in data processing systems and services, the FCC would be limited, under present law, to a rather small role. New federal or state legislation which might be thought necessary at some time in the future could enlarge the role of the FCC as well as the role of other federal or state agencies with regard to matters of privacy. The authors are not qualified to discuss the jurisdiction of various state and federal bodies to meet the problems of privacy and security in computer-based services, and will only outline some of the possibilities for legislation or other governmental action.

1. *System Licensing and Inspection.* System licensing requirements could be devised under which no person would be privileged to offer the services of a multiple access computer system until the system included hardware and programming safeguards of the kind outlined above. Licensing requirements could be combined with periodic inspection aimed at insuring that these safeguards were in fact being consistently employed.

---

6. FCC Notice of Inquiry, Docket No. 16979, para. 25 (J).

7. 47 U.S.C.A. Sec. 605.

2. *System Certification.* A voluntary program might be instituted under which a data processing firm could comply with certain standards and be subject to inspection, thereby securing the right to some sort of official certification by a governmental body.

3. *Licensing of Personnel.* One possible solution to the second aspect of the problem would be to impose licensing requirements, in the manner of other professional licensing, upon the *personnel* who are to operate these systems. Persons who wished to enter this phase of the computer industry as systems operators, programmers, etc., could be required to meet standards of both technical competence and reliability of character. Additionally, companies could be required to bond such personnel against the possibility of their unauthorized disclosure or use of data which causes damage to the user.

4. *Compulsory Insurance and/or Bonding.* Data processors offering multi-access computer services could be required to obtain insurance or bonds which would compensate any computer user who suffered damage as a result of unauthorized disclosure or use of his data, or of loss of that data. Unfortunately, to date the insurance industry does not seem to have recognized the need for underwriting some risks associated with time-sharing service operations. Basic fire and "errors-and omissions" insurance is available to such a service operator, but more complete liability coverage will be required.[8] As remote-accessed computer services become widespread, and business users demand protective coverage for the privacy and security risks to their data, the insurance carriers will undoubtedly make it available to the computer service organizations.

---

8. Robert P. Bigelow, "Legal & Security Issues Posed by Computer Utilities," *Harvard Business Review,* September-October 1967, p. 154.

5. *Criminal Sanctions* could be imposed for (a) the unauthorized disclosure or use of information contained in a multi-access computer system, and (b) the failure of any company or person to comply with rules requiring implementation of specified safeguards.

CHAPTER XI

SUMMARY

Several long-term policy problems in the computer and communications industries are explored in this book. Attention is centered on issues under study by the Federal Communications Commission in its public inquiry into the "Regulatory and Policy Problems Presented by the Interdependence of Computer and Communication Services and Facilities." These problems are primarily in six areas:

1. Should commercial offerings of data processing and information services be regulated?

2. Should communications common carriers be permitted to offer data processing and information services to the public and, if so, under what ground rules?

3. Should non-common carriers be permitted to offer message-switching services to the public?

4. Are common carrier restrictions on the use of their communication lines reasonable?

5. Are the existing common carrier communications facilities and service offerings satisfactory for data communications?

6. Is the storage and transmission of proprietary data sufficiently safeguarded by available mechanisms and practices?

The discussion begins with a general background summary which covers common carrier communications and the data processing industry in the United States. The first problem area, the possible regulation of data processing services, is discussed only briefly because of the lengthy and non-controversial discussions on this topic by the organizations responding to the Notice of Inquiry. In agreement with the clear consensus of interested parties, including the common carriers, data processing service bureaus, computer equipment manufacturers and the users of these services, the authors conclude that data processing and information services do not fall within the FCC's jurisdiction, and it is neither necessary nor desirable for these services to be regulated at this time.

In Chapter III the interest of the common carriers in, and their initial steps towards, the provision of data services to the public are described. The problems inherent in carrier provision of data services and the alternative ground rules under which carriers might be required to operate in this marketplace are discussed. It is concluded that it would be unnecessarily restrictive to completely prohibit the carriers from offering data services. However, in order to prevent carriers from obtaining unfair competitive advantage, they should be required to establish wholly separate subsidiaries for commercial offerings of data processing and information services, and the subsidiary should not be permitted to either lease communication channels from or sell data services to the parent carrier corporation.

Chapter IV describes the new technology of *computer-based* message switching and the interest of non-carriers, such as the Bunker-Ramo Corporation, in adding message switching features to existing data processing or information systems — to offer "hybrid" services. The question of whether computer-based message switching should be offered exclusively by common carriers is considered. It is observed that computer-based message switching is not a "natural monopoly" service and is sensitive to technological developments; also various users may require systems with markedly different operating characteristics. It is therefore concluded that the opportunity to provide message-switching services should not be limited to common carriers, in spite of the fact that the transfer of "hard copy" messages between firms is a service which may be required on a regular basis. An alternative solution, somewhat less satisfactory to the authors, would be to allow non-carriers to provide message-switching services only if these services are an "incidental" part of a larger primary business of offering data processing or information services.

Chapter IV also considers the nature of two new message-switching services offered by Western Union — SICOM and INFO-COM. Technologically, these services are commendable since they allow *substantial* reductions in communications cost by using time-division rather than frequency-division multiplexing to derive low-speed communication channels. However, Western Union only provides the low-cost channels in conjunction with its own terminals and SICOM or INFO-COM computer facilities. Since it seems that these facilities are separable, the SICOM and INFO-COM services appear to constitute illegal tie-in's contrary to the antitrust laws of the United States. It is recommended that Western Union be required to provide the low cost communications channels separately, as a general public offering.

In Chapter V the reasonableness of the common carrier restrictions regarding the direct attachment of customer-provided devices ("foreign attachments") to the public switched communication network is examined. The carriers claim that these restrictions are necessary to protect the "integrity" of the national network, to avoid "divided responsibility" for maintaining the network, and to preserve the carriers' freedom to introduce improvements and innovations into the network without opposition from owners of devices attached to the network. We find that these claims are overstated, and enumerate several objections to the restrictions from the data user's point of view.

Chapter VI discusses the common carriers' restrictions regarding the interconnection of private communication systems to the common carrier network. The technical arguments given by the carriers to support the restrictions are identical to those given to support the foreign attachment restrictions. The carriers also claim that if unlimited interconnection were allowed, organizations would engage in "cream-skimming"; the authors believe that this fear is substantially overstated, although it may be true in very limited instances. If the carriers are truly obtaining economies of scale, cream-skimming will be minimal. In the provision of local voice-grade communication systems (such as in-house PBX's or public mobile radio service) cream-skimming cannot occur, and here interconnection has recently been allowed.

In Chapter VII two forms of line sharing are described: "time-sharing," where several organizations alternate in the use of a single channel, and "capacity-sharing," where a firm channelizes a line and "resells" the subchannels to other organizations. Limited forms of sharing are allowed under carrier tariff provisions. The authors explain how expanded line sharing would reduce communication costs by allowing optimally sized channels to be provided, by increasing leased line utilization levels, by injecting competition into the activity of channelizing communication lines, and by eliminating the carrier practice of pricing communication channels of different bandwidths according to factors other than cost.

Chapter VIII discusses the adequacy of common-carrier facilities and services for the data communications user. It is observed that the national network was originally designed for voice communications and has been adapted, in relatively minor ways, to accommodate data. Also, the rate at which digitally oriented facilities are being installed is not commensurate with the rapid rate of growth of data communications. As a result there are shortcomings in both the technology employed and the services offered for data communications. The absence of channels with bandwidths of the same order as the data transfer rates of typical data devices, particularly in between 2,400 and 50,000 bits per second, and the need for higher quality channels, preferably based upon digital transmission methods, is discussed. The need for a message switching service in which the user would be charged according to the amount of information sent, rather than the amount of time "holding" the line, is noted. It may also be appropriate to reduce the minimum charge time for dialed calls in light of the changing average call duration due to "data" phone calls. It is suggested that "short period" leased lines are needed and can be provided either by appropriate carrier tariffs or by permitting users to "time-share" lines.

The concept of a digital network is presented, and further study of digital networks and of the appropriate rate of converting the existing network to a digital one is recommended.

Chapter IX discusses the possible entry of a new type of common carrier, which we have called the special service common carrier. Such carriers would lease to firms inter-city trunk communication channels for both voice and data communications. They would not necessarily provide end-to-end service, which could be accomplished either through telephone company or privately constructed local connecting links. A wide range of channel bandwidths would be offered, and restrictions on the use of facilities would be minimal. The arguments in favor of allowing SSCC's to enter the national telecommunications market are evaluated and public benefits found to outweigh the possible injury to existing carriers.

In Chapter X the available technical safeguards to assure the privacy of proprietary data transmitted over communication channels and/or stored at computer centers are examined and found to be sufficient. The authors believe no governmental steps are necessary to assure security of data at this time.

APPENDIX A

Before the
FEDERAL COMMUNICATIONS COMMISSION FCC 66-1004
Washington, D. C. 20554 90954

In the Matter of)
)
Regulatory and Policy Problems Presented) DOCKET NO. 16979
by the Interdependence of Computer and)
Communication Services and Facilities)

NOTICE OF INQUIRY

Adopted: November 9, 1966; Released: November 10, 1966

By the Commission: Commissioner Wadsworth absent.

I. Preliminary Statement

1. The modern-day electronic computer is capable of being programmed to furnish a wide variety of services, including the processing of all kinds of data and the gathering, storage, forwarding, and retrieval of information — technical, statistical, medical, cultural, among numerous other classes. With its huge capacity and versatility, the computer is capable of providing its services to a multiplicity of users at locations remote from the computer. Effective use of the computer is therefore becoming increasingly dependent upon communication common-carrier facilities and services by which the computers and the user are given instantaneous access to each other.

2. It is the statutory purpose and responsibility of the Commission to properly regulate interstate and foreign commerce in communications so as to make available to all the people of the United States a rapid, efficient, nationwide and worldwide communications service with adequate facilities at reasonable charges. (See Section 1 of the Communications Act of 1934, as amended.) Thus, the Commission must keep fully informed of developments and improvements in, and applications of, the technology of communications and of related fields. (See Section 218.) Moreover, the growing convergence of computers and communications has given rise to a number of regulatory and policy questions within the

purview of the Communications Act. These questions require timely and informed resolution by the Commission in order to facilitate the orderly development of the computer industry and promote the application of its technologies in such fashion as to serve the needs of the public effectively, efficiently and economically. To this end, the Commission is undertaking this inquiry as a means of obtaining information, views and recommendations from the computer industry, the common carriers, present and potential users, as well as members of the interested public. The Commission will then be in a position to evaluate the adequacy and efficacy of existing relevant policies and the need, if any, for revisions in such policies, including such legislative measures that may be required. It will also enable the Commission to ascertain whether the services and facilities offered by common carriers are compatible with the present and anticipated communications requirements of computer users. The Commission will then be in a position to determine what action, if any, may be required in order to insure that the tariff terms and conditions of such offerings are just and reasonable and otherwise lawful under the Communications Act. (See Section 201 (b) and Section 202 (a).)

II. Emerging Computer Enterprises

3. A brief review of the more important types of computer enterprises now emerging will serve to illustrate the growing convergence and interdependence of communication and data processing technologies.

4. First of all, there is the so-called in-house use of computers. Banks, aircraft manufacturers, universities and other types of institutions frequently own or lease computers primarily for their own use. In the past, the batch-processing technique has generally been employed to satisfy the needs of the in-house users. Recently, however, time-sharing systems have been installed, particularly at universities and hospitals following the example of pilot Project MAC at the Massachusetts Institute of Technology. Because more than enough capacity exists to satisfy normal in-house needs, be they mathematical computation, data processing, simulation, or storage and retrieval, the idle or excess capacity is readily salable to others. Banks and aircraft manufacturers have already made such time available to persons outside the enterprise. Economies of scale may well lead to larger and larger machines with consequent incentive for in-house computer owners to sell computer time to the general public. Efficient utilization of these computers implies organization of time-sharing systems. It likewise implies in-

creased use of communication channels obtained for the most part from communications common carriers pursuant to tariffs filed with this Commission or state regulatory commissions, depending upon the intrastate or interstate nature of the channel.

5. Secondly, several of the major computer manufacturers maintain computer service bureaus. They sell computer time to customers and usually operate on a batch-process basis. However, conversion to time-sharing is proceeding rapidly. The potential for providing the computer with general economic data to complement specific company or industry data, has led to the establishment of data banks which can be used for such purposes as economic forecasting, product marketing analysis, and more specialized uses such as legal and medical reference library services. Multiple access to such data banks is again dependent on communications links obtained from common carriers under applicable tariffs.

6. Additionally, there are hundreds of non-manufacturing firms which offer a wide range of data processing and specialized information services. These services may be provided on either a batch processing or time-sharing basis. Many of these concerns are local in scope, but others are equipped with multiple access computers and are endeavoring to develop national time-sharing systems of which communication channels will be an integral part.

7. Finally, there are some very highly specialized computer services currently being offered. An example is the stock quotation service. For a number of years, brokers and financial institutions throughout the country have been supplied with up-to-the-minute prices and quotations on securities and commodities through central real-time computers. The service enables a broker to query the computer's store of market data and receive the information on a print-out or visual display device. It has been proposed that the computers be programmed to provide capability for storing and processing buy and sell orders between individual brokers. In both instances, private line circuits leased from common carriers under applicable tariffs supply the connecting link between the computer and the brokers.

8. Other specialized computer services combining data processing and communications include a hospital information service, a coordinated law enforcement service utilizing computers to tie together the law enforcement efforts of a number of local authorities, and various kinds of reservation services.

9. Most, if not all, of the major computer manufacturers offer for sale or lease computers which can be programmed for message and circuit switching in addition to their data processing functions. There are a number of operational computerized message switching systems owned by large corporations in diverse fields. Most of these systems replaced electro-mechanical switching units provided by the communications common carriers. Motivations of increased business efficiency and maximization of the capabilities of the computer are apparently leading toward the acquisition by large corporations of computer systems. These systems permit data processing and message switching to be effectively combined with communication channels linking remote locations to form a real-time data processing and communications system.

III. Computers and the Common Carriers

10. The communications common carriers are rapidly becoming equipped to enter into the "data processing" field. Common carriers, as part of the natural evolution of the developing communications art, are making increased use of computers for their conventional services to perform message and circuit switching functions. These computers can likewise be programmed to perform data processing functions. For example, Western Union is establishing computer centers, not only for its public message and Telex systems, but eventually to provide as well a variety of data processing, storage and retrieval services for the public. The first such computer centers, planned by the company as part of its "national information utility" program, was opened March 16 of this year in New York City. This center, and others to be established in key cities, will be programmed to offer time-sharing, information processing and data-bank services. Western Union's planning looks to the establishment, through a national, regional and local network of computers, of a gigantic real-time computer utility service which would gather, store, process, program, retrieve and distribute information on a broad scale. This company will also arrange to design, procure and install all necessary hardware for fully integrated data processing and communications systems for individual customers, and provide the total management service for such systems.

11. The Bell System has not yet indicated any plan to provide a similar information service, or to offer local data processing services to the public. However, it is implementing a program to convert all central offices from electro-mechanical switching systems to electronic switching. Similar conversion

programs are being undertaken by other carriers in the industry. Interface, terminal and outstation equipments are being developed by the industry to match computer systems with communication channels. It might be observed here, that the Touch-Tone telephone instrument has significant potential as a computer input device, utilizing the telephone switched network. After a connection to a multiple access computer is completed in the regular manner, the same buttons can be pressed to enter information into the computer or to query the computer and get back a voice answer. A number of systems of this type are now in service.

12. International carriers have recently proposed new computer message switching and data processing services. One such carrier offers a service to air lines under which it switches messages between and among the various leased circuits connected to its computer. In addition, it plans to employ the same computer to store and supply up-to-the-minute seat inventory information with respect to flights of those air lines subscribing to this additional service, through communication facilities connected to air line offices and agencies on an on-line real-time basis. Other carriers plan to introduce similar service offerings.

IV. Discussion of the Problems

13. The above review, although by no means exhaustive, is illustrative of the convergence and growing interdependence of the computer and communications. This convergence takes a variety of different forms and applications thereby making it difficult to sort them into simple discrete categories. It is impossible at this time to anticipate fully the nature of all of the policy and regulatory problems that future developments may generate. Nevertheless, it is desirable to focus on those problems that are presently definable within the existing state of this burgeoning industry.

14. Communication common carriers, whose rates and services are subject to governmental regulation, are employing computers as a circuit and message switching device in furtherance of their undertakings to provide communication channels and services to the general public. There is now evidence of a trend among several of the major domestic and international carriers to program their computers not only for switching services, but also for the storage, processing and retrieval of various types of business and management data of entities desiring to subscribe therefore in lieu of such industries providing this service to themselves on an in-house basis or contracting with computer firms for the service.

15. Accordingly, we find communication common carriers grafting on to their conventional undertaking of providing communication channels and services to the public various types of data processing and information services. One such carrier has, in fact, committed its future to using its combined resources of computers and communication channels to meet the information requirements of the business community and other professional and institutional segments of our society by the establishment of a national and regional centralized information system. As a consequence, common carriers, in offering these services, are, or will be, in many instances, competitive with services sold by computer manufacturers and service bureau firms. At the same time, such firms will be dependent upon common carriers for reasonably priced communication facilities and services.

16. As previously indicated, a large number of non-regulated entities are employing computers to provide various types of data processing and specialized information services. The excess capacity of the in-house computer is made available for a charge to others; in other instances computer service bureaus sell computer time to a number of subscribers on a shared-time basis; and in still other instances, highly specialized information and data bank services are provided. At an ever increasing rate, with the development of time-sharing techniques, remote input and output devices of the users are linked to the computer by communication channels obtained from common carriers. The users located at the remote terminals are served so rapidly that each is under the illusion that he alone has access to the central processor. The flexibility of the computer makes possible, in addition to data processing services, message switching between various locations of the same customer, or between several different customers. This allows the data processing industry to engage in what heretofore has been an activity limited to the communications common carrier.

17. Common carriers have thus far taken different approaches to the question of the applicability of the regulatory provisions of the Communications Act to their computer service offerings. Notwithstanding that various aspects of such offerings appear to involve activities, such as message switching, which historically have been regarded as common carrier activities subject to regulation, no consistent policy is established and followed with respect to the filing of tariffs by carriers to cover those offerings. This is understandable considering the competitive activities of a similar nature by non-regulated entities as well as the apparent difficulties in classifying the various elements of a computer service into discrete communication and non-communication compartments.

18. From the common carriers' standpoint, regulation should extend to all entities offering like services or to none. It is urged that the ability to compete successfully depends on the flexibility required to meet the competition; and that the carriers would be deprived of this flexibility if they alone were restricted in their pricing practices and marketing efforts by the rigidities of a tariff schedule. Thus, we are confronted with determining under what circumstances data processing, computer information and message switching services, or any particular combination thereof — whether engaged in by established common carriers or other entities — are or should be subject to the provisions of the Communications Act. We expect this inquiry to be of assistance to the Commission in evaluating the policy and legal considerations involved in arriving at this determination.

V. Communication Tariffs and Practices

19. The interdependence between data processing and communication channels is bound to continue under the impetus of remote processing in combination with the growth of time-shared computer systems and services. In the past, the relationship between the relative cost of the two segments was of little concern. Data processing was expensive and in a relative sense higher than its communication counterpart. The trend toward lower EDP costs resulting from larger computer systems, has tended to shift the relative cost positions. Indeed, there is some indication that in the near future communication costs will dominate the EDP-communications circuit package. It is natural, then, that the computer industry finds its attention devoted increasingly to communication tariffs and regulations, in its search to optimize the communication segment of the package. In fact, fears are expressed that the cost of communications may prove to be the limiting factor in the future growth of the industry.

20. While the charges of the carriers are of prime importance, including the question of minimum periods of use, other tariff provisions and restrictions should also be scrutinized. Such tariff provisions as those relating to shared use and authorized use may well be in need of revision in light of the new advanced technology. Likewise, any restriction on the use of customer owned or provided equipment, including multiplexing equipment, must be reviewed for their effects on a burgeoning industry.

228 Appendix A

21. This then is another area of concern. Are the service offerings of the common carriers, as well as their tariffs and practices, keeping pace with the quickened developments in digital technology? Does a gap exist between computer industry needs and requirements, on the one side, and communications technology and tariff rates and practices on the other?

VI. The Problem of Information Privacy

22. The modern application of computer technology has brought about a trend toward concentrating commercial and personal data at computer centers. This concentration, resulting in the ready availability in one place of detailed personal and business data, raises serious problems of how this information can be kept from unauthorized use.

23. Privacy, particularly in the area of communications, is a well established policy and objective of the Communications Act. Thus, any threatened or potential invasion of privacy is cause for concern by the Commission and the industry. In the past, the invasion of information privacy was rendered difficult by the scattered and random nature of individual data. Now the fragmentary nature of information is becoming a relic of the past. Data centers and common memory drums housing competitive sales, inventory and credit information and untold amounts of personal information, are becoming common. This personal and proprietary information must remain free from unauthorized invasion or disclosure, whether at the computer, the terminal station, or the interconnecting communication link.

24. Both the developing industry and the Commission must be prepared to deal with the problems promptly so that they may be resolved in an effective manner before technological advances render solution more difficult. The Commission is interested not only in promoting the development of technology, but it is at the same time concerned that in the process technology does not erode fundamental values.

VII. Items of Inquiry

25. In view of the foregoing, it is incumbent upon the Commission to obtain information, views and recommendations from interested members of the public in order to assist the Commission in resolving the regulatory and policy questions

presented by this new technology. Accordingly, such information, views and recommendations are requested in response to the following items of inquiry:

A. Describe the uses that are being made currently and the uses that are anticipated in the next decade of computers and communication channels and facilities for:

1. Message or circuit switching (including the storage and forwarding of data);

2. Data processing;

3. General or special information services;

4. Any combination of the foregoing.

B. Describe the basis for and structure of charges to the customers for the services listed in A above.

C. The circumstances, if any, under which any of the aforementioned services should be deemed subject to regulation pursuant to the provisions of Title II of the Communications Act.

1. When involving the use of communication facilities and services;

2. When furnished by established communication common carriers;

3. When furnished by entities other than established communication common carriers.

D. Assuming that any or all of such services are subject to regulation under the Communications Act, whether the policies and objectives of the Communications Act will be served better by such regulation or by such services evolving in a free, competitive market, and if the latter, whether changes in existing provisions or law or regulations are needed.

E. Assuming that any and all of such services are not subject to regulation under the Communications Act, whether public policy dictates that legislation be enacted bringing such services under regulation by an appropriate governmental authority, and the nature of such legislation.

F. Whether existing rate-making, accounting and other regulatory procedures of the Commission are consistent with insuring fair and effective competition between communications common carriers and other entities (whether or not subject to regulation) in the sale of computer services involving the use of communications facilities; and, if not, what changes are required in those procedures.

G. Whether the rate structure, regulations and practices contained in the existing tariff schedules of communications common carriers are compatible with present and anticipated requirements of the computer industry and its customers. In this connection, specific reference may be made to those tariff provisions relating to:

1. Interconnection of customer-provided facilities (owned or leased) with common carrier facilities, including prohibitions against use of foreign attachments;

2. Time and distance as a basis for constructing charges for services;

3. Shared use of equipment and services offered by common carriers;

4. Restrictions on use of services offered, including prohibitions against resale thereof.

H. What new common carrier tariff offerings or services are or will be required to meet the present and anticipated needs of the computer industry and its customers.

I. The respects in which present-day transmission facilities of common carriers are inadequate to meet the requirements of computer technology, including those for accuracy and speed.

J. What measures are required by the computer industry and common carriers to protect the privacy and proprietary nature of data stored in computers and transmitted over communication facilities, including:

1. Descriptions of those measures which are now being taken and are under consideration; and

2. Recommendations as to legislative or other governmental action that should be taken.

26. Accordingly, there is hereby instituted, pursuant to the provisions of Sections 4(e) and 403 of the Communications Act of 1934, as amended, an inquiry into the foregoing matters.

27. In view of the scope and complexity of the matters involved, it appears desirable that interested persons be afforded an opportunity to suggest additions to and modifications or clarifications of the items of inquiry specified above. To this end, all interested persons are invited to submit appropriate recommendations in this regard on or before December 12, 1966. The Commission will thereupon issue such supplement to this Notice of Inquiry as may be warranted and will then specify a date by which written responses to said Notices shall be required.

28. All filings in this proceeding should be submitted in accordance with the provisions of Sections 1.49 and 1.419 of the Commission's Rules (47 CFR 1.49, 1.419).

FEDERAL COMMUNICATIONS COMMISSION

Ben F. Waple
Secretary

APPENDIX B

Before the
FEDERAL COMMUNICATIONS COMMISSION
Washington, D. C. 20554

FCC 69-468
29679

In the Matter of)
)
Regulatory and Policy Problems)
Presented by the Interdependence)   DOCKET NO. 16979
of Computer and Communications)
Services and Facilities)

REPORT AND FURTHER NOTICE OF INQUIRY

Adopted: May 1, 1969; Released: May 9, 1969

By the Commission: Commissioner Wadsworth absent;
Commissioner Johnson concurring in the result.

1. This Inquiry was begun by the Commission on its own motion by a Notice of Inquiry released on November 10, 1966. It was our objective to provide a public forum for the discussion, examination and resolution of a number of regulatory and policy questions that appeared to be emerging from the growing interdependence of computers and communications services and facilities. Accordingly, our Notice identified a variety of specific issues with respect to which we invited interested persons to submit information, views and recommendations.

2. In broad outline, the areas of concern addressed by our Notice included present and future computer uses of communications facilities and services; the adequacy of existing facilities and services of common carriers; the need for new and improved common carrier offerings; whether and under what circumstances the rendition by common carriers or others of data processing and other computer services involving the use of communications facilities should be free from or subject to governmental regulation; whether and under what conditions the entry into such services by common carriers and others should be controlled; and what measures, if any, are required to be taken by the computer industry, communications common carriers, or government to protect the privacy of data stored in computers or transmitted over communications facilities. The date of March 5, 1968, was fixed for the submission of such responses.

3. We received about sixty responses representing a broad cross section of interest in computers and communications. Responses came from trade associations, government agencies, professional organizations, communications common carriers, computer manufacturers, computer service organizations, computer users and others. In some instances, we were given extensive, well researched and comprehensive presentations. In other cases, respondents submitted thoughtful and provocative delineations of specific problems affecting their particular interests. Virtually all respondents acknowledged the timeliness of the Inquiry because of the currency and importance of the problems which were arising out of the interaction of computer and communications technologies. They stressed a need to deal with the adaptation of existing communications services, regulations, facilities and practices to the data communications requirements now arising out of the growing use of computers and computer technology.

4. Since the inception of the Inquiry in November, 1966, a number of developments have occurred which are relevant to, and have impact upon, the issues involved in the Computer Inquiry. We therefore are of the view that it would be timely and useful to a proper public understanding of the current status of the Inquiry and the issues involved therein to review these developments as they affect those issues and the Commission's regulatory responsibilities and required actions.

5. As indicated above, the Notice of Inquiry brought forth useful data and comment from a broad cross section of interests. Our review of the responses made it clear that our available internal resources would not be sufficient to give the responses the expert treatment and attention which was necessary to assure the comprehensive consideration and analysis demanded by the matters involved.[1] We decided that research funds available could be usefully employed for an independent and expert study which would evaluate the responses and make recommendations to the Commission. We contracted with the Stanford Research Institute to conduct such a study.

---

1. In its response, the Association for Computing Machinery, a professional organization in the field of computer technology, offered its assistance to the Commission. We accepted the ACM's offer, and it developed and presented to the Commission and its staff a series of seminars dealing with the technical aspects of computer use and the nature of the growing involvement of communications with computers. The seminars provided a most useful background for our consideration of the issues involved in the Inquiry.

6. SRI's study has now been completed. Its results are encompassed in a series of seven reports which have been published and are available to the public. The reports consist of summaries and analyses of the responses in the Inquiry; information added by SRI to the material in the responses; a discussion of decision-making and research techniques SRI proposes as additional regulatory tools; and SRI's conclusions and recommendations as to the disposition or further regulatory treatment of the issues in the Inquiry. Copies of the SRI study may be purchased from Clearing House for Federal Scientific and Technical Information of the Department of Commerce or may be inspected at the offices of the Commission in Washington, D.C.[2]

7. The Commission has not yet fully evaluated the findings and recommendations of SRI in the context of our regulatory responsibilities and the objectives and issues of the Computer Inquiry. In connection with this evaluation and the determinations yet to be made by us as to the further regulatory actions that may be required, we will welcome and are inviting comments on the SRI reports from respondents to this Inquiry or any other persons who may be interested.

8. Although our evaluation of the responses to the Inquiry and the SRI reports remains to be completed, we are in a position to make certain observations concerning the results thus far accomplished by the Inquiry and the posture of the issues involved therein. *First,* it is evident that the responses as well as the SRI study provide the Commission and all segments of the computer and communications industries with a more informed basis upon which to gain an understanding of the problems that arise out of the growing interdependence of computers and communications. *Second,* both the responses and the SRI study serve as a significant point of departure for the more definitive identification and further consideration of the specific issues which confront the Commission and the industry and which must be resolved if communications and computer technologies are to be amalgamated in a fashion which will most effectively serve the public interest. *Third,* as will be noted hereinafter, the Inquiry has already generated constructive actions by the communications common carriers in response to certain of the specific problems which have been crystallized by the responses to the Inquiry. *Fourth,* the responses to the Inquiry and the SRI study

---

2. By publishing the SRI reports, the Commission is not necessarily accepting, approving or rejecting any of SRI's conclusions and recommendations. As an independent, research organization, SRI bears full responsibility for the content of the reports.

have enabled us to relate constructively various proceedings already instituted by the Commission with respect to the tariff rates, regulations, and practices of the common carriers to the problems arising out of the interaction of computers and communications. *Fifth,* the Inquiry has pointed up those problems involved in the computer use of communications which need to be made the subject of further proceedings by the Commission, or which require the development of more information and experience before we can decide what measures can or should be taken to assure that the public interest will be served in this important field.

9. One subject which was already under examination by the Commission in a pending proceeding but which, nevertheless, evoked substantial comment from a number of respondents, concerned the tariff provisions of the telephone companies forbidding attachment of customer-owned devices and the interconnection of private communications systems with the common carrier switched network. The then existing tariff limitations against attachments and interconnection were held to be unlawful by the Commission in its decision in the Matter of Use of Carterfone Device in Message Toll Telephone Service, Docket No. 16942, 13 FCC 2d 420, June 26, 1968, petition for reconsideration denied, 14 FCC 2d 571, September 11, 1968.

10. Subsequently, the telephone companies filed revised tariffs which were allowed to become effective on January 1, 1969. These revisions, in effect, opened up the switched network to the interconnection of customer-owned devices and systems subject, however, to various conditions and limitations specified in the tariffs. At that time we made no determination as to the lawfulness of the revised tariffs, but we did recognize that there were a number of objections to various features of the new tariffs which warranted further consideration. Among other things, it was claimed that certain regulations of the new tariffs unduly restricted or interfered with the use of the switched communications network for computer purposes. Some of the questions are of a substantial nature and remain to be resolved. Nevertheless, the lifting of the ban on interconnection of customer-owned equipment and systems will expand greatly competitive opportunity for the manufacture and sale of peripheral and special communications equipment and systems for use in connection with the switched network.

11. By our Memorandum Opinion and Order of December 24, 1968, we instructed the Chief of the Common Carrier Bureau to convene a series of technical and engineering conferences, to be participated in by all interested

236   Appendix B

parties, for the purpose of considering the various questions. It is expected that the conferences will determine what further tariff modifications might be warranted and implemented without formal action by the Commission, or whether formal hearings should be held on unresolved issues. The Chief of the Common Carrier Bureau has begun to bring together representatives of all interests concerned with these problems.[3]

12. Thus, the problems arising out of restrictions on attachments and interconnection of customer-owned communications equipment and systems which may affect computer uses as well as other uses of the switched network are included within the scope of proceedings now underway. Computer interests are participating in these proceedings, and all those having an interest in these matters should communicate their views to the Chief of the Common Carrier Bureau.

13. One important consequence of the growing dependence of computer technology and use upon communications is that new questions have been raised regarding the rate structures and practices of communications common carriers. Respondents in the Inquiry expressed various views regarding the adequacy or suitability of the carriers offerings for data transmission. For example, there is considerable complaint regarding the three-minute minimum rate schedule applicable to message toll telephone service. Because large volumes of data can be transmitted over the switched network in bursts which consume seconds rather than minutes, the three minute minimum is regarded as uneconomical. In response, AT&T will shortly introduce a one-minute initial period rate schedule on a trial basis in a limited number of locations. The offering will be available to both voice and data users of the switched network at substantial savings.

14. Computer technology has cast an entirely new light existing restrictions upon customer sharing of communications facilities. Here again our Inquiry has revealed that, so far as computer use of communications services is concerned, current tariff restrictions upon sharing are regarded as defeating the efficient and economic use of common carrier facilities and running counter to the developing needs of data users. Data users are not the only ones who have objected to limitations upon sharing of communications facilities. A partial response to the

---

3. The Commission will use the resources of a professional organization if suitable arrangements can be made, to assure maximum expert technical advice and objectivity in dealing with the problems.

concerns expressed by the respondents in the Inquiry is found in AT&T's recently filed tariff liberalizing the sharing of private line circuits of voice grade or less bandwidth.[4] Certain aspects of sharing regulations of the carriers are under examination in the *Telpak* Sharing case, Docket No. 17457, which is currently in a decisional status;[5] other aspects will be the subject of hearings in the private line rate investigation instituted on July 16, 1968, in Docket No. 18128. We intend to look to the responses of those who may communicate their views regarding the SRI reports to see if there is need for other proceedings or actions with respect to tariff provisions which affect sharing of communications facilities and services.

15. The most notable regulatory problem raised by the respondents and discussed by SRI in its study arises out of the rendition of data processing services by common carriers, and the provision of what are claimed to be communications services by unregulated data processing organizations. This problem breaks down into a number of issues. The first of these is whether communications common carriers should be permitted to sell data processing services and, if so, what safeguards should be imposed to insure that the carriers will not engage in anti-competitive or discriminatory practices. On the other side, we have the problem of the extent to which unregulated data processing organizations should be permitted to sell communications as a part of a data processing package not subject to regulation. This situation leads to the difficult problem of separating by definition what are communications services and what are data processing services. There does not appear to be any public need or demand for regulation of what is generally understood to constitute data processing. However, serious differences arise over what to do when data processing is part and parcel of a mixed offering which includes communication type functions, most notably message switching.

16. SRI contends that for the immediate future we should limit our regulatory action to assuring that in leasing facilities to Western Union for its use in furnishing data processing services, no preference is given to Western Union by

---

4. A new limited trial offering of a wide-band private line service, providing, among other features, for unlimited customer sharing was recently filed by AT&T to be effective July 1, 1969. The Commission has not yet taken any action with respect to this tariff.

5. A recommended decision was issued in this matter by the Chief of the Common Carrier Bureau on April 25, 1969.

AT&T compared to the terms upon which such facilities are supplied by AT&T to its other customers. SRI also advises that we keep a close watch on how other carriers operate their data services. Many of the respondents take a more urgent view of this problem, a number of them advocating that carriers be either totally barred from providing data processing services or subjected to rigorous safeguards designed to separate unregulated data processing services from regulated communications activities.

17. An aspect of this arose in connection with our consideration of Western Union's SICOM and INFO-COM tariffs. We allowed these tariffs to go into effect but without prejudice to our further examination of the question of regulation of computer functions and communications services. (In the Matter of Western Union Telegraph Company Tariff FCC No. 251 Applicable to SICOM Service 11 FCC 201.) We will shortly initiate an appropriate proceeding as a basis for definitive decisions as to what requirements shall be imposed upon carriers with respect to data services; and whether computer services which involve data communications should be subject to regulation whether engaged in by carriers or others, and the specific form of any such regulation.

18. Privacy and security of data during both transmission and storage in computer memory are extremely important to many of those who answered the Computer Inquiry. We intend to give further consideration to the needs which may exist in this area and to the regulatory actions which may be required. We believe it would be best, however, to obtain more information regarding present and future needs and the technical, operational and economic implications involved in meeting those needs. Until we have given further analysis to the problem, we are leaving open for the time being the question of specific regulatory or other actions which might be usefully taken to assure privacy and security of data.

19. Responses to the Inquiry and the SRI study have also pointed out an increasing number of technological problems relating to the adequacy of existing communications facilities to meet the needs of the growing computer technology. Thus, they raise questions as to the need for a digital transmission network designed specifically for data communications requirements, the need for greater transmission speed and accuracy, the need for greater variety of bandwidth configurations, the need for more sophisticated equipment to assure security of data, and the need for development of domestic satellite networks for data communications requirements and many others. Although the data produced by

the Computer Inquiry and the SRI study provide a most valuable point of departure, more definitive and comprehensive technical data relative to these needs must be developed.

20. We are in an era of rapidly advancing technologies and an ever-present demand for expanded, improved and new communications capabilities. In these circumstances, it is inevitable that difficulties and conflicts will arise as computer and communications technologies encounter each other as has happened during the past decade. The regulatory processes of government, of course, stand ready to assist in the resolution of these problems and this Inquiry is one example of regulatory initiative. However, of major importance is the establishment and maintenance of an appropriate dialogue between computer interests and communications common carriers so as to give each a better understanding of the existing and prospective needs, capabilities and limitations of the other. By this process, constructive accommodation can be reached for the benefit of all thereby minimizing the need for regulatory action and maximizing the effectiveness of voluntary, responsible actions within the private sector.

21. The Computer Inquiry has been a valuable first step toward making available to all concerned relevant and reliable information on a continuing basis. As an ongoing process, we intend to acquire data and information related to technological developments and trends in data communications. SRI underscores the need for systematic collection of such data as vital to the Commission, industry and computer users in understanding and anticipating problem areas. We are already taking steps to supplement the data accumulated in the Computer Inquiry in a number of areas. In addition, we propose to enlist the cooperation and participation of industry and professional organizations as part of the continuing effort to maintain an effective program for the public exposure of relevant data and the treatment of specific problems. The specific framework for these activities will be determined after opportunity for consultation. Here again, we will welcome specific suggestions from interested persons.

* * *

In summary, the Computer Inquiry has been an effective vehicle for the identification and better understanding by Government, industry and computer users of the problems spawned by the confluence of computer and communication technologies which has taken place in the past decade. Certain of these problems have been, or are being, effectively dealt with in other proceedings before the Commission or by voluntary action taken by the industry. The treatment of other problems requires further proceedings or the acquisition of more information and experience before we can decide what further actions, if any, may be needed. Finally, we intend to establish procedures for the continued collection and evaluation of reliable data concerning existing and prospective developments in the data communications field.

We propose to keep this Inquiry open for the filing by interested persons of:

(a) Comments regarding the SRI study, its information, conclusions and recommendations;

(b) Comments and recommendations regarding the methods which will best promote the continuous availability of relevant and reliable technical and other information related to the interdependence of computers and communications.

Comments, proposals, and recommendations should be filed with the Commission not later than June 24, 1969. Upon consideration of these submissions, we will take such further actions as may then be indicated.

FEDERAL COMMUNICATIONS COMMISSION

Ben F. Waple
Secretary

APPENDIX C

The chart on the following page summarizes the issues addressed in the FCC Computer/Communications Inquiry, and the positions taken by virtually all of the participants, in their Responses filed in March 1968. Several non-substantive or late filings are omitted. Positions taken by the respondents with respect to these issues are indicated by "Y for "Yes," "N" for "No," and "C" for "Conditional Yes." The numbers following indicate the page numbers in the respective Responses. In addition, a number of respondents discussed certain issues without taking clear positions on them, and these discussions are not shown on the chart. It should also be noted that many of the respondents addressed only a subset of the issues in the Inquiry, limiting themselves to those with which they were particularly concerned and/or avoiding those issues on which they did not wish to take a public stand.

242 Appendix C

	Regulation of EDP	Carrier Provision of EDP		Non-Carrier Message Switching			Carrier Tariff Restrictions				Adequacy of Carrier Communication Service					Privacy
	Regulate EDP Services	Permit Carriers to Offer EDP Services	Permit Only thru Subsidiary	Forbid Message Switching by Non-Carriers	Permit Unrestricted	Permit thru Primary Business Test	Abolish "Foreign" Attachment Rule	Relax Interconnection Prohibition	Relax Sharing/Resale Prohibition	Reduce MTT Minimum Charge Time	Need Additional Bandwidths	Need Widespread Switched Broadband	Need Higher Quality Lines	Provide Additional Specifications	Introduce T-carriers, ESS more rapidly	Government Action Required for Privacy
Assoc. of Am. Railroads	N-15				Y-15											N-20
Am. Banking Association	N-19															N-13
ADAPSO	N-7													Y-19	Y-19	
Aeronautical Radio, Inc.						Y-9										
Aerospace Industries Assoc.	N-2															N-20
Aetna Life & Casualty Co.																N-16
Am. Inst. of Indust. Engin.	N-9			Y-114												Y-52
Am. Newspaper Publish. Assoc.																
Am. Petroleum Institute				Y-22												N-156
A.T.&T.											Y-18		Y-19	Y-17	Y-61	
Am. Trucking Association											Y-22			Y-15		
BEMA	N-67			Y-11		Y-81	Y-17	Y-17	Y-23	Y-149	Y-11		Y-151	Y-148	Y-55	N-156
Bunker-Ramo Corporation	N-21			Y-39		Y-3	Y-10	Y-10	Y-10	Y-55	Y-10			Y-60		
California PUC										C			Y-33			
Central Info. Proc. Corp.	N-3			Y-9			Y-3	Y-3	Y-3	Y-135	Y-16			Y-17		N-6
Collins Radio	N-4	N-5		Y-5			Y-17 / Y-13	Y-17	Y-17 / Y-14	Y-45	Y-15 / N-44		Y	Y-15		Y-23
Computing & Software Inc.	N-2			Y-5		Y-3	N-29	N-29	N-29							
Comsat Corporation	N-11										Y-143		Y-141		Y-22	Y-27
Control Data Corporation	N-10	N-15						Y-15	Y-15	Y-17	Y-55		Y-19	Y-19	Y-55	
Credit Data Corporation							Y-22	Y-25	Y-25	Y-26						
Eastern Airlines																
EDUCOM	N-11			Y-86		Y-85	Y-121	Y-119	Y-132	Y-141	Y-138	Y-156	Y-49	Y-138	Y-141	N-145
Electronic Industries							Y-10	Y-18	Y-18							N-26
Electrospace Corporation	N-10					Y-14	Y-18	Y-50	Y-50	Y-24	Y-45		Y-69	Y-42	Y-54	Y-52
Federal Reserve System	N-36	Y-40			Y-44		Y-50	Y-53	N-63					Y-72		
General Electric	N-44	Y-49					N-54									Y-61
General Service Adminis.																Y-77
G.T.&E. Service Corp	N-133	Y-134			Y-13	Y-121	Y-160	Y-18	Y-18	N-23	N-24				Y-158	N-172
Humble Oil													Y-156			
I.B.M. Corporation	N-15			Y-14			Y-18	Y-18	Y-18	N-24						
I.E.E.E.																
I.T.T. Corporation																

Appendix C 243

Organization	Regulation of EDP: Regulate EDP Services	Carrier Provision of EDP: Permit Carriers to Offer EDP Services	Carrier Provision of EDP: Permit Only thru Subsidiary	Non-Carrier Message Switching: Forbid Message Switching by Non-Carriers	Non-Carrier Message Switching: Permit Unrestricted	Non-Carrier Message Switching: Permit thru Primary Business Test	Carrier Tariff Restrictions: Abolish "Foreign" Attachment Rule	Carrier Tariff Restrictions: Relax Interconnection Prohibition	Carrier Tariff Restrictions: Relax Sharing/Resale Prohibition	Adequacy: Reduce MTT Minimum Charge Time	Adequacy: Need Additional Bandwidths	Adequacy: Need Widespread Switched Broadband	Adequacy: Need Higher Quality Lines	Adequacy: Provide Additional Specifications	Adequacy: Introduce T-carriers, ESS more rapidly	Privacy: Government Action Required for Privacy	C
Justice Department	N-65		N-84	C-84		Y-90		Y-29	Y-37	Y-44							N-10
Law Research Service	N-7							Y-8	Y-8	Y-8		Y-9		Y-10			
Lockheed Aircraft Corporation	N-12	N-13						Y-62 / Y-31	Y-62 / Y-3	Y-32			Y-33	Y-34			Y-37
McGraw-Hill	N-27	Y-28															
Microwave Communications, Inc.	N-18		Y-19					Y-27	Y-26	Y-28		Y-30		Y-30			N-31
Nat'l Assoc. of Manufacturers	N-37		Y-70				Y-46	Y-72	Y-72	Y-91		Y-95	Y-94	Y-95			N-96
Nat'l Assoc. of Reg. Util. Com.	N-22	Y-25					Y-3	Y-47	Y-47	N-48		Y-5			Y-5		N-53
Nat'l Comm. for Util. Radio								Y-1	Y-1								
N. Am. Computer & Comm. Co.																	
Nat'l Retail Merchants Assoc.																	
Randolph Computer Corp.																	
RCA Communications																	
Republic Sys. & Programming																	
Rixon Electronics	N-30																
Soc. Inter. de Telecom. Aeron.																	
Spindletop																	
Tech. Communic. Corporation						Y-DE4	Y-9	Y-G13 / Y-12	Y-G13 / Y-12	Y-G23	Y-G15 / Y-13	Y-H19 / Y-14	Y-H14	Y-H111	Y-15	Y-H120	N-J34
Towson Labs																	
Union Pacific Railroad					Y-9		Y-6	N-G6 / Y-22	N-G6 / Y-21	N-G11 / Y-24		Y-8 / Y-H14		Y-13			N-12
Unitab Co.					Y-CDE43												
United Airlines				Y-13													Y-J1
UNIVAC Div. of Sperry-Rand	N-DE3	Y-C28															
University Computing Co.	N-7	N-9															
Univer. of Maine	N-11																
U.S. Indep. Telephone Assoc.	N-16	N-10															
VIP Systems	N-CDE46	Y-F1															
Western Union Telegraph Co.	N-19																
Xerox Corporation																	

BIBLIOGRAPHY

Contents

I.	Computer/Communications Policy Issues	245
	A. General Discussion of the Issues	245
	B. Interfacing with the Public Telephone Network	246
	C. Security and Privacy	247
II.	Studies by the President's Task Force on Communications Policy	248
	A. Task Force Staff Papers	249
	B. Task Force Consultants' Studies	249
III.	Remote-Access Computing and Information Systems	250
	A. The Technology of Multi-Access Computing	250
	B. The "Computer Utility" Concept	251
	C. Future Computer Services	252
IV.	Data Communications Technology	253
	A. Data Communications Network Design	253
	B. Transmission Theory and Engineering	254
	C. Switching Systems	254
	D. Evolving Requirements for Data Communications Services	254
	E. Performance of the Telephone Network for Data Transmission	255
	F. New Communications Technology and Services	256
	G. Digital Transmission	257
V.	Regulation of Communications Common Carriers	258
	A. Regulatory Theory	258
	B. The Regulatory Process	258
VI.	Reference Documents	259
VII.	Periodicals	261

I. COMPUTER/COMMUNICATIONS POLICY ISSUES

A. General Discussion of the Issues

Baran, Paul, "The Future Computer Utility," *The Public Interest,* Summer 1967. A well-written discussion of several emerging policy questions.

"Behind the Communications Mess," *Business Week,* Nov. 18, 1967, pp. 66-74.

The Impact of Government on Information Processing: The FCC Inquiry, Diebold Research Program Document MB11. New York: The Diebold Group, Inc., 1968. A review of the specific issues in the FCC Computer Inquiry, including excerpts reflecting the positions of the major respondents.

Irwin, Manley R., "The Computer Utility," *Datamation,* XII, No. 11 (1966), pp. 22-27.

——————, "The Computer Utility: Competition or **Regulation?**" *Yale Law Journal,* LXXVI, No. 7 (1967), pp. 1299-1320.

——————, "New Policy for Communications," *Science and Technology,* April 1968, pp. 76-84.

——————, "Time-Shared Information Systems: Market Entry in Search of a Policy," *Proceedings of the 1967 Fall Joint Computer Conference,* pp. 513-520. Washington, D.C.: Thompson Book Company, 1967. Discussion of events and issues leading to the FCC Computer Inquiry.

Noel, Walter M., Jr. and V. Reed Manning, *Implications of Regulatory Developments in U.S. Telecommunications,* Arthur D. Little, Inc., Service to Management Report. Cambridge, Mass.: May 1969.

Simonson, W.E., "Data Communications: The Boiling Pot," *Datamation,* XIII, No. 4 (1967), pp. 22-25.

Stanford Research Institute, *Reports to the Federal Communications Commission on Docket No. 16979.* Menlo Park, Calif.: Feb. 1969.

Vol. 1. *Policy Issues Presented by the Interdependence of Computer and Communications Services*

Vol. 2. *Analysis of Policy Issues in the Responses to the FCC Computer Inquiry*

Vol. 3. *Decision Analysis of the FCC Computer Inquiry Responses*

Vol. 4. *Patterns of Technology in Data Processing and Data Communications*

Vol. 5. *Digest of the Responses to the FCC Computer Inquiry*

Vol. 6. *A Preface to a Theory of Regulation*

Vol. 7. *A Dynamic Financial Model of a Utility*

(Available from the U.S. Department of Commerce Clearinghouse for Federal Scientific and Technical Information, Springfield, Va., Accession Nos. PB 183 612, PB 183 613.)

Strassburg, Bernard, (Chief, FCC Common Carrier Bureau), "The Computer as an 'Information Utility' and its Regulatory Implications for the Federal Communications Commission," Address before the Institute on Management Information and Data Transfer Systems at the Session on Communications Carrier and Management Information Systems, Oct. 21, 1965, 11 pp.

—————, "Marriage of Computers and Communications — Some Regulatory Implications," Address before the Association for Computing Machinery, Washington, D.C., Chapter, Oct. 20, 1966, 14 pp. One of the earliest indications of FCC interest in the interdependence of computers and communications.

B. Interfacing with the Public Telephone Network

Courtney, Jeremiah, (interviewed by Robert E. Tall), "What the FCC *Carterfone* Decision Means to the Communications Industry," *Communications Magazine,* Oct. 1968, pp. 8-12.

Cox, Kenneth A. (FCC Commissioner), Untitled address before the National Retail Merchants Association, Montreal, Canada, October 8, 1968. Discusses the problem of attaching customer-provided devices and networks to the public telephone system, as raised in the *Carterfone* case. Also discusses the issues raised in the FCC Computer Inquiry.

―――――, "CATV and Interconnection of Telephone Devices: Where Do We Go From Here?" Address before the Eightieth Annual Convention of the National Association of Regulatory Utility Commissioners, Chicago, Ill., Nov. 13, 1968.

―――――, "The Telephone Industry and the FCC," Address before the Fifteenth Annual Meeting of the National Telephone Cooperative Association, Las Vegas, Nev., Jan. 30, 1969. Discussion of the policy problems associated with the attachment of customer-provided devices and networks to the public telephone system.

Hrusoff, R.R., "Telephone Attachments and Recording Devices (Legal and Technical Difficulties)" *Public Utilities,* May 12, 1966, pp. 33-37.

Patton, Phillips B., "How *Carterfone* Should Advance Telephone Service," Address before the Third Annual Special Seminar of the International Communications Association, Pittsburgh, Pa., Jan. 21, 1969. 24 pp. Interesting discussion of the history of the foreign attachment rule, and the nature and implications of the recent tariff relaxations.

Romnes, H.I., (Chairman of the Board, AT&T), "Dynamic Communications for Modern Industry," Address before the Annual Meeting of the American Petroleum Institute, Chicago, Ill., Nov. 13, 1967. Statement of the AT&T position on the attachment of customer-provided devices and networks to common carrier communication channels.

C. Security and Privacy

Bigelow, Robert P., "Legal and Security Issues Posed by Computer Utilities," *Harvard Business Review,* Sept.-Oct. 1967, pp. 150-161.

Dennis, R.L., *Security in the Computer Environment.* Santa Monica, Calif.: Systems Development Corporation, Aug. 18, 1966.

Glaser, E.L., "A Brief Description of Privacy Measures in the Multics Operating System," *Proceedings of the 1967 Spring Joint Computer Conference,* pp. 303-304. Washington, D.C.: Thompson Book Company, 1967.

Graham, Robert M., "Protection in an Information Processing Utility," *Communications of the ACM, XI,* No. 5 (1968), pp. 365-369.

Harrison, Annette, *The Problem of Privacy in the Computer Age: An Annotated Bibliography,* RAND Report RM-5495-PR/RC. Santa Monica, Calif.: The RAND Corp., Dec. 1967. 125 pp.

Hoffman, Lance, J., "Computers and Privacy: A Survey," *Computing Surveys,* I, No. 2 (1969), pp. 85-103.

Miller, A.R., "National Data Center and Personal Privacy," *Atlantic Monthly,* Nov. 1967, pp. 53-57.

Peters, Bernard, "Security Considerations in a Multi-Programmed Computer System," *Proceedings of the 1967 Spring Joint Computer Conference,* pp. 283-286. Washington, D.C.: Thompson Book Company, 1967.

Petersen, H.E., and Rein Turn, *System Implications of Information Privacy,* RAND paper P-3504. Santa Monica, Calif.: The RAND Corp., April 1967. (Also in *Proceedings of the 1967 Spring Joint Computer Conference,* pp. 291-300. Washington, D.C.: Thompson Book Company, 1967.)

Ware, Willis H., *Security and Privacy in Computer Systems,* RAND paper P-3544. Santa Monica, Calif.: The RAND Corp., April 1967. 28 pp.

Westin, Alan F., *Privacy and Freedom,* New York: Atheneum, 1967. 487 pp.

II. STUDIES BY THE PRESIDENT'S TASK FORCE ON COMMUNICATIONS POLICY

President's Task Force on Communications Policy, *Final Report,* Washington, D.C.: U.S. Government Printing Office, 1969. 414 pp. plus app.

A. Task Force Staff Papers

A Survey of Telecommunications Technology, Part One, (Staff Paper 1). Available from the U.S. Dept. of Commerce Clearinghouse for Federal Scientific and Technical Information (CFSTI), Springfield, Va., Accession No. PB 184 412.

A Survey of Telecommunications Technology, Part Two, (Staff Paper 1). CFSTI Accession No. PB 184 413.

Domestic Applications of Communications Satellite Technology, (Staff Paper 4). CFSTI Accession No. PB 184 416.

The Domestic Telecommunications Carrier Industry, Part One, (Staff Paper 5). CFSTI Accession No. PB 184 417.

The Domestic Telecommunications Carrier Industry, Part Two, (Staff Paper 5). CFSTI Accession No. PB 184 418.

The Role of the Federal Government in Telecommunications, (Staff Paper 8). CFSTI Accession No. PB 184 423.

Bibliography, (Staff Paper 9). CFSTI Accession No. PB 184 424.

B. Task Force Consultant's Studies

Breyer, Stephen G. (Harvard Law School), *Computers, Communications, and Regulation,* 1968. (Unpublished). 114 pp.

Irwin, Manley R. (Whittemore School of Business and Economics, University of New Hampshire), *Vertical Integration in the Communications Industry: A Public Policy Critique,* 1968. (Unpublished). 203 pp.

Sheppard, Harrison J. (Antitrust Division, U.S. Department of Justice), and Kenneth G. Robinson, Jr. (University of North Carolina School of Law), *Report to the President's Task Force on Communications Policy in Re the Western Union Telegraph Company,* 1968. (Appendix B of Staff Paper 5, above.) 163 pp. plus exhibits.

Trebing, Harry M. (Director, Institute of Public Utilities, Michigan State University), and William H. Melody (Economic Studies Division, FCC Common Carrier Bureau), *An Evaluation of Domestic Communications Pricing Practices and Policies,* 1968. (Appendix A of Staff Paper 5, above.) 275 pp.

III. REMOTE – ACCESS COMPUTING AND INFORMATION SYSTEMS

A. The Technology of Multi-Access Computing

Auerbach Corporation, *Jointly Sponsored Study of Commercial Time-Sharing Services* (two volumes), Final Report 1570–TR–1. Philadelphia, Pa.: Dec. 1968. Volume 1 discusses the time-sharing services market, user characteristics, technology, and future potential. Volume 2 summarizes the business and technical aspects of 19 major time-sharing services firms.

Corbató, F.J., and V.A. Vyssotsky, "Introduction and Overview of the MULTICS System," *Proceedings of the 1965 Fall Joint Computer Conference,* pp. 185-196. Washington, D.C.: Thompson Book Company, 1965. A concise statement of the definition and design requirements of a "computer utility," and an overview of the second time-sharing system developed by MIT's Project MAC.

Crisman, P.A., ed., *The Compatible Time-Sharing System: A Programmer's Guide,* 2nd ed. Cambridge, Mass.: MIT Press, 1965. A reference manual describing the capabilities of CTSS, a widely-known time-sharing system developed by MIT's Project MAC.

Karplus, Walter J., ed., *On-Line Computing: Time-Shared Man-Computer Systems.* New York: McGraw-Hill Book Company, 1967. 337 pp.

Knight, Kenneth E., "Changes in Computer Performance," *Datamation,* XII, No. 9 (1966). Traces the development of digital computers from 1944 to 1962, developing mathematical cost/effectiveness trends.

————, "Evolving Computer Performance 1963-67," *Datamation,* XIV, No. 1 (1968). Continues the development of cost/effectiveness trends for more recent computer mainframes, begun in his Sept. 1966 article.

Martin, James, *Design of Real-Time Computer Systems.* Englewood Cliffs, N.J.: Prentice-Hall, Inc., 1967. 629 pp. A comprehensive and readable tutorial which has become a "classic" computer industry reference.

B. The "Computer Utility" Concept

Barnett, C.C., Jr., et al, *The Future of the Computer Utility*. New York: American Management Association, 1967. 158 pp.

Canning, Richard G., ed., "Computer Utilities: No Easy Road Ahead," *EDP Analyzer,* Oct. 1967.

——————, ed., "The Lure of the Computer Utility," *EDP Analyzer,* Sept. 1967.

David E.E., Jr., and R.M. Fano, "Some Thoughts About the Social Implications of Accessible Computing," *Proceedings of the 1965 Fall Joint Computer Conference,* pp.243-247. Washington, D.C.: Thompson Book Company, 1965.

Denz, Robert F., "The Case Against the Computer Utility," *Proceedings of the 22nd National Conference, Association for Computing Machinery,* 565-572. Washington, D.C.: Thompson Book Company, 1967.

Duggan, Michael A., "Computer Utilities – Social and Policy Implications: A Reference Bibliography," *Computing Reviews,* IX, No. 10 (1968), pp. 631-644.

Fano, R.M., "Computer Utility and the Community," *IEEE International Convention Record,* Pt. 12, 1967, pp. 30-37.

——————, "MAC System: The Computer Utility Approach." *IEEE Spectrum,* Jan. 1965, pp. 56-64.

Greenberger, Martin, "The Computers of Tomorrow," *Atlantic Monthly,* May 1964. To the best knowledge of the authors, the term "computer utility" was coined by Dr. Greenberger in this prophetic article.

Gruenberger, F., ed., *Computers and Communications – Toward a Computer Utility*. Englewood Cliffs, N.J.: Prentice-Hall, Inc., 1968. 232 pp.

McFall, Russell W., (President and Chairman of the Board, Western Union Telegraph Company), "The Age of the Communicator," Address before the International Communications Association, Montreal, Canada, May 2, 1966.

───────── , "New Partners in Progress: Communications and Computers," *Computers and Automation,* Oct. 1966. An Address before the 21st National Conference of the Association for Computing Machinery, Los Angeles, Calif., Aug. 30, 1966.

───────── , "The Place of Western Union in the Information Revolution," *Signal,* Dec. 1965, pp. 13-15.

Parkhill, Douglas F., *The Challenge of the Computer Utility.* Reading, Mass.: Addison-Wesley Publishing Company, 1966. 207 pp. A well-written discussion of some of the first computer/communications systems, the technology and economics of such systems, and the implications for the future.

Selwyn, Lee L., "Information Utility," *Industrial Management Review,* (MIT Sloan School of Management), Spring 1965.

C. Future Computer Services

American Bankers Association, *Proceedings of the National Automation Conference.* New York: American Bankers Association (Published annually).

Anderson, Allen H., et al, *An Electronic Cash and Credit System.* New York: American Management Association, 1966.

Brown, George W., James G. Miller, and Thomas A. Keenan, eds., EDUNET: *Report of the Summer Study on Information Networks.* New York: John Wiley and Sons, Inc., 1967. 440 pp.

Bush, Vannevar, "As We May Think," *Atlantic Monthly,* Vol. 176 (July 1945), pp. 101-108.

"Electronic Money and the Payments Mechanism," *1967 Annual Report of the Federal Reserve Bank of Boston.*

Greenberger, Martin, ed., *Computers and the World of the Future.* Cambridge, Mass.: MIT Press, 1964. Collected papers presented by a number of eminent authorities at an MIT computer conference.

Information. San Francisco: W.H. Freeman and Co., 1966. First published as a collection of articles in the September 1966 issue of *Scientific American.* Discussion by noted authors of computer and communications technology, and applications in education, science, business, and industry.

Licklider, J.C.R., *INTREX, Report of a Planning Conference on Information Transfer Experiments.* Cambridge, Mass.: MIT Press, 1965.

——————, *Libraries of the Future.* Cambridge, Mass.: MIT Press, 1965, 219 pp.

IV. DATA COMMUNICATIONS TECHNOLOGY

A. Data Communications Network Design

Chaney, W.G., "Basics of Data Transmission," *Communications Magazine,* six-part series, March-August 1969. Reprints available from the publisher.

Gentle, Edgar C., Jr. ed., *Data Communications in Business.* New York: American Telephone and Telegraph Co., 1965. An elementary but useful discussion of data communication concepts.

IBM Corporation, *Data Communications Primer,* Form C20-1668. One of the best introductions to data communications available.

Martin, James, *Telecommunications and the Computer.* Englewood Cliffs, N.J.: Prentice-Hall, Inc., 1969. An excellent tutorial on telecommunications technology and the use of common carrier facilities for data transmission.

——————, *Teleprocessing Network Organization.* Englewood Cliffs, N.J.: Prentice-Hall, Inc., 1969. A detailed description of the principles of designing and optimizing data communication networks.

254 Bibliography

B. Transmission Theory and Engineering

American Telephone and Telegraph Co., Long Lines Department, *Principles of Electricity Applied to Telephone and Telegraph Work.* New York: AT&T, 1961. A text used by AT&T to train their engineering and plant personnel in the technology of telephone transmission facilities.

————, *Transmission Systems for Communication.* By Members of The Technical Staff, Bell Telephone Laboratories, 1964. 784 pp. Available from Graybar Electric Co., 420 Lexington Avenue, New York, N.Y.

Angelakos, D.J., and T.E. Everhart, *Microwave Communications.* New York: McGraw-Hill Book Company, 1967. 272 pp.

Bennett, William R., and James R. Davey, *Data Transmission.* New York: McGraw-Hill Book Company, 1965. 356 pp.

Gatland, Kenneth W., ed., *Telecommunication Satellites: Theory, Practice, Ground Stations, Satellite Economics.* Englewood Cliffs, N.J.: Prentice-Hall, Inc., 1964.

Raisbeck, Gordon, *Information Theory.* Cambridge, Mass.: MIT Press, 1964.

C. Switching Systems

McKay, K.G., "Networks," *Science and Technology,* April 1968, pp. 45-50.

"No. 1 Electronic Switching System," AT&T Monograph No. 4853. Papers from the *Bell System Technical Journal,* XLIII, Sept. 1964, pp. 1831-2609. Technical description of the Bell System's new computer-controlled circuit switching equipment.

Rubin, M., and C.E. Haller, *Communications Switching Systems.* New York: Reinhold Publishing Company, 1966.

D. Evolving Requirements for Data Communications Services

Andrews, G.E., and F. Kennedy, *The Data Communications Market in the United States,* Arthur D. Little, Inc., Service to Management Report. Cambridge, Mass.: Sept. 1966, 58 pp.

McPherson, John C., "Data Communication Requirements of Computer Systems," *IEEE Spectrum,* Dec. 1967, pp. 42-45.

Mills, Richard G., "Communications Implications of the Project MAC Multiple-Access Computer System," *IEEE International Convention Record, Part I,* 1965, p. 238.

Oran, Frank G., "Communications for Time Sharing at Bell Telephone Laboratories," *IEEE Transactions on Communications Technology,* Feb. 1968, pp. 12-18.

E. Performance of the Telephone Network for Data Transmission

Alexander, A.A., R.M. Gryb, and D.W. Nast, "Capabilities of the Telephone Network for Data Transmission," *Bell System Technical Journal,* May 1960. Study of data communication error rates on voice-grade telephone channels.

Bodle, D.W., and P.A. Gresh, "Lightning Surges in Paired Telephone Cable Facilities," *Bell System Technical Journal,* XL, No. 2 (1961), p. 547.

Elliott, E.O., "A Model of the Switched Telephone Network for Data Communications," *Bell System Technical Journal,* XLIV, No. 1 (1965), p. 89.

Fennick, J.H., and I. Nasell, "The 1963 Survey of Impulse Noise on Bell System Carrier Facilities," *IEEE Transactions on Communications Technology,* Aug. 1966, p. 520.

Hinderliter, R.G., "Transmission Characteristics of Bell System Subscriber Loop Plant," *IEEE Transactions on Communications and Electronics,* Sept. 1963, p. 464.

Kelly, J.P., *Test of Data Transmission Via Switched Message Networks for SAGE and BUIC,* MITRE Technical Memorandum 3380. Bedford, Mass.: The MITRE Corp., March 1964. A detailed study of transmission characteristics and error rates on the switched telephone network.

Morris, R., "Further Analysis of Errors Reported in 'Capabilities of the Telephone Network for Data Transmission'," *Bell System Technical Journal,* XLI, No. 4 (1962), p. 1399.

Nasell, I., "The 1962 Survey of Noise and Loss on Toll Connections," *Bell System Technical Journal,* XLIII, No. 2 (1964), p. 697.

―――――――, "Some Transmission Characteristics of Bell System Toll Connections," *Bell System Technical Journal,* XLVII, No. 6 (1968), p. 1001.

―――――――, C.R. Ellison, and R. Holmstrom, "The Transmission Performance of Bell System Intertoll Trunks," *Bell System Technical Journal,* XLVII, No. 8 (1968), p. 1561.

O'Neil, D.R., *Error Control for Digital Data Transmission Over Telephone Networks,* MITRE Report No. ESD–TR–65–87. Bedford, Mass.: The MITRE Corp., May 1965. (Available from U.S. Department of Commerce Clearinghouse for Federal Scientific and Technical Information, Springfield, Va., Accession No. AD 616 678.)

Townsend, R.L., and R.N. Watts, "Effectiveness of Error Control in Data Communication Over the Switched Telephone Network," *Bell System Technical Journal,* XLIII, No. 6 (1964), p. 2611.

F. New Communications Technology and Services

Baran, Paul, *LSI: The Basic Module for New Communication Networks,* RAND paper P-3904. Santa Monica, Calif.: The RAND Corp., Sept. 1968.

Davis, Robert L., "New Wideband Data Communications Services," *Datamation,* XIV, No. 9 (1968),pp. 62-66.

Electronic Industries Association, *Trends in Communications Research and Development.* Washington, D.C.: Feb. 27, 1968.

James, Richard T., "High-Speed Information Channels," *IEEE Spectrum,* April 1966, pp. 79-95.

Johnson, John Paul, "A Conversation with Russell W. McFall [President and Chairman of the Board, Western Union Telegraph Co.] ,"*Communications Magazine,* May 1969, p. 24.

Johnson, Nicholas (FCC Commissioner), "CATV: Promise and Peril," *Saturday Review,* No. 11, 1967. A thoughtful discussion of the potential of broadband transmission.

Romnes, H.I. (Chairman of the Board, AT&T), "Computers and Communications," Address before the 1968 Spring Joint Computer Conference, Atlantic City, N.J. Discussion of AT&T's plans for providing new data transmission services.

Western Union Technical Review, April 1967. Special issue on new and proposed Western Union communication services: ISCS, INFO-COM, SICOM, Hot/Line, Dial-Pak (Broadband Exchange), and domestic satellite systems.

G. Digital Transmission

Baran Paul, *On Distributed Communications.* Eleven-volume study for the U.S. Air Force; RAND Memoranda RM-3420-PR, RM-3638-PR, RM-3097-PR, RM-3762-PR, RM-3763-PR, RM-3764-PR, RM-3765-PR, RM-3766-PR, RM-3767-PR. Santa Monica, Calif.: The RAND Corp., 1964. A lengthly and authoritative study of an all-digital telecommunications system for voice and data transmission.

Franklin, R.H. and H.B. Law, "Trends in Digital Communication by Wire," *IEEE Spectrum,* Nov. 1966, pp. 52-58.

Harley, G.C., and J.H. Dejean, "Potentialities of an Integrated Digital Network," *Telecommunications,* Feb. 1968, pp. 15-21.

Hill, W. Jack, "Multiplexing: The Science of Mixing Voices," *Electronics World,* Feb. 1968, pp. 43-46.

Hoth, D.F., "Digital Communication," *Bell Laboratories Record,* XLV, No. 2 (1967), pp. 39-43.

Mayo, J.S., "Pulse Code Modulation," *Scientific American,* March 1968, pp. 102-108.

Pierce, John R., "Some Practical Aspects of Digital Transmission," *IEEE Spectrum,* Nov. 1968, pp. 63-70.

Travis, L.F., and R.E. Yaeger, "Wideband Data on T-1 Carrier," *Bell System Technical Journal,* XLIV, No. 8 (1965), p. 1567.

Ulstad, M.S., "A High-Performance Digital Data-Transmission System," *IEEE Transactions on Communication Technology,* Feb. 1968, pp. 115-119.

V. REGULATION OF COMMUNICATIONS COMMON CARRIERS

A. Regulatory Theory

Averch, Harvey, and Leland Johnson, "Behavior of the Firm Under Regulatory Constraint," *American Economic Review,* LII, No. 5 (1962).

Bonbright, James C., *Principles of Public Utility Rates.* New York: Columbia University Press, 1961.

Kaysen, Carl, and Donald F. Turner, *Antitrust Policy: An Economic and Legal Analysis.* Cambridge, Mass.: Harvard University Press, 1959. 345 pp.

Shepherd, William G., and Thomas S. Gies, eds., *Utility Regulation: New Directions in Theory and Policy.* New York: Random House, 1966. 284 pp.

Trebing, Harry M., *Performance Under Regulation.* East Lansing, Mich.: Michigan State University Press, 1968. 169 pp.

Wilcox, Clair, *Public Policies Toward Business,* 3rd ed. Homewood, Ill.: Richard D. Irwin, Inc., 1966. 882 pp.

B. The Regulatory Process

Arpaia, Anthony F., "The Independent Agency — A Necessary Instrument of Democratic Government," *Harvard Law Review,* LXIX (1956), pp. 483-506.

Beelar, Donald C., "Cables in the Sky and the Struggle for Their Control," *Federal Communications Bar Journal,* XXI, No. 1 (1967), pp. 26-41. History of the controversies over the nature, extent, and control of microwave communications in the United States.

Booz, Allen and Hamilton, *Organization and Management Survey of the Federal Communications Commission,* Report to the U.S. Bureau of the Budget, 1962.

Doyle, Steven E., "Do We Really Need A Federal Department of Telecommunications?" *Federal Communications Bar Journal,* XXI, No. 1 (1967), pp. 3-16.

Drew, Elizabeth Brenner, "Is the FCC Dead?" *Atlantic Monthly,* April 1967, pp. 29-36. Informed and witty article on the organization and responsibilities of the FCC.

Federal Communications Commission, *The FCC in Fiscal 1969: A Summary of Activities.* (Published annually).

Friendly, Henry J., "A Look at the Federal Administrative Agencies," *Columbia Law Review,* LX, No. 4 (1960), pp. 429-446.

Goulden, Joseph C., *Monopoly.* New York: G.P. Putnam's Sons, 1968. 350 pp. A dramatized discussion of the history and practices of the American Telephone and Telegraph Company.

Hecter, Louis J., "Problems of the CAB and the Independent Regulatory Commissions," *Yale Law Journal,* LXIX, No. 6 (1960), pp. 932-977.

Landis, James M., *Report on Regulatory Agencies to the President-Elect,* Submitted by the U.S. Senate Subcommittee on Administrative Practice and Procedure. Washington, D.C.: U.S. Government Printing Office, 1960.

Parks, Norman L., "The Sovereign State of Bell," *The Nation,* Oct. 30, 1967, pp. 430-435.

_____, "Who Will Bell the Colossus?" *The Nation,* Oct. 23, 1967, pp. 391-393.

United States v. Western Electric Company, Inc., and American Telephone and Telegraph Company (Consent Decree), 13 RR 2143; CCH 1956 Trade Cases, sec. 68,246 (D.C.N.J., 1956).

VI. REFERENCE DOCUMENTS

Auerbach Data Communication Reports. Auerbach Info Inc., 121 N. Broad Street, Philadelphia, Pa. A subscription service providing descriptions of data communication equipment and services.

The Communications Act of 1934, With Amendments and Index Thereto. Washington, D.C.: U.S. Government Printing Office (Updated as required; price 50 cents).

Defense Documentation Center, *a DDC Bibliography on On-Line Computer Systems,* Vol. 1, Sept. 1968; 163 citations, all abstracted. Available from U.S. Dept. of Commerce, Clearinghouse for Federal Scientific and Technical Information, Springfield, Va. This volume is supplemented by an Unclassified-Limited volume, with 52 citations, all abstracted. Accession Nos. AD 675 050, AD 840 090.

Federal Communications Commission, *Statistics of the Communications Common Carriers.* Washington, D.C.: U.S. Government Printing Office (Published annually).

IBM Corporation, *Data Communications Glossary,* Form No. C20-1666.

Selected Articles from the Lenkurt Demodulator. San Carlos, Calif.: Lenkurt Electric Co., Inc., 1966. 717 pp. A highly readable collection of articles on telecommunications technology.

Time-Sharing Industry Directory. Time-Sharing Enterprises, Inc., 251 W. DeKalb Pike, King of Prussia, Pa. 19406. A subscription service providing information on commercially available computer time-sharing and remote access services.

Executive Office of the President — Office of Telecommunications Management, "Chronological Resume of Some Significant Incidents in U.S. Telecommunications from 1866." (Unpublished).

Part 1, to 1961 (Jan. 1962), 41 pp.
Part 2, Oct. 1961 to Dec. 1962, 21 pp.
Part 3, Jan. to Dec. 1963, 19 pp.
Part 4, Jan. to June 1964, 9 pp.
Part 5, July to Dec. 1964, 17 pp.
Part 6, Jan. to June 1965, 22 pp.
Part 7, July to Dec. 1965, 24 pp.

United States Independent Telephone Association, *Independent Telephone Statistics.* Washington, D.C.: USITA (Published annually).

Withington, Frederic G., *The Computer Industry — 1969-1974.* Arthur D. Little, Inc., Service to Management Report. Cambridge, Mass.: June 1969 (Prepared annually).

VII. PERIODICALS

Bell Laboratories Record. Bell Telephone Laboratories, Inc., Mountain Ave., Murray Hill, N.J. (Published monthly).

Bell System Technical Journal. Bell Telephone Laboratories, Inc., Mountain Ave., Murray Hill, N.J.

Communications of the ACM. Association for Computing Machinery, 1133 Avenue of the Americas, New York, N.Y. 10036. (Published monthly).

Datamation. F.D. Thompson Publications, Inc., 35 Mason St., Greenwich, Conn. Most widely read periodical in the computer industry. (Published monthly).

EDP Analyzer. Canning Publications, Inc., 134 South Escondido Avenue, Vista, Calif. (Published monthly).

Federal Communications Bar Journal. Federal Communications Bar Association, 1343 H Street, N.W., Washington, D.C.

IBM Systems Journal. IBM Corporation, Armonk, N.Y. (Published quarterly).

Industrial Communications. Washington Radio Reports, Inc., 388 National Press Bldg., Washington, D.C. 20004. A telecommunications industry weekly newsletter.

The Lenkurt Demodulator. Lenkurt Electric Co., Inc., San Carlos, Calif. (Published monthly; free upon request).

Signal Magazine. Journal of the Armed Forces Communications and Electronics Association, 1725 Eye Street, N.W., Washington, D.C. (Published monthly).

Telecommunications Reports. 1204-1216 National Press Bldg., Washington, D.C. 20004. The authoritative newsweekly of the telecommunications field.

Telephony Magazine. Telephony Publishing Corp., 608 South Dearborn, Chicago, Ill.

Western Union Technical Review. Western Union Telegraph Co., 60 Hudson Street, New York, N.Y. (Published quarterly).

INDEX

A

Above 890 case, 109, 116-117, 139, 181
Acoustic coupling devices:
  Carterfone device, 85
  description, 99-100
  use on dial telephone network, 100-103
Advanced Computer Utilities Corporation, joint venture with Western Union, 29
Advanced Research Projects Agency (see ARPA computer network)
Aeronautical Radio, Inc. (ARINC), 51, 63, 69, 123
AIRCON (see RCA Global Communications, Inc.)
American Telephone and Telegraph Company (AT&T):
  antitrust consent decree, 5, 20, 32-33, 38n
  Bell System organization, 4-5, 139
  Business Information System (BIS), 31
  Dataphone 50 service, 155
  dial telephone service, 22n, 155
  Electronic Switching System (ESS):
    arranged with Data Features (ESS-ADF), 56-57
    performance, 56-57, 91, 170
    reliability, 56-57
    suitability for EDP, 30-31
  growth, 2
  international telephone service, 7
  Picturephone, 174
  private line service, 104n, 124n, 155
  private microwave, opposition to, 181
  provision of EDP services, 30-33
  Series 8000 service, 155, 157-158, 162n
  Series 11,000 service:
    description, 125
    distinguished from TELPAK, 125
  size, 5, 187
  T-carrier system (see Digital transmission)
  TELPAK service: 155, 158-159, 181-182
    sharing of, 122, 136n, 142-144
  TWX service: 50, 51n, 54-55, 155, 160
    pricing, 165, 170
    purchased from AT&T by Western Union, 28-29, 65-66
  WATS service, 155
Analog transmission (see also Modems), 170-172
ARINC (see Aeronautical Radio, Inc.)
ARPA computer network: 67, 129-134, 168
  effect of line sharing restriction, 133-134
  motivation for, 132-133
Asymmetric channels (see Data communications)
Auerbach Corporation, xi
"Authorized user" tariff provisions (see also Line sharing), 122-124
AUTODIN (Automatic Digital Network), 68
Automatic Electric Company (see GT&E)
Average-cost pricing: 55
  definition, 109
  geographic averaging, 109-110, 189
  technological averaging, 110-111
Averch, Harvey, 38, 143

B

Baran, Paul, 153, 168, 173
Beelar, Donald C., 113, 181
Bell System (see AT&T)
Bell Telephone Laboratories, Inc. (see AT&T, Bell System organization)
Bigelow, Robert P., 215
Bolt Beranek and Newman, Inc., 132
Booz, Allen & Hamilton, 153, 157
Bunker-Ramo Corporation: 52
  message switching services, 21-22
  opposition to SICOM service, 73
  Telequote stock quotation service, 11, 123
  TOPS (Tele-Center Omni-Processing System), 52

Index

Business Equipment Manufacturer's Association (BEMA): 17
    opposition to SICOM service, 73

C

Carrier system, 180
Cathode-ray tube display (see CRT terminal)
Carterfone case, ix, 85-87, 89n, 94, 101, 106
CATV, 175, 182, 202
Channelizing equipment (see Multiplexing)
Channels, communication (see Data communications)
Charging by the bit, 50, 166-167
Circuit Switching:
    advantages over Message Switching, 49-50
    commercial services:
        Broadband Exchange (see Western Union)
        Dataphone 50 service (see AT&T)
        Dial telephone service (see AT&T)
        Telex (see Western Union)
        TWX (see AT&T)
Common carriage, definition (see also Communications common carriers), 1
Communication channels (see Data communications)
Communication system design, 126n, 127
Communications Act of 1934, 1-3, 17, 19
Communications common carriers:
    data transmission services (see Data communications)
    history, 1-2, 187
    industry structure, 4-10
    international, 7
    new carrier entrants (see Special service common carriers)
    obligations of, 3
    provision of EDP services: viii-ix, 19-20, 26-43, 209
        as regulated services, 37-38
        as unregulated services 35-37
        through restricted separate subsidiary, 40-42
        through separate subsidiary, 39-40

"Rate base" regulation:
    definition, 2, 147
    effects upon carrier investment decisions, 115-116, 140
    effects upon carrier risk-taking, 71-72
    effects upon multiple regulated services, 37-38
    effects upon unregulated services, 35-37
tariffs:
    definition, 3
    restrictions contained in, ix-x
telegraph company:
    domestic (see Western Union)
    international, 7
telephone companies:
    Bell System (see AT&T)
    Independents, 6, 187
    interconnection with Bell System, 107-108
Communications Satellite Corporation (see Satellites, communication)
Compatible Time Sharing System (CTSS), 59-60
Computer-based information services, 11-12
Computer equipment:
    cost-effectiveness trends, 151
    suppliers, 10
Computer industry, structure 10-12
Computer Sciences Corporation, attempted merger with Western Union, 28
Computer utility, vii, 16-19, 20, 26-29
COMSAT (see Satellites, communication)
Concentrator, data transmission, 45
Conditioning, channel, 169n
Connect time, 167, 170, 207
Continental Telephone Corporation, 30
Corbató, F.J., 13n, 61
Cost accounting, common carrier (see also Price discrimination):
    cost allocation, 35-37
    cost separation, 35-37
    cross-subsidization among communication services, 37-38, 144
    joint costs, 35-36

Cream skimming (see also Average-cost pricing):
  by Bell System, 113
  by private microwave systems, 112, 114
  by special service common carriers, 112-114, 189-190
  definition, 111
  geographic, 114
  technological, 115-116
Credit Data Corporation, 11
Crosstalk, 88-89, 93, 145-146, 163
CRT terminal, 158, 160-161

D

Dalcode (see also Multiplexing), 74-75, 77, 160
Data communications (see also Digital transmission):
  asymmetric channels, 160-161, 207
  available common carrier channel bandwidths: 24, 154-163
    compared with EDP equipment operating speeds, 156-159
    high speed, 163
    low speed, 160-161
    medium speed, 161-162
  channel reliability, 24, 163-164
  error rates, 163-164, 169, 206
  growth rates, 153-154, 200
  maximum theoretical data transmission speed, 156
  price structure (see also Average-cost pricing), x, 164-170
  systems cost trends, 150-151
  terminal devices, 157-163
Data processing, regulation of, 16-19
Data set (see Modem)
Dial network
  definition, 22n
  pricing, 164, 170
Digital transmission:
  description, 170-172, 204-206
  digital communication network (see also University Computing Company), 24, 170-174

  pulse code modulation (PCM), 171
  T-Carrier system, 81n, 141, 163, 172
Data Transmission Company (see University Computing Company)
Drew, Elizabeth Brenner, 4

E

EDUCOM (Interuniversity Communications Council), 67
Electronic Switching System (ESS) (see AT&T)
Error rates (see Data communications)

F

FCC (see Federal Communications Commission)
FDM (see Multiplexing)
Federal Communications Commission (FCC):
  computer/communications inquiry, vii, xi
  formation, 1-3
  resources, 3-4
  responsibilities, 3-4, 74
Foreign attachments (see also *Carterfone* case; *Hush-a-Phone* case; Modems): 83-103
  Data Access Arrangement, 89
  divided responsibility, 91-92
  equipment innovation, 92-93, 97-98
  network control signaling, 87, 89-91, 94-95
  residential telephones, 93-94, 102
  system integrity, 88-91, 102
  tariff restrictions, 22-23, 83-84, 87
  tariff revisions:
    description, 87, 101-102
    impact, 96-103
Frequency-division multiplexing (see Multiplexing)

G

General Telephone & Electronics Corporation (GT&E): 20, 172

Automatic Electric Company, 6, 139
  General System, 6
  GT&E Data Services Corporation, 30, 40
  Lenkurt Electric Co., Inc., 6, 139
Graham, Robert M., 212n
Greenberger, Martin, 16n

H

Honeywell DDP-516 computer, 132
Hush-a-Phone case, 84, 101

I

Impulse noise, 163-164
Independent telephone companies (see Communications common carriers)
INFO-COM (see Western Union)
Information utility (see Computer utility)
INTELSAT (see Satellites, communication)
Interconnection (see also *Carterfone* case):
  definition, 101
  economic considerations (see Average-cost pricing; Cream skimming)
  farmer service lines, 107
  network control signaling unit (see also Foreign attachments), 105
  private microwave systems, 112-113, 116-118
  special service common carriers, 112, 118-120, 206
  tariff restrictions, 104-105
  technical considerations (see also Foreign attachments), 106-109
Interconnection restrictions, 23
Interdata Communications, Inc., 201
Interface Message Processor (IMP), 131-133, 168
International Business Machines Corporation (IBM):
  development of System/360, 71-72
  proposed primary business test, 63
  test of data transmission via satellite, 9-10
International Communications Association (ICA), 126
International Telephone and Telegraph Corporation (ITT): 7, 20

ITT Data Services, 30
Interstate Commerce Commission, regulation of communications by, 1
Irwin, Manley R., vii-x, 64
ISCS (see Western Union)

J

Johnson, Leland, 38, 143
Johnson, Nicholas, 115n, 150
Joint costs (see Cost accounting)
Joint user (see Line sharing, tariff revisions)
Jones, Malcolm M., 212n

K

Kappel, Frederick, 154
Knight, Kenneth E., 151

L

Large scale integration (LSI), 171
Leased lines (see Data communications)
Lenkurt Electric Company (see GT&E)
Line sharing (see also "Authorized user" tariff provision; Price elasticity of communications demand):
  capacity sharing (see also Line sharing, channelizing):
    carrier opposition, 145-147
    effects upon equipment supply, 139-142
    motivation for, 137-144
    price discrimination, 142-145
  channelizing of communication lines (see also Multiplexing):
    bandwidth optimization, 138
    definition, 122, 135-136
  economic impact:
    long-run effects, 149
    short-run effects, 148-149
  Joint user (see Line sharing, tariff revisions)
  line brokerage (arbitrage), 127, 141, 144, 147, 152
  motivation for, 121-122

resale of communication lines, 122, 129, 137
Series 11,000 service (see AT&T)
tariff restrictions, 23, 52, 122-123
tariff revisions, 53, 123-124, 137, 202
TELPAK sharing (see AT&T)
time-sharing of communication lines:
   definition, 122, 126
   motivation for, 126, 127
   "short period" leased lines, 126
   use in message switching systems, 52, 128-129
Long Lines Department (see AT&T, Bell System organization)

M

MAC, Project (MIT): 13, 59-60, 167
   message switching capability, 59
Mallard, Project, 81
Marginal-cost pricing, 191-192
Martin, James, 156n, 164n
McFall, Russell W., 20, 26-27, 29-30
McPherson, John G., 12n
Melody, William H., 143-144
Message switching, commercial services:
   AIRCON (see RCA)
   INFO-COM (see Western Union)
   ISCS (see Western Union)
   PMS (see Western Union)
   SICOM (see Western Union)
   TCCS (see Western Union)
   TOPS (see Bunker-Ramo Corp.)
Message switching, description of: 44-51
   advantages over circuit switching, 46-49
   digital network, 172-174
   distributed network, 44, 46, 52-53
   duplexed equipment, 48-49
   hybrid systems, 45-46, 59-63, 67-68, 80
   paper tape systems, 44-45
   pricing (see Charging by the bit)
Message switching, use of ESS for (see also AT&T), 56-57
Message switching services, regulatory status of:
   current regulations, 52-53

   policy alternatives: 54-73, 80-81
      primary business test, 59-65
      restricted to common carriers, 54-59
      unregulated specialized services, 65-73
Message switching systems, types of:
   industry-oriented, 51, 67-73
   private, 50-51, 66-67
   public, 50, 65-66
Microwave Communications, Inc. (MCI) (see also Special service common carrier), 25, 178, 183-185, 188, 190, 192-198, 201-202
Microwave Transmission Corp. (see University Computing Company)
Microwave transmission:
   characteristics, 179-181
   digital transmission, 204-205
   economies of scale, 116-117
   history, 179-180
Mills, Richard G., 167n
Modems (see also Acoustic coupling devices):
   data transmission speeds, 155n, 156
   description, 23, 83-84
   digital transmission, not needed for, 172
   growth rate, 200
   impact of revised foreign attachment rule:
      acoustic coupling, 99-101
      modem technology, 97-98
      tone-generation terminals, 98-99
   restricted by foreign attachment rule, 83
MULTICS (MIT), 60-61, 212n
Multiplexing (see also Line sharing, channelizing): 45, 170, 172
   frequency-division (FDM), 45, 74-75, 78, 136, 156, 172
   time-division (TDM): 45, 136, 160, 172, 205-206
   used in SICOM service, 74-79, 82

P

Parity bit, 164
Parkhill, Douglas F., 16n
PCM (see Digital transmission)
Picturephone, 31

Polling, 167
President's Task Force on Communications Policy, vii, xi-xii, 186, 190-191
Price discrimination, 55n, 142-145
Pricing, communications (see Average-cost pricing; Data communications; Marginal-cost pricing)
Price elasticity of communications demand, 150-151
Primary business test, 63-65, 80
Privacy:
   legal controls: 213-216
      bonding, 215
      certification, 215
      criminal sanctions, 216
      licensing of personnel, 215
      system licensing, 214
   system safeguards: 210-212
      audit trail, 212
      cryptographic devices, 211
      duplexed EDP equipment, 212
      "execute only" memory, 212
      file "dumps", 212
      hardware traps, 212
      limited access to terminals, 211
      password, 210
      "public" files, 211
      "read only" memory, 212
Private microwave systems (see also *Above 890* case; Microwave transmission): 116-118, 177-178, 181-182
   AT&T opposition, 112-113
   AT&T TELPAK response, 143
Private Wire Service (PWS) (see Western Union)
Project MAC (see MAC)
Public exchange network, definition, 22n
Public Message Service (PMS) (see Western Union)
Pulse code modulation (see Digital transmission)

R

"Rate base" regulation (see Communications common carriers)

RCA Global Communications, Inc.: 7, 51
   AIRCON service, 51, 128
Real-time system, definition, 12
Regenerative repeaters, 171, 205
Regulation:
   of common carriers (see Communications common carriers, "rate base" regulation of)
   of data processing, 16-19
   of teleprocessing services, viii
Remote-access data processing: 16
   conversational time-sharing systems, 12-13
   data collection systems, 13-14
   definition, 12
   information distribution systems, 14-15
   inquiry systems, 13
   remote batch processing systems, 14
   remote document production systems, 14
Risk aversion of regulated firms, 71-72
Roberts, Lawrence G., 133
Romnes, H.I., 31-32

S

Satellites, communication: 7-10
   Communications Satellite Act of 1962, 7-8
   Communications Satellite Corporation (COMSAT), 7-9, 25, 110n
   data transmission via, 9-10
   domestic, 8-9, 25
   International Telecommunications Satellite Consortium (INTELSAT), 8
Scantlin Electronics, Inc.: 123
   opposition to SICOM service, 73
Security (see Privacy)
Service bureau, data processing, 10-11
Servan-Schreiber, J.J., vii
Seven-Way Cost Study, AT&T, 37, 191
Shannon-Hartley law, 156
Sharing, communications line (see Line sharing)
Short-period leased lines, 165-166
SICOM (see Western Union)
Software house, 10-11

Index   269

Special service common carriers (see also Average cost pricing; Cream skimming; Microwave Communications, Inc.; University Computing Company; Interconnection), 25, 111
  arguments against, 187-190
  competitive benefits, 186-187
  competitive response of existing carriers, 190-192
  description, 177-179
  distinguished from present carriers, 176-177
  FCC's MCI decision, 192-199
  impact upon existing carriers, 200
  long-range potential, 200-203
  service characteristics, 183-186
  use of microwave, 177-178, 182
Stanford Research Institute (SRI), 174
Store-and-forward system (see Message switching system)
Stromberg-Carlson, 172
Switched network, definition, 22n

T

T-carrier system (see Digital transmission)
TCCS (see Western Union)
TDM (see Multiplexing)
Telephone company (see AT&T)
Teletype Corporation (see also AT&T, Bell System organization), 5, 31
Teletypewriter, 79, 158, 160
Telex (see Western Union)
Telex Computer Communications Service (TCCS) (see Western Union)
Thermal noise, 163
Time-division multiplexing (see Multiplexing)
Time-sharing, computer, 16-19
Time-sharing, of communication lines (see Line sharing)
TOPS (see Bunker-Ramo)
Touch-Tone telephone, 31
TransAmerican Microwave, Inc., 204
TWX (see AT&T)

U

Ultronic Systems, Inc., 123
United States Independent Telephone Association (USITA), 6
United Utilities, Inc., 6, 30
UNIVAC:
  418 computer, 73
  1108 computer, 48n
University Computing Company, proposed data network: 28, 175n, 203-209
  computer switching, 207
  digital transmission, 204-206
  local distribution, 206
  outlook for UCC proposal, 208-209
  services to be provided, 207

V

"Value of service" pricing, 55n, 142-145
Voice telephone network, definition, 22n
Vyssotsky, V.A., 13n

W

Western Electric Company 5, 96, 139, 186
Western Union Computer Utilities, Inc., formation of, 29
Western Union International, Inc., 7
Western Union Telegraph Company (WU): 6-7, 18, 20, 61
  attempted merger with Computer Sciences Corp., 28
  Broadband Exchange, 155
  comments on cost separation, 36
  INFO-COM (Information Communications Service), 27-28, 45-46, 50-51, 68, 73-79, 82, 128, 160
  ISCS (Information Services Computer System), 48, 66
  legal citation service, 27
  long-range plans, 26-30
  microwave, early use of, 179
  opposition to AT&T line sharing liberalization, 135

PICS (Personnel Information Communication System), 11, 27
PMS (Public Message Service), 6-7, 62n, 65-66
provision of private computer systems, 33
PWS (Private Wire Service), 74, 76-79
SICOM (Securities Industry Communications Service): 27-28, 45-56, 73-79, 82, 160
   antitrust aspects, 75-77, 82
   FCC decision, 75-79
   nature of the service, 74-75
TCCS (Telex Computer Communications Service), 44n
Telex service, 50, 54-55, 65-66, 155, 160
   pricing, 165

On Your Own: Professional Growth
through Independent Nursing Practice

On Your Own:

PROFESSIONAL GROWTH THROUGH INDEPENDENT NURSING PRACTICE

Edited by: Mary Lee Lynch,
RN, MSN, CS

Wadsworth Health Sciences Division
Monterey, California

Wadsworth Health Sciences Division
A Division of Wadsworth, Inc.

© 1982 by Wadsworth, Inc., Belmont, California 94002.
All rights reserved. No part of this book may be reproduced, stored in a retrieval system, or transcribed, in any form or by any means—electronic, mechanical, photocopying, recording, or otherwise—without the prior written permission of the publisher. Wadsworth Health Sciences Division, Monterey, California 93940, a division of Wadsworth, Inc.

Printed in the United States of America

10 9 8 7 6 5 4 3 2 1

Library of Congress Cataloging in Publication Data:

On your own.

    Bibliography: p.
    Includes index.
    1. Nursing—Practice.   2. Nurse practitioners—United States.
I. Lynch, Mary Lee. (DNLM: 1. Nurse practitioners. WY 128 058)
RT86.7.05   610.73'06'8   81-19702
ISBN 0-8185-0507-9   AACR2

Sponsoring Editors: Ed Murphy/Jim Keating
Production: Ron Newcomer & Associates, San Francisco, California
Manuscript Editor: Antonio Padial
Interior Design: Otto Spek
Cover Design: Al Burkhardt
Illustrations: Irene Imfeld
Typesetting: Graphic Typesetting Service, Los Angeles, California

                    Allen County Public Library
                        Ft. Wayne, Indiana

Preface

Since 1965, when the first nurse practitioner programs started, we have seen the rise of a new kind of independent nurse practitioner. These nurses, whether alone or in groups, have set up independent practices to offer patients nursing care as an entity in itself. For the first few years the emphasis was on primary medical care and pediatrics. In recent years we have seen the emergence of the psychiatric/mental health clinician, the nurse generalist, the oncology specialist, and other nurse clinicians establishing themselves in private practices. After identifying, assessing, and solving patient problems within hospital and community health settings, these practitioners have found it expedient and professionally rewarding to become independent practitioners.

How does one go about establishing a private nursing practice? Where does one start looking for information? How does the nurse break from the traditional model and establish a private practice? Will clients seek help from a nurse? Can the nurse make an adequate living? What are the legal, business, and ethical implications of such a practice? All of these questions, and many more, surface as the nurse attempts to sort out priorities and directions.

The limited, current literature presents a variety of practices and their particular problems, advantages, or disadvantages. Very little, however, has been written to answer the basic questions and give needed information to the aspiring independent practitioner. This book puts it all together: it attempts to answer basic questions and at the same time present some model practices.

There is little valid research about the need for and acceptance of this type of health-care delivery. But if we share experiences and research we can formulate and substantiate the need for new and independent nursing functions. This kind of nursing needs a firm scientific base, differentiated from medical practice, yet able to meet the needs of our clients. We need to emphasize research and the sharing of experiences if we are to meet some important goals: improved health care, a firm professional base for nurses, and acceptance by the community and other health professionals.

I canvassed the country for practitioners whose experiences could benefit others. Despite the problems, I have assembled a group of authors whose experiences illustrate both the difficulties and rewards of independent practice. Although it was impossible to include practitioners from all fields, one can generalize from their experiences. All practitioners are responsible for and accountable to their clients; these nurses are evidence that today there are alternatives for the nurse who seeks more satisfaction and responsibility.

Section I treats the changing role of nurses, business, legal, and research considerations of independent practice. These areas are basic to the beginning practitioner, yet information is scarce. This section addresses nurses' urgent need for practical information by giving realistic advice from professionals who know the problems. Section II is a narrative account by independent nurses in several kinds of practice. They describe their practices in detail and address the issues they see as vital. These narratives are a rich source of inspiration and information. It is interesting to draw comparisons among them; yet there is a core of agreement and shared experience that define private practice in a universal way.

This book in no way belittles traditional hospital nursing. Certainly there will always be a need for the hospital nurse. But the educationally prepared, innovative, and independent practitioner needs access to alternative types of practice. Every day we lose qualified nurses because they have become disenchanted with the system. These nurses are moving into other fields in search of more autonomy and higher incomes. Independent practice is an answer, but certainly not the only one. Many practitioners are creating jobs in industry, in the courts, in colleges as counselors, and in other nontraditional settings. For whatever reasons, the professional nurse of today is rebelling against a subservient role.

There is much to rebel against: the frustration of not being able to use one's skills fully, the suppression of initiative and leadership, and the inflexible bureaucratic structure. There is change, but it is slow. Too often the nurse is still seen as a "handmaiden to the physician." As in the past, nurses continue to allow themselves to be dictated to by physicians and administrators. Representation on health committees and policy-making boards is minimal despite the fact that nurses make up the largest component of the health-care-delivery system. Nursing, which is predominantly a woman's field, is affected by the stereotyping of women in general.

Recently we have seen an increasing concern for cost containment of health care and also a more knowledgeable public interested in quality care. The National Joint Practice Commission and the Institute of Medicine have published reports emphasizing the need for the nurse practitioner to be designated as a primary provider of health care. Although these reports differ on the need for medical supervision, and although some disagree with these reports, they are recognition of the contribution the nurse practitioner can make. What is needed now is more meaningful communication between nurses and physicians rather than fruitless debate on such controversial topics as physician supervision of nurses and the role of the nurse practitioner.

Resistance to the new roles of nurses was and is a major barrier encountered by most of our authors. This resistance, either overtly or covertly expressed, can be seen as a defense against real or implied threats to the medical profession. Unfortunately, there is resistance on the part of some nurses also. The challenge is to turn intergroup competition into intergroup collaboration. This collaboration is rooted in plain good sense and an obvious truth: No health professional is independent in the sophisticated, interdependent world of modern medicine.

Change is always inevitable and always threatening. As a group, nurses are not accustomed to using their potentially immense political power. To effect constructive changes in the health-care system and the federal, state, and local laws affecting nursing, nurses need to become aware of that power and organize. Of equal importance is education of the public to the role of the nurse practitioners and the range of their skills. Professional organizations can help by waging effective public relations campaigns to educate the consumer. Nurses themselves can promote this changing image by becoming involved in the community.

Two important issues facing the nurse practitioner are the variety of nurse practice acts and reimbursement for services. The state of Maryland recently set precedent by passing a law requiring every health insurer in the state to reimburse for the services of nurse midwives. An important provision of the bill is that insurance companies cannot require the nurse midwife to be under the supervision of the physician. Perhaps this law will be the start of a nationwide reimbursement system that will cover all nurse practitioners.

The emergence of nurses as private practitioners is a timely answer to the needs of both consumers and practitioners in the health-care-delivery system. An added advantage to private practice is the possibility for variety and mobility—a relevant issue in today's changing and mobile society. Private practice, however, can be rewarding as well as frustrating. Each practitioner must weigh the advantages and disadvantages before making a final commitment. The independence of private practice may look less inviting when clients, families, friends, and other professionals start making after-hours demands. Greater independence means greater responsibility.

For the nurse willing to accept the challenge, the opportunities are available. Nurses practicing in independent settings share an important characteristic: a dedication to treating the whole person within his or her cultural setting. Emphasis is on case finding, health teaching, health counseling, prevention, and rehabilitation. Accountable to themselves and their clients, nurses can develop new methods in the delivery of health services that are creative, genuinely responsive to public demands, and personally and professionally fulfillling.

Mary Lee Lynch

Contents

Section I—Business/Professional Considerations 1

1 **How to Start a Private Practice** 2
   Elsie Simms, RN, PhD

   The Market Survey
   The Marketing Mix
   Financing
   Business Decisions
   Summary

2 **Ethical Considerations for the Nurse as Primary Provider** 29
   Martha Henderson, MSN

   Formulating One's Ideal
   The Process of Ethical Reflection
   Code for Nurses
   ANA Code for Nurses with Interpretive Statements for Nurses as Primary Providers
   Summary

3 **Legal Planning for Private Practice** 52
Phyllis M. Gallagher, RN, JD

Defining the Practice
Choosing a Legal Structure
Minimizing the Risks of Malpractice
Organizing for Political Support

4 **Research Aspects of Independent Practice** 72
Marilyn T. Oberst, RN, EdD

The Practice Base: Intuition vs. Science
Research and Independent Practice
Research Possibilities in Independent Practice
Ethical Considerations
Some Difficulties and Solutions
Summary

5 **Obtaining Hospital Visiting Privileges** 98
Donna Lee Wong, RN, MN, PNP

The Beginning
Applying for Visiting Privileges
Advantages of Obtaining Visiting Privileges

Section II—Clinical Practice 115

6 **Hanging Out a Shingle:
The Joys and Sorrows of Private Practice** 116
Karlene Kerfoot, MA, RN

Defining Private Practice
Preparing for Private Practice
Visibility and Credibility
Specific Skills for Private Practice
Selection, Transfer, and Termination of Clients

Maintaining a Practice
Difficulties and Rewards

7 A Joint Psychiatric Practice 139
Margaret T. Campbell, RN, MSN, and Owen E. Clark, MD

Background
Initial Planning
Defining Focus and Scope
Setting the Stage
Community Liaison
The Treatment Program
Financial Considerations
Nonprofit Status
How It All Worked Out
General Discussion and Considerations
Summary

8 Multidisciplinary Group Practice 166
Keville Frederickson, RN, EdD

Initiating Practice
Professional Differences
Therapeutic Modalities

9 Nursing on the Prairie 181
Elaine R. Nielson, RN

Appendix I 193
Appendix II 196
Index 197

On Your Own: Professional Growth through Independent Nursing Practice

SECTION

I

Business/Professional Considerations

CHAPTER

1

How to Start a Private Practice

Elsie Simms, RN, PHD

How does one start a private practice? Where does one go for information and counseling? How can a nurse who goes into private practice be assured a good chance of success? How does one acquire a clientele? How does one find financial backing to begin and expand services? How does one advertise the services? What skills are required to maintain the practice once it is established? Any nurse starting a private practice must consider these questions.

Nurses who take the risk of starting nurse-managed health-maintenance clinics provide a much-needed service. Many Americans receive less than optimum health care, and nurses have the skills to offer this care. Many nurses take too narrow a view of their options. They think only of working for wages in large organized agencies and do not consider venturing into independent practice. With an understanding of the skills and knowledge required to begin and maintain a small business, nurses could supply much of the health care needed by the underserved public.

This chapter answers some questions for those just beginning an independent practice and suggests how to find answers to the questions that arise as a private practice grows. The beginning entrepreneur can look for those answers in many places. There are some very useful books on small-business management. A few of these are listed at the end of this chapter. The beginner can get good advice from sources as diverse as federal government documents, the local banker, and the neighborhood accountant. Some information, such as a lawyer's advice, can cost a considerable amount of money, whereas equally valuable information is totally free.

The Market Survey

Before considering anything else, the beginner should survey the market. A market survey is the assessment of an area and its people to determine the probable consumer response to the service one plans to offer. A good survey reveals the number of consumers within the area; their social, psychophysiological, and demographic characteristics; and their attitudes toward health services. Ideally, the market survey is conducted before the offered services are determined. The survey predicts which services the community will welcome, leads to informed decisions about what services to offer, and enhances the new venture's chance for success.

Market surveyors can obtain data from federal and local government departments. The city, county, or state regional planning and development department is a good source of current and pertinent information. Telephone directories are a source of numbers and addresses. If more-detailed information is needed, the government documents division of the nearest state university library may oblige. Data may not be available for the precise geographic area in which the nurse practitioner plans services. The practitioner must then interpolate the available information to determine the demographics of the chosen area. The area chosen may be as small as a circle with a ten-block radius in a large city or as large as a county in a Western state. The geographical size of the area to be served is determined by its population and by the number and kinds of services planned.

Defining Objectives

It is usually best to start a market survey by defining the objectives of the new enterprise. After these objectives, or goals, are determined, information is gathered to facilitate decisions on how to attain the objectives. There is no pat formula for the final decision—that is part of the risk—but decision makers should use the information gathered to generate several plans of action. After thinking through each plan to its probable outcome, decision makers choose and act on one or more of these plans. This method of decision making allows the entrepreneur to reevaluate the effectiveness of various actions as the business progresses and new information becomes available.

The following list shows how one might go about collecting information for a market survey.

Main Objectives:
1. To offer a needed health-care service to the public
2. To maximize profit/risk ratio yet maintain a socially responsible stance

Supporting Objectives:
1. To offer the services needed by a defined population
   Required information:
   a. A list of possible services
   b. The transportation system in the area
   c. The health practices and problems of the population
   d. The attitudes of the population toward health care
   e. The availability of other health-care services in the area
2. To define a population that needs and will use a nurse practitioner service
   Required information:
   a. The type of housing in the area

  b. The age (by percentage groups in ten-year increments) of the population in the area
  c. The population mobility
  d. The availability of other health-care services in the area
  e. The religious and ethnic makeup of the population
  f. The health practices and problems of the population
  g. The acceptability of the service to the population of the area
  h. The birth rate in the area
3. To determine the ability of the chosen population to support the service financially
 Required information:
  a. The occupations of people in the area
  b. The percentage or number of individuals in each occupation
  c. Income by percentage of population
  d. Attitudes toward payment for health-care services in the area
4. To choose the site for the service
 Required information:
  a. The cost to rent or purchase proposed sites
  b. The transportation system in the area
  c. The availability of referral services in the area
  d. The availability of the proposed sites
5. To determine how much money and time must be invested
  a. The cost to rent or purchase proposed sites
  b. The cost of furnishings and supplies
  c. The break-even point

Of course the survey should reflect the characteristics of the service area and the service plan. This list should serve only as a departure point for those planning a survey.

 As a next step in the survey, the nurse or nurses should list the services they could offer the community. The list may include (1) health assessments for all age groups; (2) maintenance nursing care, such as injections from a prescription, dressings, and irrigations; (3) emergency first-aid care; (4) in-service programs for hospitals, parent-teacher groups, or nurses'-association groups; (5) teaching and counseling in prenatal and postpartum preparation, diet control, wound self-care, pre- and postsurgical care and activities, and the like; (6) writing grants and conducting research for hospitals or other health agencies; (7) group counseling and teaching sessions, such as classes for recovering postcoronary–artery bypass patients, postmastectomy classes, relaxation training groups, teenage pregnancy groups, parenting groups, weight-control groups, diabetic groups, grief and loss counseling, and psychotherapeutic groups. These examples are only a beginning list of the many services a nurse could offer the community. Once the list is relatively complete, the entrepreneur is ready to test the market response to the services contemplated.

Survey Data and Their Implications

Demographic Data. What clues do demographic data offer the nurse practitioner who wants to start a business? Demographic information provides the physical and economic profile of a proposed market, including information about housing. Do the people in the area live in apartments, private homes, or tenements? The answer tells the would-be provider what potential clients can afford. The survey should also reveal the age distribution of potential clients. If there are many young adults, there will probably be a high demand for child health care and first-aid emergency care. There will also be a market for parenting classes. If the population is largely retired, the health-maintenance demands will be great. The survey should also reveal how frequently people move in and out of the community. If, as near military bases or colleges, the young-adult population is highly mobile, many individuals will not have an established health-maintenance source. If, however, the population is relatively stable, it may be very difficult to attract clientele unless they are underserved or highly dissatisfied with existing health-maintenance services.

It is important to know the major sources of income of the community. If the major employer is an industrial plant that provides free health services, it may be difficult to attract clients. If the major industry is mining or agriculture, a different service or mixture of services will be needed for each type of major income source. In many communities, large industrial companies offer complete health insurance. In such a community, a nursing practice could survive only if nursing services were covered by the third-party payer. Payment to nurses for health-maintenance services by insurance companies or federal health insurance would provide a mechanism for establishing nursing practices in many areas of the nation that are now severely underserved.

The transportation system of a community should be studied before selecting a service site. If many families use private automobiles, parking becomes an important consideration. If there is a widely used, inexpensive public transportation system, the clinic should be near central routes. If neither public nor private transportation is available in the target community, it may be wise to provide a shuttle service once or twice a week and to locate the clinic within walking distance of the majority of the clients.

Income levels, as revealed by a market survey, have important implications for the nurse practitioner planning a service. People in the lower-income ranges spend a greater percentage of their incomes and have more illnesses and accidents than people in middle-income brackets. However, it may be more difficult to collect fees for services from such a population. Surveys of national spending among households of various incomes reveal that those in the lower third of the income range spend approximately one-fourth of their income on health care, whereas the middle- and upper-income groups spend approximately one-third or more of their income for health care.

If there is a considerable amount of poverty or borderline poverty in the area being surveyed, the practitioner may be able to obtain private or government grants for establishing a health-services clinic. Middle-income populations tend to pay bills more regularly but also are more careful to seek assistance for health maintenance. Upper-income groups tend to have established health-service providers and will not use a new source unless a unique and needed service is offered. A new service may possibly attract clients from the upper-middle and upper economic brackets if the community is in a growth area. If the community is shrinking, physicians may be the first to leave the area.

Other demographic information of value to the practitioner who contemplates starting a private enterprise include (1) the availability and cost of clinic space, (2) the probable physical safety and the safety of collected funds in a given area, (3) the availability of referral services such as physical therapy, respiratory therapy, pharmacies, and medical practitioners.

Behavioral Data. Behavioral data also offer important insights. This portion of the market survey includes such information as the religious and ethnic makeup of the community and the health practices traditional to those groups. Most of this information can be obtained from the state and local public health department and from hospitals serving the area. An interview with the nursing-service directors of these agencies may reveal the general answers to questions concerning health practices and problems of the people in the area of interest. If further information is required, the social or human services department is a ready source of information about community health practices and problems.

Are the dietary practices of the community creating nutritional deficiency and health problems? Certainly obesity should be considered in this context as well as infant and childhood deficiency problems. Are support systems for the sick efficient and extensive? Are people of the ethnic group in the area concerned about maintaining health, or do they seek outside assistance only after the support system has been exhausted and the client is severely ill and debilitated? What is the level of hygiene acceptable to the community? What is the general level of health in the community? Is the attitude of people fatalistic, or do they believe people control their own fates? Is cash or credit the acceptable method of payment for services, or does tradition rule that health care is a charity service? Is the community bilingual? Before embarking on a business enterprise, the wise practitioner seeks answers to these questions. The success and acceptability of a new private practice depends on how well it accommodates these behavioral characteristics.

An area frequently ignored in market surveys for health services is the different concepts of time among ethnic groups. Usually, White Anglo-Saxons are future oriented, but many cultural and ethnic groups are present oriented. A few cultural groups are eternity oriented. These characteristics should determine whether clients drop in or make appointments. These characteristics also have a bearing on whether discrete, well-defined services would be better received than holistic services. They may also determine whether you will need the sanction of the ethnic group's religious leader to be acceptable to the community.

Extremely important to the large majority of nurses contemplating a private practice is whether there is a cultural expectation that health-service providers be either male or female. Some cultures, such as middle-American Whites, have difficulty viewing females as knowledgeable and competent health providers, whereas many European and Spanish-American cultures expect the major health-service provider to be female. Certainly many cultural and ethnic groups expect perinatal services to be provided exclusively by a female. Some groups are more concerned with the age of the health provider and will not consult a very young practitioner even if it is evident that a younger individual is competent.

Conducting the Survey

There are marketing-research agencies that will provide a complete survey and analysis for a fee. A beginning service, especially an individual or a two- or three-party partnership often cannot afford a professional survey. The rule of thumb is to spend no more than 5% of the initial investment on data collection and analysis. It is very possible for individuals to do their own marketing survey and stay within the 5% guideline limits. This survey does not require the large samples or control needed for a sociological research project, but accurate, current data are important. Sources of required information for supporting objectives listed on pp. 4–5 are as follows:

1. a. A list of possible services—from the nurse's own knowledge and expertise. Koltz's[1] list of typical services may stimulate

creative thinking (for information on this book, see Further Readings on page 27).

b. The transportation system—an area, street, or highway map from the local chamber of commerce or the highway department. Frequently, these two sources can supply traffic-count information. The local transportation system company or government department will supply route maps for public conveyances and usage counts. The state labor-department census discloses how many people own automobiles.

c. The health practices and problems of the population—the Health Systems Agency and the state Comprehensive Health Planning Council gather and supply most of this information. It may be wise to conduct a telephone or personal survey of local public-health nursing departments, Visiting-Nurse Services, and hospitals. The questions should be prepared beforehand. For instance, one might ask how many major surgeries are performed each month, whether existing facilities offer presurgery teaching and postsurgery instruction in preparation for discharge, and whether there is a health-maintenance clinic in the area. The questions should be as specific as possible and pertinent to the contemplated practice.

d. The attitudes of the population toward health care—the local public-health nursing department can supply this information. A question about attitudes should be included in the telephone survey.

e. The availability of other health-care services in the area—the Health Systems Agency can supply this information. Another ready source is the Yellow Pages of the telephone directory or a service map of the area.

2. a. The type of housing in the area—the local chamber of commerce, the state labor-department census, or the U.S. Department of Commerce are the best sources. Other possible sources for this information are the public-utilities companies in the area. Every ten years, the U.S. Department of

Commerce publishes the Standard Metropolitan Statistical Area study, which divides an area into census tracts and includes such information as population; its age composition, personal income, employment; and amount of trade. These publications are usually available in the government-documents division of the library.
- b. The age (by percentage groups in ten-year increments) of the population in the area—same source as 2a.
- c. The population mobility—the city or county planning division or the commerce department of the public utilities company.
- d. The availability of other health-care services in the area—same source as 1e.
- e. The religious and ethnic makeup of the population—same source as 1d.
- f. The health practices and problems of the population—same source as 1c.
- g. The acceptability of the service to the population of the area—this question should be included in the telephone survey of the public-health department, visiting-nurse service, and hospital nursing-service directors. The beginner may also want to do a mailed survey similar to the one described in Chapter 4 of Koltz's[1] book.
- h. The birth rate in the area—the county or state public-health department, Division of Vital Statistics.
3. a. The occupations of people in the area—the state labor-department census or the local chamber of commerce.
  - b. The percentage or number of individuals in each occupation—same source as 3a.
  - c. Income by percentage of the population—same source as 3a.
  - d. Attitudes toward payment for health-care services in the area—the accounting department of the local hospitals and visiting-nurse services.
4. a. The cost to rent or purchase proposed sites—local realty agents who deal in commercial sites.

      b. The transportation system—same source as 1b.
      c. The availability of referral services—same source as 1e.
      d. The availability of clinical supplies—the Yellow Pages of the area telephone directory and purchasing departments of the area hospitals.
      e. The centrality of the proposed sites—the area map, which is marked and coded with the population and transportation information previously collected.
5. a. The cost to rent or purchase proposed sites—same source as 4a.
    b. The cost of furnishings and supplies—the Yellow Pages of the area telephone directory will give a listing of office furniture suppliers. Most retail outlets will supply brochures and price lists of various lines of furnishings.
    c. The break-even point—described later in this chapter.

Analyzing the Survey

Informed decisions about the marketing mix can be made after this survey is completed and analyzed. Analyzing a market survey is similar to analyzing a nursing assessment. Lebell[2] states that this process requires "no greater analytic sophistication than required or justified in most practical situations." Certainly market analysis is no more difficult than the analysis required in a nursing process.

    One method of organizing the material is to divide the area map into sections around major population and transportation centers. It requires approximately 4,500 people to support a drugstore, and there is about one physician for every 1,500 people in most metropolitan areas. A generous estimate for the population necessary to support a private nursing practice would be between 2,500 and 5,000. One can reduce or expand the area around the transportation centers on the map to include that number of people. One can select four or five areas and then list the nursing care needs, the referral sources, the supply availability, and the desirability and costs of various office locations for each of these areas. After organizing survey information in this manner, one may evaluate each area. Chapter 3 of Lebell's[2] book contains a detailed description of one technique to aid in such an evaluation.

The Marketing Mix

The marketing mix is a combination of the product, the promotion, the price, and the distribution. In a health-care facility, medical services are the product. Promotion is any advertising of the offered services, and price is a monetary value assigned to those services. Distribution is the method of dispensing services, whether from a fixed facility or facilities, at the homes of clients, on an emergency basis, and the like.

The marketing mix reflects choices about which services to offer, how to present the offer to the public, how to select the appropriate market for the offered services, and how to seek compensation for services given. These decisions are made by determining the total amount expended and acquired by the enterprise and then allocating this dollar amount among the variables of the mix.

Product and Distribution

After analyzing the market survey to determine the health and spending patterns of the client population, the practitioner can make decisions about what services to offer and where the clinic should be. Of course, the decision about site should come only after analysis of transportation patterns and the cost of various locations. Also affecting the decision about clinic site is the availability and cost of space and the local availability of consumable supplies such as linens, dressing materials, solutions, and the like.

Price

Price setting is an essential component of the marketing mix and will affect the consumers' response to the service offered. Determining the price of a service is a complex decision. Economic theory dictates that pricing should be determined to maximize revenue. Accounting theory dictates that pricing should be based on cost accounting, which is the process of determining the cost of producing a unit of service and adding to that cost a reasonable percentage of profit. In practice, few prices reflect either theory precisely, but both theories must be considered. An entrepreneur wants to maximize revenue, and thereby profit, but social responsibility usually also enters into the pricing decision. Certainly pricing usually should follow a reasonable profit-over-cost principle.

Many Americans, however, equate high cost with good quality. For example, few women are convinced by the overwhelming evidence that there is little or no difference in the ingredients or proportion of ingredients in cosmetics and believe that a $5.00 lipstick from an exclusive cosmetic distributor is of better quality than the 50¢ lipstick from the local variety store. Nurses have long been victimized by not understanding this price-quality equation by the public and frequently find that their services are undervalued because they are reasonably priced. A prudent pricing policy reflects this American bias.

Complicating the pricing decision is the sometimes severe competition for the health-care market. Probably competition will weigh heaviest in the pricing decision. If the prices charged by similar health-care services in the area allow for full utilization by the public of the practitioner's time and permit a reasonable profit over expenses, then it would be wise to set comparable prices. In the discussion later in this chapter about the break-even point, other pricing considerations will become apparent.

Promotion

After the pricing decision is made, promotion or marketing decisions can be considered. Marketing is a method of creating consumer demand for the product. It is a form of selling that conditions the public to use the services offered. Advertising informs and educates the consumer about the benefits of the service.

How are new services promoted? There is a time-honored way: (1) attract attention, (2) arouse interest, (3) create desire, (4) convince the public they will be satisfied with the service, and (5) get the public to use the service. A goal of advertising may be to inform the public that nurses can offer services ordinarily sought from other, more expensive sources. In other words, the advertising strives to change the buying habits of the public.

Ethical barriers to advertising are being challenged by many professionals. Still, the professional who advertises services runs the risk of offending potential clients as well as colleagues. Caution in advertising is advised. Several acceptable advertising methods are available to the nurse(s) promoting a new private practice. Business cards, ads in the Yellow Pages, and fliers are very acceptable advertising. It is common practice for professionals to announce new services in the newspaper. This announcement should include the address, hours of service, and telephone number of the new practice as well as the types of services offered and the qualifications of the practitioner. It is also most appropriate to call on hospital nursing-service administrators and other health-service practitioners who may refer clients to the new service. It is wise to leave fliers describing new services with these other professionals.

One can get ideas for formats for ads from the Yellow Pages, the business section of any newspaper, or the numerous fliers most nurses receive from professional organizations. Printers and typesetters can show brochure samples and assist in producing the flier. Most printers also can refer clients to graphic artists who will make up a sample flier for a fee. If a local community college has a commercial art department, one can go there for further examples and assistance.

Behavioral studies reveal that most individuals respond to the following consumer appeals: (1) comfort or the possession of good health, (2) curiosity or "try-it-you-might-like-it," (3) fear of becoming ill or increasingly uncomfortable, (4) security from real or imagined physical, emotional, or financial threats to well-being, (5) affection from or for the provider, (6) usefulness of the service, or (7) pride and approval. The best advertisement is a satisfied consumer. No practitioner should forget that word-of-mouth can make or break a practice.

Location affects promotion. The decision of how much to spend on advertising can hinge on community attitudes toward advertising and, of course, on what has already been spent on the product (often a function of location) and distribution.

Financing

The simplest method of financing a new venture is through private funds: one's own funds or the funds of friends, relatives, or partners. One may decide to incorporate and sell shares in the new business. Some resourceful nurses may seek a private or government grant to begin the practice and then move to a fee-for-services practice when the grant elapses. One may choose from among many other alternatives. Financing, however, is obtained most frequently through a loan from a commercial investor such as a bank, savings and loan association, or commercial loan association.

One's own funds require neither repayment nor interest payments. However, it is important to be aware that such funds can be invested elsewhere to provide at least bank-interest rates of return. Relatives' or friends' funds usually require repayment with interest, although the interest is usually considerably less than one would pay a commercial lending agency. Grants usually do not need to be repaid, but the time and work required to plan, write, and defend a grant are time and work that can possibly be put to more profitable use. Periodic reports and final evaluations and continuation requests are also very time consuming for grant holders. Commercial loans, of course, require (1) some time and work to plan the loan request and (2) repayment of principal and interest.

The Small Business Administration (SBA) will make loans directly to small businesses, but more frequently loans through this government office are made in cooperation with a bank or other lending institution. The Small Business Administration supervises and licenses lending institutions that provide long-term loans to business enterprises. When seeking to finance a new business through the SBA, one should first attempt to acquire a loan from a reputable bank that has handled one's business and loans in the past. If, after presenting a well-thought-out business plan, one is refused by the bank and can find no alternative private funding with reasonable terms, then the best next step is to apply to the regional SBA for assistance. For brochures and publications about small businesses, write the Superintendent of Documents, Printing Office, Washington, D.C., 20402. Two helpful publications are *A Small Business of Your Own* and *A Survey of Federal Government Publications of Interest to Small Business.* The government documents section in the nearest university or college library usually has copies of most of these materials.

However one decides to procure the financing to begin the business, it is very important to shop for the money carefully. Each quarter-percent jump in interest rate, say on $1,000 over 15 or 20 years, increases costs by between $365 and $500; if one borrows $10,000, this difference is $3,000 to $5,000. One must also consider how loan repayments are calculated; this factor can change the total cost of the money borrowed and the amount of time one has the use of the money.

Besides the cost of the loan, one should evaluate the quality and qualifications of the lending institution. One must remember that lending institutions are selling the use of money. The reputation of the seller is important. A nurse should seek out a progressive banker who is interested in service to people and new ventures. If, as is likely, the nurse cannot find a banker who is familiar with health-care delivery business, it is to the nurse practitioner's advantage to educate the banker about opportunities in health care. The banker should be confident, interested, and involved in the community to be served. It is important to know beforehand if the officers of the lending institution will be accessible for conferences after the enterprise is launched. It is also important to know if the institution will disclose the credit ratings of future suppliers and customers.

Here are some other questions the new entrepreneur should ask before deciding on a lender.

- *Are the deposits insured by a federal agency?* Federally insured deposits assure the depositor that funds will be safe.

- *Will the lender finance new equipment?* The assurance of reasonable term financing for equipment may allow the entrepreneur to purchase equipment at a considerable savings even if sales or bargains on such equipment occur when cash is low.
- *Will the lender discount accounts receivable?* The service of discounting accounts receivable will remove the necessity of contracting a collection agency and provide the entrepreneur with a more stable cash flow. After first buying unpaid client accounts for less than full value, the lender assumes the responsibility of collecting the full fee from the client. The percentage of reduction is the lender's fee for this service.

No borrower should hesitate to ask questions or to take notes during interviews with lenders. The borrower is shopping for the product (service) offered by the lenders. It is their role to sell themselves and their services. However, the lenders are entitled to expect the loaned commodity to be returned in good order. Therefore, the lending institution will also require information from the entrepreneur. This information is best presented in a profile of the proposed business.

The Business Profile

The business profile allows the bank to judge the viability of the proposal. The profile includes the following information: the borrowers' credit ratings, reputations for repaying past loans, personal monetary assets, education and past experience that enable them to offer the service, licenses held, and the amount of life insurance held. The profile should include a carefully considered and well-written description of the purpose and organization of the business. A statement of how the business is expected to generate adequate income to repay the loan within the specified loan term is essential to the business profile. Part of this description should be a projected balance sheet. Graphs can be used to show expected growth rates.

The findings of the marketing survey should be part of the business profile. The profile should include all data that support the decision to begin a private practice, such as population figures, health-status findings, costs of renting and obtaining necessary services, and information concerning the need for and acceptability of the service. The last part of this report should include proposed fees for specific services.

The Break-Even Point

To summarize the report, one should calculate a break-even point. The calculation is relatively simple. It requires approximately 15 minutes to perform most simple services, such as drawing blood, applying a dressing, or giving an injection. Therefore, each 15-minute period can be considered a "production unit." As a first step in determining the break-even point, one lists all projected expenses. The list is divided into fixed (overhead) expenses and variable expenses. Fixed expenses include rent or monthly mortgage payments, insurance, utilities (such as heat, lights, and water), accountant and janitorial fees, and all partners' minimum living costs as salaries. These expenses remain relatively fixed regardless of the number of clients treated. Variable expenses include collection costs, cotton balls, pledgets, dressings, linens, disposable syringes, solutions, and all other disposable items required in the practice. Variable expenses fluctuate with the number of clients seen.

The second step in the calculation is determining the average variable expense for each 15-minute period (the production unit). One arrives at this figure by dividing the year's variable expenses by the yearly production units. The usual number of yearly production units is approximately 10,000, but the figure varies depending on whether one works an eight- or ten-hour day and a five- or six-day week. The third step is arriving at an acceptable charge to the public for a production unit of service. This method is less precise than using cost accounting to estimate the charge since it is determined by what is being charged in the community for similar services offered by other practitioners, such as physicians, psychologists, respiratory therapists, or physical therapists.

The final step in calculating the break-even point is to subtract the average variable expense for each production unit from the income for each production unit. The remainder is then divided into the total annual fixed expense to determine how many production units will be required each year to break even. This number will indicate how many clients one must serve and how many hours of service one must give before any profit is realized. If breaking even requires more production units than it is possible to expect in a year, it is time to rethink expenses or incomes.

It is important to know the break-even point because from it one may deduce how much money will be needed to finance the new enterprise until it is self-sufficient.

Business Decisions

Partnership or Corporation?

The decisions of where to locate, how to advertise, and how to finance the practice have been discussed. Some important choices remain. One is whether to maintain the clinic alone, form a partnership, or incorporate the practice. One must weigh the advantages and disadvantages of partnerships and corporations.

A partnership is formed by contract between two or more individuals. Total liability for the entire business is assumed by each partner, and this liability extends to personal property not connected to the business per se. The contract between the partners limits the individual partner's right to sell their interest in the partnership. The division of the profits of the business is stipulated in the contract, but there is no legal limitation on how much the partners may draw from the assets. The cost to begin a partnership is limited to the cost of legal advice concerning the contract. There are no extra taxes or fees levied against the partnership. However, the death of a partner dissolves the business contract and the partnership. Partnerships are frequently incapacitated during probate if a partner dies without a will.

To incorporate a business, the entrepreneur files an application with a designated state official. When the application is approved, the state issues a charter of incorporation, which evidences the fact that the corporation has been authorized to issue capital stock and to conduct business. Capital stock is simply an offer to sell a fixed portion of the business (corporation) to the public. The corporation is strictly regulated by both federal and state laws. A corporation is a legal entity and therefore can be held accountable for liabilities, but the individual stockholder has limited personal liability. Withdrawal of assets is limited by law to portions of the profit distributed as dividends. It is very easy for the individual stockholder to transfer or dispose of the stock if a buyer can be found. Stock can also serve as collateral to secure loans, whereas a partnership contract cannot. However, a corporation, as a legal entity, must pay taxes on its profits before dividends are distributed. Then the individual stockholder must pay income taxes on his or her personal income, which includes the dividends paid on the stock owned. This in effect, results in double taxation on the profits of a corporation. Because corporations are legally controlled both by state and federal government agencies, corporations must file frequent reports and entail more paperwork than partnerships.

Before one decides how to register the business, one should consult a lawyer who is familiar with state corporate law and small-business formation. Most lawyers charge for each consultation; practitioners who anticipate frequent consultations may wish to explore the possibility of a retainer-fee system.

Accounting Systems

An accountant's advice is invaluable when planning and setting up an accounting system and when preparing state and federal tax forms. As the practice grows, an accountant can also give advice about methods to maximize the profitable management of assets. Many accounting firms offer either a retained or a per-consultation service.

Although the accountant will help one select and understand the best bookkeeping system for one's practice, it may be beneficial to become acquainted with some accounting principles. The most fundamental principle in accounting, the accounting equation, may be manipulated as any algebraic formula is manipulated. This equation in its simplest form is as follows:

$$\text{Assets} = \text{Liabilities} + \text{Ownership}$$

Assets are recorded as the dollar value of owned things such as land, buildings, office and clinic furniture, business automobiles, office and clinic supplies, cash and accounts receivable, copyrights, and prepaid liability insurance. Liabilities are the amounts owed to creditors, such as loans payable, accounts payable, wages payable, and taxes payable. Ownership or the owner's equity is the value of the assets less the value of the liabilities. The sources of the owner's equity are invested money and business earnings.

This equation is the basis of a business balance sheet. The two sides of the equation must balance or be equal. The purpose of an accounting system is to (1) balance assets with ownership and liabilities and (2) create a record on which to base business decisions, pay taxes, and establish a credit rating.

The examples below are based on a fictitious company established by three partners to provide nursing services in a clinic. The P.R.N. Nursing Company was formed by Sue Pernoiss, R.N., Jane Roe, R.N., and Norman Nurse, R.N.

THE P.R.N. NURSING CO.
Balance Sheet

July 31, 19___

ASSETS		LIABILITIES & PARTNERS' EQUITY	
Cash	$7500.00	Accounts Payable	$1750.00
Furniture & Equipment	1750.00	S. Pernoiss' Equity	2500.00
		J. Roe's Equity	2500.00
		N. Nurse's Equity	2500.00
Total	$9250.00	Total	$9250.00

The accounting data are often recorded in a ledger with separate accounts for each area of transactions. These accounts are usually kept in loose leaf binders or files. A separate account is kept for each element of the ownership: the liability and the assets. It is simpler to draw up periodic balance sheets if the accounts are arranged and numbered in the ledger as they are in the balance sheet. Thus the accounts for the P.R.N. Nursing Company may appear as follows:

ASSETS:
Cash	1
Furniture & Equipment	2
Office & Clinic Building	3
Accounts Receivable	4
Supplies	5

LIABILITIES:
Accounts Payable	10
Taxes	11
Building Mortgage	12

PARTNERS' EQUITY:
Sue Pernoiss	20
Jane Roe	21
Norman Nurse	22

These accounts are typically recorded on the left and right—or, respectively, debit and credit—side of an accounting "T". These two terms are not the same as *debt* or *income*. *Debit* and *credit* refer only to the left and right sides of an accounting form. Convention and consistency dictate that increases in asset accounts be recorded on the left or debit side of the account and that increases in liability and owners'-equity accounts be recorded on the right or credit side of an account. For example, P.R.N. account sheets might look like this:

CASH   account #1

7/15/79	N. Nurse	2,500.00		
7/16/79	J. Roe	2,500.00		
7/16/79	S. Pernoiss	2,500.00		

FURNITURE & EQUIPMENT   account #2

| 7/17/79 | Reception rm. | 1,750.00 | | |

ACCOUNTS PAYABLE   account #10

| | | | 7/17/79 | S&T Furniture Co. | 1,750.00 |

EQUITY: NORMAN NURSE   account #22

| | | | 7/15/79 | Deposited in practice account | 2,500.00 |

EQUITY: SUE PERNOISS   account #20

| | | | 7/16/79 | Deposited in practice account | 2,500.00 |

Before transactions are posted they are usually recorded in a journal. All transactions are listed in the journal in chronological order. A journal entry includes a date, the account titles of the accounts credited or debited, and the amount of the transaction. The purpose of the journal is to reduce the possibility of error, to show the offsetting debit and credit entries for each transaction, and to give a chronological record of all transactions of the business. A journal might look like this:

LF	DATE	ACCOUNT DEBITED	ACCOUNT CREDITED	DEBIT AMOUNT	CREDIT AMOUNT
	7/15/79	Cash #1	Equity: N. Nurse #22	2,500.00	2,500.00
		Deposited in practice account			
	7/16/79	Cash #1	Equity: S. Pernoiss #20	2,500.00	2,500.00
		Deposited in practice account			
	7/17/79	Furniture & Eqpt.#2	Acc't Payable #10	1,750.00	1,750.00
		Purchased on account			

These three records—the journal, the ledger, and the balance sheet—make up the basic bookkeeping requirements for a business account. Transactions should be recorded in the journal as soon as possible after they have occurred. It is preferable to post the journal entries in the ledger accounts daily. Posting is the process of recording in the ledger accounts the debits and credits as they are indicated in the journal. As an entry is transferred from the journal to the ledger account, the journal is annotated with the ledger folio (LF), or ledger page number on which the transaction was posted. This annotation provides a cross-reference between the journal and the accounts and helps the bookkeeper to keep track of how much of the journal has been posted.

Periodical trial-balance sheets check the mathematical correctness of the ledger and assist the accountant in preparing various statements. A trial balance is simply listing and adding all debits and credits, which should be equal. Such trial balances do not necessarily assure the complete accuracy of the accounts because it is possible to enter the same amount twice or to fail to enter both the debit and credit amounts. Nevertheless, the trial balance is useful for many ongoing decisions. The accounts should be audited to assure complete accuracy.

This entire bookkeeping process may be computerized. An accountant can advise when computerized accounts are appropriate for a practice. Certainly, computerization can reduce much of the record keeping.

Collections

The new nurse practitioner must decide how best to collect from clients. Many professional practices require clients to pay cash at the time of services, but this approach may not be appropriate for some practices. Some entrepreneurs may want to use a collection agency or may wish to sell the accounts receivable to a bank. Some collection agencies charge per account collected; others charge a retainer fee or on a percentage basis. Of course it is quite possible to do one's own billing and collecting, but this work is tedious and routine.

Monthly billing can consume expensive secretarial time and postage. Furthermore, collection may not be the most profitable use of the practitioner's skills and time.

Licenses

Every state has some mechanism for regulating business. Usually the state requires those establishing or conducting a business to acquire a license. The license must be purchased and is renewable periodically. Regulatory agencies are listed in the telephone directory under the name of the state and usually have such names as *Department of Business Licensing and Regulations* or *Industrial Commission*. A letter or telephone call to this regulatory agency will reveal how to apply for any business licenses required by the state. It is also wise to check with city and county government agencies to determine if city or county business licenses are required as well. Any lawyer knowledgeable about small-business operation will know licensing requirements.

Finally, after all of this information gathering and analyzing, you are ready to see your dream come to fruition! With a will to succeed in your venture, go forth: purchase your business regulation licenses, acquire funding, arrange for office space and equipment, buy insurance, prepare news releases, prepare advertising fliers and business cards, contact referral services, draw up contracts with legal, accounting, and collecting services, hang out your shingle, and open your doors to a private nurse practice that will serve your public!

Summary

The practitioner contemplating an independent practice must make many decisions. A market survey can help the decision maker choose which service to offer, where and how to offer it, how to promote it, and how much to charge for it. The market survey is used to prepare a business profile, a document that can help the beginner finance the venture. Lawyers and accountants can help the beginner shop for money, establish the business legally, and set up accounting systems. When the new enterprise opens its doors, the practitioner will reap rewards for careful planning.

Author's Biography

Dr. Elsie Simms received a baccalaureate degree in nursing at the University of Colorado and a bachelor's degree in general business from Idaho State University. She received her master's degree in nursing at Indiana University and her doctorate from the University of Arizona.

Presently, she is director of undergraduate programs, College of Nursing, University of Akron, Akron, Ohio.

REFERENCES

1. Koltz CJ Jr: *Private Practice in Nursing, Development and Management.* Germantown, Md, Aspen Systems Corp, 1979.
2. Lebell D: *The Professional Services Enterprise: Theory and Practice.* Sherman Oaks, Calif, Los Angeles Publishing Co, 1973.

FURTHER READINGS

Boone LE, Kurtz DJ: *Contemporary Business.* Hinsdale, Ill, Dryden Press, 1976.
Editors of the Financial Post: *Running Your Own Business,* ed 3. Maclean Hunter Ltd, 1974.
Ford I: *Buying and Running Your Own Business.* London, Business Books, 1970.
Frantz FH: *Successful Small Business Management.* Englewood Cliffs, NJ, Prentice-Hall, 1978.
Fregly B: *How to Be Self-Employed: Introduction to Small Business Management.* Palm Springs, Calif, ETC Publications, 1977.
Hailes WD Jr, Hubbard RT: *Small Business Management.* Albany, NY, Delmar Publishers, 1977.
Higdon H: *The Business Healers.* New York, Random House, 1970.
Koontz H, Pulmer RN: *A Practical Introduction to Business.* Homewood, Ill, Richard D Irwin, 1975.
Lasser (JK) Institute, New York: *How to Run a Small Business.* New York, McGraw-Hill, 1974.
Mallonee JD: *Introduction to Business Dynamics.* Boston, Holbrook Press, 1976.
Newcomer HA: *Private Business Enterprise.* Columbus, Ohio, Merrill, 1974.
Putt WD: *How to Start Your Own Business.* Cambridge, Mass, MIT Press, 1976.

Rost OF: *Going into Business for Yourself.* New York: McGraw-Hill, 1945.
Stegall DP, Steinmetz LL, Kline JB: *Managing a Small Business*, rev ed. Homewood, Ill, Richard D Irwin, 1976.
Tungate LA: *Financial Management for the Small Businessman.* Boston, Chapman and Grimes, 1952.
Winter EL: *Your Future in Your Own Business.* New York, R Rosen Press, 1966.

ANNOTATED REFERENCES

Hollingsworth AT, Hand HH: *A Guide to Small Business Management: Text and Cases.* Philadelphia, W B Saunders Co, 1979.
   A particularly good section on site and location analysis. Discusses some of the personal assets essential to the entrepreneur.
Koltz CJ Jr: *Private Practice in Nursing, Development and Management.* Germantown, Md, Aspen Systems Corp, 1979.
   A description of the development and management of one private nursing practice with a short questionnaire to assess community need for health services that could be supplied by a nurse. Also provides a list of possible services and the pricing system used by the author.
Lebell D: *The Professional Services Enterprise: Theory and Practice.* Sherman Oaks, Calif, Los Angeles Publishing Co, 1973.
   Written specifically for the small individual practice with a realistic approach. Chapter 3 contains a description of a simple method of data analysis and planning.
MacFarlane WN: *Principles of Small Business Management.* New York, McGraw-Hill, 1977.
   Particularly helpful are the chapters on marketing procedures: Chapter 17, "How to Use Marketing Research," and Chapter 18, "The Pricing of Products and Services." This book also contains an appendix of addresses of Small Business Administration field offices and lists of useful free and for-sale publications as well as other information sources.
Mancuso JR: *How to Start, Finance and Manage Your Own Business.* Englewood Cliffs, N.J.: Prentice-Hall, 1978.
   The author gives an excellent detailed description of what to include and how to write a business plan. He also includes sample business plans and resource information in six fine appendixes.
Tate CE: *The Complete Guide to Your Own Business.* Homewood, Ill, Dow-Jones-Irwin, 1977.
   This text is geared to small manufacturing businesses larger than most private practices, but it does give much detailed, current information on business practices and serves as an ongoing reference.

CHAPTER

2

Ethical Considerations for the Nurse as a Primary Provider

Martha Henderson, MSN

One of your patients is a 70-year-old woman who is lucid, active, healthy, and obviously enjoying her retirement. On a routine visit she reports slight postmenopausal bleeding, and you discover a small pelvic mass. Your back-up physician, who knows the patient superficially, advises, "Ignore it; she is too old for a work-up. A gynecologist would not treat her anyway." What would you do?

This chapter will discuss the importance of ethical thinking in the nurse's professional practice and suggests ways of dealing with the kind of dilemma described above. First, there will be a brief introduction to ethics in health care and consideration of the ethical foundations of nursing practice. Then the American Nurses Association (ANA) Code for Nurses will be discussed as a guide for ethical reflection on the actions of the nurse who is a primary provider.

Ethics illumines what the best course of action in a given situation might be. It is derived from a branch of moral philosophy and deals with issues of human conduct.[1] Bioethics, or health-care ethics, deals with questions in the area of health care. These questions may concern public, sociopolitical issues within the health-care delivery system, such as allocation of resources; professional issues, such as standards for new nursing roles; or private ethical issues, such as justice in dealing with individual clients. The specific focus of this chapter is the nurse with a private practice and the ethical implications of that new role, with its new autonomy, authority, and responsibility. As nurses broaden their professional perspective and practice, it is appropriate to reexamine how they make decisions regarding the private- and public-health issues of those seeking care.

Formulating One's Ideal

Before one examines specific ethical dilemmas and decisions, a return to the roots of the profession may be appropriate—a look at the 'raison d'être' of nursing. Ethics involves clarifying one's own ideal: Who ought I to be, as a nurse? What is my commitment to my patient? The fundamental issue of who the nurse is precedes the issue of what the nurse does.

A nurse is one who nourishes or nurtures another's growth. Health is the state at which an organism functions optimally. To heal is to restore to health or make sound or whole. By putting these concepts together, one sees the nurse as a facilitator of healing—a person who helps remove obstacles to another's optimal functioning. Each person has the potential for being whole and realizing her or his unique gifts and strengths. The result is healthful relationships with oneself and others. As weaknesses are discovered or imposed, help from a caring nurse can create an environment conducive to healing. A blight may be removed, or accepted and integrated into one's life in such a way that optimal functioning can be regained. Nurses have the opportunity to become partners in healing as they bring their individual gifts, strengths, and ways of caring to their nursing tasks. They can participate in and respond to another's healing as he or she moves in a unique way toward wholeness.

The Process of Ethical Reflection

As nurses try to maintain a commitment to their ideal of nursing, ethical dilemmas will arise. Each nurse must decide the best course of action in each situation. Ethical decision making appeals to certain moral principles that help guide behavior. One such principle may be that honesty in communication is imperative for a trusting relationship. While such principles may be helpful, they may also conflict with other principles. Ethical dilemmas arise when conflicting obligations exist or when there is disagreement over the right or best course of action. These dilemmas necessitate careful ethical reflection; that is, thoughtful reasoning combined with appropriate use of general guiding principles. The process of ethical reflection entails stating the problem or issue, collecting pertinent data, assessing the values and relevant moral principles in question, generating alternative choices of action, and considering the consequences of each action. In this way, the rationale for an ethical decision can be articulated. As nurses learn to engage in such critical reflection and decision making, they begin to appreciate the ethical nature of their practice.

A part of ethical reflection is the clarification of values; that is, identifying the values one affirms. Values may derive directly from parental, societal, or religious influences or beliefs; for example, *abortion is wrong*. Values may be synonymous with moral principles; for example, *always be honest*. Values clarification makes one conscious of values held and allows one to examine them for consistency with one's actions and with the moral principles one affirms. Clarifying one's values can facilitate ethical reflection—that is, an objective, systematic evaluation of a dilemma, during which one considers all data, relevant principles, and choices and arrives at an enlightened, wise decision.

Code for Nurses

In addition to personal guiding principles, there are helpful professional guidelines for ethical behavior given in the ANA Code for Nurses. This document states that a code "provides one means for the exercise of self-regulation . . . and indicates a profession's acceptance of the responsibility and trust with which it has been invested by society."[2] This code can act as a common ground for discussion of professional responsibility and ethical behavior in nursing practice.

In the following discussion, the code is reviewed and evaluated for applicability to the nurse functioning in the expanded role of primary provider in a variety of settings. When applied to nurses, the term *primary provider* means nurses through whom clients enter the healthcare system and who have their own patient population. Ethical issues will be raised, and several examples of ethical reflection or guiding principles will be given. Of course, the examples merely suggest the range and scope of ethical considerations; they are not meant to be comprehensive. The interpretative comments are meant to be broad enough to have relevance for a variety of nurses—that is, "primary providers", nurse practitioners, clinical specialists, and nurse psychotherapists; however, these comments may reflect the clinical bias of the author, who is a geriatric nurse practitioner.

ANA Code for Nurses with Interpretive Statements for Nurses as Primary Providers

1. The nurse provides services with respect for human dignity and the uniqueness of the client unrestricted by considerations of social or economic status, personal attributes, or the nature of health problems.

One ethical issue suggested in statement 1 can be stated thus: Is nondiscriminatory access to health care a right for everyone in this country? Suppose Mrs. Jones, a poor woman, comes to your general practice with a nonacutely ill child. Your partner does not want the practice to accept Medicaid patients because the government will only reimburse 80% of the fee. (The predicament is exaggerated if the woman does not have Medicaid, or Medicaid will not reimburse nurse-practitioner services.) The only other health facility, a clinic in the next county, uses temporary physicians of various specialties. Mrs. Jones does not have transportation. A moral principle for you is that all human beings deserve equal opportunities, especially in health care. Conflicting data are that you do not want to alienate your partner, you must meet expenses, and seeing Mrs. Jones may prompt other Medicaid patients to come to you.

Alternative solutions and consequences may include the following:

• Because of the expense and your partner's stance, you decide to refer the child to the clinic and try to arrange transportation. Because you are concerned about the quality of care clients receive there, you speak with the clinic board members about hiring appropriately trained and stable providers. You work on getting a transportation service.

• You see the child and later try to persuade your partner to see some poor patients, because access to care is a problem. Your income will decline, so you must somehow adjust your lifestyle. If you are unsuccessful in persuading your partner to see poor patients, you may negotiate to continue seeing some poor clients and receive less income. Or you may get another loan and open an independent practice. Though you feel satisfied that patients are getting needed care, you also may feel overworked and underpaid! (The author has found that, when financially feasible, a practice with clients of varying backgrounds and characteristics challenges the skills and flexibility of the provider, gives more provider satisfaction, and maintains a higher quality of care than a segregated practice.)

While choosing either of the above alternatives, you may also get involved in working for full third-party reimbursement for all providers, including nurse practitioners.

Another ethical issue implied in statement 1 is related to the first.

Can all clients receive the kind of care they need regardless of socioeconomic status, geographic location, or age? There is an insidious tendency for nurse practitioners who can receive third-party reimbursement to become "doctors" for the disadvantaged; for example, the situation in rural or isolated settings may demand that the nurse practitioner manage critically ill, unstable patients independently. However, client needs and provider competencies, not socioeconomic or geographic factors, should determine the appropriate use of health resources. Although nurse practitioners certainly can provide the bulk of primary care, available, competent medical back-up is imperative.

The health-care system is full of examples of discrimination against minorities and disadvantaged groups, such as the mentally ill, the poor, and the elderly. Discrimination may take the form of inaccessibility to service, long waits in clinics, service from a different provider at every visit, cursory care, and service by an unqualified provider. What would you do if you practiced in a prison hospital that hired inadequate numbers of physicians who were poorly qualified or disrespectful of clients? Nurses can challenge or help change these inequities in policy by becoming more involved in political and legislative work on health-care issues. In their own practices, nurses as primary providers should be exemplary in providing care to all their patients regardless of individual background or characteristics, with "respect for human dignity and the uniqueness of the client."

2. The nurse safeguards the client's right to privacy by judiciously protecting information of a confidential nature.

An ethical question raised by statement 2 is: How much personal client information can be shared with the health team in order to assure quality and continuity of care without breaching a confidence? For example, should interactions between the client and the staff, which the client has shared with you in confidence, yet are contributing to a client's depression, be discussed with those involved? While there must be professional judgment here, "the rights, well-being and safety of the individual client should be the determining factors in arriving at this decision."[2(p6)] One may consult with the client regarding this matter. Also remember that it is more difficult to safeguard the confidentiality of material after it is written down than before. If it is appropriate to share some private information with members of the team, it may be helpful to discuss the necessity of confidentiality with them. Sometimes troublesome issues may be discussed with colleagues or consultants without revealing the client's name.

A related ethical question is: With whom should client information be shared? Dilemmas often arise when an elderly or young client's family desires to know diagnostic or prognostic information. For example, the discovery of an 80-year-old patient's malignancy may tempt the provider to reveal facts to family before the truth is told to the patient. The questioning of a patient's ability to understand or deal with unfavorable news must be examined carefully. Fletcher and Yarling stress the moral right of clients to information about themselves.[3,4(pp40-50)] Although the practitioner may not always be obligated to provide all information to a client, in certain cases (for instance, in the care of a terminally ill patient), the nurse is obligated to provide it if requested by the patient.[4(p45)] It may be helpful to remember that information belongs to the patient. Unless there are clear contraindications, such as indisputable incompetence, the client should decide with whom information is shared.

3. The nurse acts to safeguard the client and the public when health care and safety are affected by the incompetent, unethical, or illegal practice of any person.

Ethical dilemmas may arise when there is a conflict between protecting one's client and respecting the autonomy of other professionals. The trustful relationship between nurse and client necessitates that the nurse sometimes act as client advocate, especially in the case of a child client with limited power. If the nurse as primary provider is aware that the care of another is adversely affecting the client, the nurse is obligated to act. If, for example, physical therapy is exacerbating a client's problem, discussing the matter with the therapist may facilitate a solution. If not, the nurse must report to some responsible administrative person or a professional licensing group through a procedure that ideally has been preestablished. Sometimes consultation with the client may reveal the correct course of action. If ambiguity exists, the nurse should seek professional colleagues' opinions for validation and support. Of course, the nurse must investigate the competence of other professionals before referring a patient to them.

The issue of safeguarding a client against incompetent, unethical, or illegal practice of another may particularly affect the nurse as primary provider if the nurse has a professional-legal relationship with a physician who is a consultant in medical matters. Reconsider the opening example of the 70-year-old woman with vaginal spotting and a pelvic mass. The physician says to ignore the problem, but respect for the patient's right to participate in decisions regarding her care dictates that she be given relevant information. In his essay, "Respect for Persons as a Moral Principle," MacLagan alludes to the collaborative nature of client-provider relationships: "It is of the essence of proper respect that we encourage others to be co-agents."[5] The dilemma involves going against the wishes of the physician in respecting the dictates of your own conscience. (Assume in this situation that referrals are possible without the physician's consent. In some situations, unapproved referral is impossible.) You may handle this dilemma by explaining to the patient the positive findings and offering her the option of a gynecological consultation. When you explain your action, the physician may not agree, but nevertheless support you, or he or she may be nonsupportive. Another alternative is to call a gynecologist and ask for a recommendation. If the gynecologist recommends a consultation, you may share this information with the client and physician, and then support the client's decision. You may have to call more than one gynecologist to get an opinion you trust.

Although nurses in independent practice may not directly confront the painful dilemmas occasioned by overlapping areas of expertise and function with the physician, these continue to be issues for nurses who are legally restricted from final authoritative decisions in some aspects of medical care, such as treatment. Some guidelines regarding this area of potential conflict may be helpful. Too much emphasis cannot be given to the initial wise choice of physician and to defined functions and administrative back-up. Many problems can be avoided if the nurse and doctor have mutual respect for each other's competence and authority and if functions are clear from the beginning. There must be administrative understanding and support of the nurse in this expanded role. It may also be helpful to have nurse and physician consultants to act as peer reviewers. It is clear no nurse should follow medical directives she considers unsafe. It is also clear that repeated suggestions for treatment with which she does not agree necessitate renegotiation of functions and authority or resignation. Confrontation, when appropriate, can foster new learning and respect among professionals and lead to improved patient care.

4. The nurse assumes responsibility and accountability for individual nursing judgments and actions.

The mark of a true professional is that she or he assesses a situation accurately and carefully, makes a decision and implements it, and then takes responsibility for the consequences.

Imperative to professional maturity is knowledge of the laws affecting one's practice. Although ethical standards may call for excellence in judgment not legally required, state practice acts do define laws and guidelines for professional practice. These guidelines should be helpful in specifying minimal standards for safe practice. The present lack of standardization among laws regarding expanded roles means that nurse practitioners must look closely at local requirements for licensure or legal sanction to practice one's profession. The ambiguous use of the terms *licensure* and *certification* have confused the legal and professional arenas, hindered the process of identifying standards of competence, and delayed national recognition of nurse practitioners. State practice acts discuss licensure; that is, the minimal requirements for the practice of one's profession. For health professionals, requirements usually include graduating from an approved training program that includes clinical work and passing a standardized test to prove minimal competence. The profession designs and administers a national standardized test. If private practices by nurses are to be safe, identifiable to the public, relatively uniform and viable across the states, and reimbursable, it seems wise to require a licensing procedure that is state administered yet nationally recognized and standardized. Certification in specialties, a process of recognizing excellence, can continue to be offered through the national nursing association.

Even if legal guidelines for practice as a nurse practitioner become clear and standardized, there will continue to be ethical dilemmas that require individual judgment and accountability. Although the law cannot tell a practitioner whether to stop treating a dying patient, the nurse may face this situation. In this case, taking into account the patient's and family's wishes is part of ethical decision making. Professional practice necessitates, as a minimum, knowing the law and, beyond that, making ethical judgments that go beyond the law in fulfilling a primary responsibility to act in the patient's best interest.

Dilemmas may also arise when a nurse practitioner asserts authority in nursing matters against the advice of others, such as administrators, physicians, and other nurses. For example, the author helped a client return home to die with dignity, with the support of the family, but against the wishes of the nursing-home administrative staff.

It is inevitable that sooner or later one makes a wrong decision. Mistakes can lead to dilemmas. Although loss of trust by the patient or family may be feared, and although it is tempting to deny the mistake, conscientious follow-up and corrective action are imperative. For example, a nurse psychotherapist may err by holding a separate conference with the family of a paranoid client, who becomes more paranoid as a result. Admitting that action and arranging a joint conference may be necessary to reestablish trust. Another example is a nurse's decision not to hospitalize a psychiatric patient who then commits homocide or suicide. The nurse may be called to court to state the rationale for the judgment and defend the action. The nurse must learn to live with the results of the decision. The day is past when nurses can claim authority but abdicate responsibility.

5. The nurse maintains competence in nursing.

This statement raises several ethical issues for nurses who are primary providers. How can competence be guaranteed to clients when there is such lack of agreement in defining and measuring competence in expanded roles and lack of uniformity in educational preparation? In spite of excellent preparation, the nurse practitioner inevitably will notice gaps in knowledge when he or she begins to practice. Although the nurse has a moral obligation to deliver competent care, a dilemma arises because the nurse is also obliged to develop self and relationships outside work. How does one fill the learning deficits, keep up with changes in information, and maintain a heavy work load and a personal life? Some alternatives for addressing the competence issue are suggested.

Obviously, some arrangement must be made for ongoing learning. Although maintaining clinical practice is essential and guarantees some learning, specific time for study may be set aside at work or in off-hours. Part-time work may permit more unpressured study. When deficiencies arise at work, one may look up information immediately or consult with a more knowledgeable person. Journal reading helps keep one abreast of new and relevant information.

While individual study is necessary, continuing education with peers provides not only new information but also the opportunity to share resources, hear case studies, and exchange support. Often state conference groups can facilitate workshops for nurses with common interests. Peer-review teams of colleagues may enable the nurse to evaluate and improve his or her work in a new way. Conferences with professionals who have expertise in other fields can also broaden the nurse's knowledge.

6. The nurse exercises informed judgment and uses individual competence and qualifications as criteria in seeking consultation, accepting responsibilities, and delegating nursing activities to others.

Again, the responsible nurse adheres to the ethical standard for ensuring a high quality of care for the client. As a mature adult, experienced clinician, and educated professional, the nurse must use judgment in initially assessing the fit between a client's needs and the nurse's own competence to meet them.

The first decision must be whether the client and nurse provider are suited for each other. Consideration must be given to the needs and characteristics of the client: expressed need, age, rapport with nurse, special circumstances (such as transportation), and unexpressed but exhibited problems. The nurse's qualifications also need to be assessed: clinical competence in area in which client is seeking help, educational preparation and credentials, rapport with client, availability.

If there is a mismatch between client and provider, the best decision is to refer the client to the appropriate person. If for some reason referral is impossible, the nurse may elect to accept, along with the client, the responsibility for care, but the nurse's limitations must be kept in mind throughout the helping process.

If need and resources seem to fit, the nurse and client enter into a helping relationship, in which the nurse continually evaluates his or her competence to provide high-quality care. Services are provided in such a way that the nurse demonstrates clinical competence, respect for the client's role in the healing process, and creativity in problem solving.

At a point in the care of some clients, the nurse must decide if consultation or referral is appropriate in order to provide the best care to the client. For example, nurses who want to provide continuous care may experience a dilemma when they recognize their own inadequacy of knowledge or skills in a particular area. Consultation should be discussed with the client beforehand.

The nurse must negotiate for consultant services from providers in his or her own specialty as well as in related fields. The nurse must consider what types of consultation are likely, where the consultations will occur, how binding consultant suggestions will be on the nurse, and how fees will be determined and paid. The nurse needs to anticipate needed services and build professional, mutually helpful relationships with colleagues.

Consultant recommendations are discussed with the client and evaluated in light of the total management of the client. The nurse's judgment and competence are invaluable in incorporating relevant suggestions into the client's care.

The nurse may have opportunities to delegate certain aspects of care to others. "Others" may include patients as they learn more about their conditions and are ready to assume more responsibility for their own care. An aide or outreach worker may assume some functions previously performed only by the nurse. Whenever another assumes responsibility for health care, the nurse, as the primary provider, is responsible for teaching that person and assessing his or her competence. The nurse is also responsible for ongoing evaluation of the person to whom duties are delegated. Flexibility is imperative. For example, diabetics, normally quite able to give themselves insulin injections, may lose this ability during acute illness. Then the nurse may teach a family member to assume this task.

7. The nurse participates in activities that contribute to the ongoing development of the profession's body of knowledge.

As nurses claim their areas of special practice, they are in unique, advantageous positions to learn and write about how health is maintained, how healing takes place, and how the nurse practitioner participates in this process. Though clinical research is greatly needed to document and develop theories of practice, primary-provider nurses need to share their ideas and practice models even in the formative stages. As nurses communicate with each other and share useful observations, a climate of mutual discovery can evolve.

The topics for investigation are numerous. One example is exploration of alternative forms of healing. Is it ethical to use traditional treatment modalities that can cause severe complications? Instead of perpetuating a system that provides symptomatic relief from pain or stress, how can the causes be eliminated or reduced? How effective are drugless methods of coping with pain (for example, touch, visualization of pleasant images, recalling supportive words from the past, physical distraction such as pacing, or finding meaning in unavoidable suffering)?[6]

Many ethical issues surround research. Who is qualified to conduct research? On whom? Under what conditions? To what end?

The ANA Commission on Nursing Research has published a helpful guide.[7] The underlying ethical principle to be remembered is that respect for individual patient rights should be foremost in every researcher's mind. This respect entails honoring the relationship of trust between client and nurse, respecting confidentiality, and ensuring informed and voluntary consent at all stages of the research.[8]

The following situation serves as an example of an ethical dilemma in research. A nurse is investigating drugless methods of coping with pain. A subject in the experimental group has responded to an analgesic in the past but wants to try a new approach. Once she is into the study, the patient experiences some increase in pain and asks for an analgesic. Although the nurse is committed to the principle of the client's right to withdraw from a study, the nurse feels the patient could give the other method more time and energy, eventually benefit by mastering her pain, and contribute helpful knowledge to other patients. How can the nurse react to the request?

- After confirming the patient's wish to take the medicine, the nurse prescribes it. The nurse may feel frustrated because the patient can no longer participate in the study, and the patient may realize the effect her decisions have in her care as the nurse honors her right to withdraw from the study.
- The nurse may negotiate with her to try the new coping method a little longer, but promise the medicine if she still wants it. She may be willing to continue in the study or she may perceive the nurse as slightly coercive and drop out.

8. The nurse participates in the profession's efforts to implement and improve standards of nursing.

The subject of standards is at the same time a personal and professional issue. The nursing profession has a responsibility to society to define what specialized nurses do and to require some demonstration of competence in knowledge base and performance of functions. Educational programs must be standardized so there is some uniformity in knowledge and clinical competence of graduates. These issues are addressed most appropriately by national councils and specialty groups. These groups need input and support from clinicians.

An ethical question implied by statement 8 is: How can the public be assured of competent care by the new primary providers unless there is a clearly defined, enforceable standard? This question poses an ethical dilemma for the profession as it tries to clarify educational standards for advanced practitioners and to deal with the licensure question previously discussed. Because educational programs and undefined professional titles abound, the question arises: Should a master's degree be required of nurses assuming more clinical responsibility and authority than usual? While the principle that the public has a right to competent care can be affirmed, so can the principle that persons (nurses) have a right to advance themselves professionally even if circumstances do not permit graduate education. There is no guarantee that a master's degree or a written test ensures clinical competence. Are the kind of conceptual thinking and research skills taught in a master's program necessary for an advanced clinician?

There are several possibilities for preparing nurses to become primary providers. Preparation may occur in continuing-education or master's degree programs. Although variations in programs may cause problems in standards setting, many nurses could take advantage of these programs and meet the needs of underserved populations. If a master's degree were required, educational standards could be more easily addressed and the public could perhaps find new nursing roles easier to understand. However, some nurses may not be in a position to attend graduate school, while others may need assistance to obtain both BS and MS degrees. Because of the desirability the degree confers, it may be difficult to keep nurses with master's degrees in the clinical settings where new providers are desperately needed, such as rural and inner-city areas. Also research is needed on the relationship between educational preparation and clinical competence in these new roles before an informed decision can be made for the future.

A national test for minimal competence is certainly a step in the right direction toward standard setting. The relationship between test scores and clinical performance also needs investigation. Standards for performance in the clinical arena need to be formulated.

There are other questions about the education and professional standards of nurses as primary providers. These questions raise ethical issues: How are nurses selected to enter advanced educational programs? How can such important qualities as empathy, conscientiousness, honesty, and responsibility be fostered, measured, and required? Who teaches advanced practitioners, and how are teachers' qualifications measured?

9. The nurse participates in the profession's efforts to establish and maintain conditions of employment conducive to high-quality nursing care.

Only as nurses learn to care about and for themselves can they care for others. Just as an anemic person cannot donate blood, neither can a nurse who is deficient in self-esteem and energy give the caring and clinical expertise necessary to facilitate healing. Traditional health-care settings have oppressed nurses and caused declines in self-esteem. Wilma Scott Heidi, a nurse and former president of the National Organization for Women, challenges nurses to say no to their oppression and claim their rightful, respectful place in the health-care system.[9] Nurses' Coalition for Action in Politics, N-CAP, is the professional organization that ensures nursing is represented in political matters affecting the profession and other important health-care interests.

If nurses as primary providers are self-employed, they are relatively free to create their own conditions of employment. If nurse providers are employees of a larger system, ethical dilemmas may arise about role definition and function. Suppose a nurse sees his or her role as meeting several needs in one setting—for example, patient work-ups and management, coordinating staff development, and home visiting. Imagine that the medical director wants the nurse to do only work-ups and management of patients in the clinic. The ethical dilemma involves conflicting obligations to organizational goals and to personal and professional goals for improved service.

Berlin says that essential to human rights is the freedom to be one's own master—a subject, not an object; a decider, not a thing to be manipulated by others.[10] Nurses must claim this freedom. Professionalism implies a high degree of autonomy and responsibility. Therefore, nurse practitioners should take the primary role in defining how their skills and time can be best used. Nurses should not sacrifice role ideals for competing organizational expectations, though they must learn to deal with conflicting obligations. Some general guidelines may be helpful. Nurses as primary providers must take special care in initial negotiations regarding their job. Functions (including on-call responsibilities), salary, benefits, (including time for continuing education), lines of authority, formal relationships to physicians and administration, and relationship of nursing staff to other staff must all be spelled out. A job description must be negotiated by the nurse practitioner and those to whom he or she is responsible until all agree.

Because every circumstance cannot be anticipated, the nurse and administrator and/or physician need to be able to renegotiate points of difference or ambiguity at any time. Relationships with other staff will need special care and should be characterized by openness to information and idea exchange. The position and responsibilities of the new provider need to be explained formally as well as informally so the staff know with what authority the nurse clinician functions. A back-up physician must be available at all times if the nurse is responsible for medical diagnosis and treatment. Although the nurse specialist may be very sensitive to rising costs of health care and want to help lower costs to the public, nurses must still regard themselves and their services as worthy of reasonable compensation. The amount will vary, depending on the setting and other factors such as education, experience, and responsibility. If a group practice splits profits, nurses should receive equal shares, rather than be restricted to fixed salaries.

High-quality practice necessitates time and space for nurse practitioners to make their unique contributions, whether they be in areas of preventive health care, counseling, patient education, staff development, group work, or home visiting. These new providers have a responsibility for developing innovative solutions that get to the root of health-care problems, instead of perpetuating bandaid treatment of immediate needs. These solutions may consist of preventive measures. For instance, one may anticipate depression in an elderly person who experiences several losses. A nurse can encourage that person to join a support group rather than wait and treat with drugs a full-blown depression later. In their new roles, nurses may become advocates for clients; for instance, a nurse could play a part in getting food stamps for a poor mother and teaching her what to buy rather than assuming an old role—treating the baby's malnutrition. Patient education may lead to solutions of health-care problems. Groups of clients with similar chronic illnesses can meet to support each other, share ideas, and learn to be more self-reliant in monitoring their health problems. Only if nurse practitioners have the authority to decide priorities can they approach creative solutions to traditional health-care difficulties and thus provide the highest quality of care.

10. The nurse participates in the profession's effort to protect the public from misinformation and misrepresentation and to maintain the integrity of nursing.

The public holds many misconceptions about health. For example, many people think that mental illness is incurable and that a psychiatric label is precise and fixed. Nurses must continually correct misinformation and educate regarding health concepts and practices.

A major issue suggested by statement 10, which may be peculiar to nurses as primary providers, centers on how nurses can maintain the integrity of nursing within the context of a new role. Involved in this issue is the definition of appropriate intervention. For example, although the nurse practitioner may be tempted to expedite a client visit by prescribing a medication, such as a tranquilizer, the nurse may feel the best care is a nursing intervention, such as listening and counseling or referring the patient to a mental-health professional. New skills, knowledge, and authority must be carefully integrated with sound nursing approaches. While nursing actions such as counseling and health teaching may sometimes lengthen a visit and limit the number of patients seen, they may be the treatment of choice. The integrity of nursing lies in the wise use of the nurse's and client's resources and joint decision making about goals and appropriate intervention.

Another ethical issue may center on the exclusivity of the new provider's practice; that is, autonomy and independence may become so important that they sometimes appear isolationist. Although nurses are rightfully redefining themselves as humanists "who can best guide the total care of the whole patient" (W. S. Heidi, unpublished paper, 1978), they must take care they avoid the trap of feigned self-sufficiency. Nursing has always been interdependent with other health and health-related professions. Appropriate consultations and referrals still seem necessary to maintain the integrity of nursing, even in a new context.

A last ethical issue is a professional one: What is the relationship between traditional nurses and nurses who define themselves in an expanded role as primary providers. Can the profession tolerate diversity and maintain a unifying concern for improved health care? Nurses as primary providers may be tempted to pull away from professional ties because of special interests and desire for greater autonomy. Traditional nurses may see nurse practitioners as "sold out to medicine" or unable to relate to "real nursing" issues. Perhaps nursing should reexamine its preoccupation with distinctions between medicine and nursing and refocus on the client's health-care needs and the nurse's role in meeting them. Fragmentation among nurses significantly weakens the fabric of professional solidarity and power. Only as nurses unite in their commonalities and support diversity among themselves can they address larger health issues with a powerful voice.

11. The nurse collaborates with members of the health profession and other citizens in promoting community and national efforts to meet health needs of the public.

Nurses as new primary providers may face a dilemma in defining the scope of their role: "Should I be interested primarily in my own professional security within the present individualistic, crisis-oriented health-care system; or should I join others in advocating system-wide changes, which will produce a system more responsive to broadly defined health needs?"

As new providers, nurse practitioners have a unique opportunity and responsibility to add a new perspective to solutions of national problems of health care. They can collaborate with others who share particular concerns. The danger is that they will perpetuate the status quo—giving crisis care, seeking their own preservation at all costs, and being used as a stop-gap measure by physicians and administrators for problems in the health-care system. The alternative is to join with others who can think creatively and plan for systemic change in a health-care system that does not meet the public's needs adequately or justly.[11-13] Interdisciplinary brain-storming and research efforts need nursing input to address future health problems and trends, such as the increase in the elderly population.

As nurses claim their rightful place professionally, they need to be represented at policy-making levels: on state and national legislative committees, health-systems agencies, and boards of county commissioners. Only as nurses present their views publicly will consumers of health care recognize nurses as advocates of improved and equitable services.

A concern for nurses who want to be primary providers is simply being able to offer their services to the public. In order for nurses to be viable as primary providers, they must address legal, financial, and public-image issues in addition to the competence matters already discussed. Nursing, medical, and pharmacy practice acts need to be updated to allow for expanded functions by the nurse. Third-party reimbursement needs to be broadened to cover nurses as primary providers in any setting. Potential clients need to be educated regarding the preparation and services of nurse specialists. While these issues are confronted on a professional level, the individual nurse in practice can investigate and address these matters locally and on a state level.

Summary

This chapter has attempted to make nurses who are primary providers aware of ethical issues they may confront in their new roles. Although no simple solutions or absolutes exist in dealing with the complex dilemmas that arise, careful ethical reasoning usually can eliminate ambiguity and facilitate wise decisions. Consultation with ethicists, ministers, or other respected colleagues may help clarify which decisions are correct. When dilemmas are pressing and ambiguity persists despite conscientious ethical reflection, the nurse may focus again on an individual vision of ideal nursing. Commitment to the health and wholeness of the client is paramount. Legalistic adherence to a code cannot replace the nurse's unique capacity to assess and respond to each situation in a caring and responsible way.

Inherent in the pioneering role of the nurse practitioner is the willing assumption of greater responsibility and the challenge of clarifying the nurse practitioner's identity, function, and effect on the health-care system. An important ethical question is whether nurses as primary providers will allow themselves to be used to perpetuate an outdated, problematic system or use their creativity, concern for clients, and political power to be part of new solutions to pressing health-care problems.

As providers in new roles and settings, nurses have the freedom and power to act according to their best judgment. They need to realize their capacity for "assertive nurturance"—a phrase and concept formed by Wilma Scott Heidi. Assertive nurturance calls for the use of a combination of masculine and feminine characteristics to provide integrated health care.[11] Nurses want to "nourish" clients toward wholeness and health. The time has come for them to do that assertively.

Author's Biography

Martha Henderson received her BSN and MSN degrees from Duke University and her certificates as a family and adult/geriatric nurse practitioner from the University of North Carolina (UNC). She has held various positions as a clinician and faculty member, including geriatric coordinator of the Nurse Practitioner Program at UNC. She is active in a North Carolina Nursing Association committee that is planning workshops on ethical issues in nursing. Because of her interest in holistic healing and ethics in health care she recently completed a Master of Divinity degree from Yale Divinity School. Currently she is director of clinical services and is a geriatric nurse practitioner at Carol Woods Retirement Community, Chapel Hill, North Carolina.

REFERENCES

1. Aroskar M, Davis A: *Ethical Dilemmas and Nursing Practice.* New York, Appleton-Century-Crofts, 1978, p 1.
2. *Code for Nurses with Interpretive Statement.* American Nurses Assoc, 1976.
3. Fletcher J: Medical diagnosis: Our right to know the truth, in *Morals and Medicine.* Boston, Beacon Press, 1954, pp 37–38.
4. Yarling R: Ethical analysis of a nursing problem: The scope of nursing practice in disclosing the truth to terminal patients. *Supervisor Nurse,* May 1978.
5. MacLagan WG: Respect for persons as a moral principle. *Philosophy: Royal Instit Philosophy* **35**:289–305, 1960.
6. Copp L: Nursing as ministry, in *Ethical Issues in Nursing: A Proceedings.* St. Louis, The Catholic Hospital Assoc 1976, pp 65–70.
7. *Human Rights Guidelines for Nurses in Clinical and Other Research.* American Nurses Assoc, 1976.
8. Aroskar M: Nursing relationships: Ethical issues and educational experiences. Paper presented at Ethics in Nursing Workshop, sponsored by the Institute of Society, Ethics and the Life Sciences, Bronxville, NY, June, 1977.
9. Heidi WS: Nursing and women's liberation: A parallel. *Am J Nurs* **73**: 24–27, 1973.
10. Berlin I: Two Concepts of Liberty (1958), in Katz J (ed): *Experimentation with Human Beings.* New York, Russell Sage Foundation, 1972.
11. Ehrenreich B, Ehrenreich J: *The American Health Empire: Power, Profits, and Politics.* New York, Harper & Row, 1971.
12. Illich I: *Medical Nemesis: The Expropriation of Health.* London, Calder and Boyers, 1975.
13. Milio N: *The Care of Health in Communities: Access for Outcasts.* New York, Macmillan, 1975.

FURTHER READING

Am J Nurs (special supplement on ethics), May 1977.
Bandman E, Bandman B (eds): *Bioethics and Human Rights: A Reader for Health Professionals.* Boston, Little, Brown, 1978.
Bergman R: Ethics: Concepts and practice. *Int Nurs Rev,* Sep-Oct 1973, pp 140–141.
Bliss A: Nurse practitioner: Victor or victim of an ailing health care system. *Nurse Practit,* May-June 1976, pp 10–14.
Curtain L: Nursing ethics: Theories and pragmatics. *Nurs Forum* **17**:5–11, 1978.
———. A proposed model for critical ethical analysis. *Nurs Forum* **17**:13–17, 1978.
Frankena WK: *Ethics,* ed 2. Englewood Cliffs, NJ, Prentice-Hall, 1963.
Gorovitz S, Jameton A, Macklin R (eds): *Moral Problems in Medicine.* Englewood Cliffs, NJ, Prentice-Hall, 1976.
Jameton A: The nurse: When roles and rules conflict, in Institute of Society, Ethics and the Life Sciences: *Hastings Center Report* **7**:22–23, 1977.
Murphy M: Making ethical decisions—systematically. *Nurs '76* **5**:13–14, 1976.
Nucholls K: Who decides what the nurse can do? *Nurs Outlook,* Oct 1974, pp 629–630.
Ramsey P: *The Patient as a Person.* New Haven, Conn, Yale University Press, 1970.
Tate BL: *The Nurse's Dilemma: Ethical Considerations in Nursing Practice.* Geneva, International Council of Nurses, 1977.
Veatch R: Models for ethical medicine in a revolutionary age, in Institute of Society, Ethics and the Life Sciences: *Hastings Center Report* **2**:5–7, 1972.

CHAPTER

3

Legal Planning for Private Practice

Phyllis M. Gallagher, RN, JD

Part of planning a private practice is considering a number of legal issues, some of which will be discussed in this chapter. The checklist below, although not a comprehensive catalogue of the legal issues that might arise in this context, will help the nurse contemplating private practice to survey its legal aspects.

Because laws relating to the practice of nursing vary from state to state, this discussion, by necessity, will be rather general. When it seems useful, California law will be used to illustrate issues. The purpose here is not to resolve, but merely to raise, issues of law relating to the private practice of nursing so that nurses considering private practice can make more effective use of legal services. The issues raised here must be resolved by reference to local law.

Planning Checklist
1. Defining the practice
2. Choosing a legal structure
3. Minimizing the risks of malpractice
4. Organizing for political support

Defining the Practice

Initially, the most important legal planning by the nurse will be defining the practice. The nurse must decide what services to offer clients and make certain that those services are within the scope of nursing practice as defined by state law. This planning step may seem obvious, but it is crucial to the success of any practice to define beforehand the scope of services.

This planning is crucial, not only because a nurse who delivers services beyond those permitted by law risks serious trouble,[1(p53)] but also because the economic success of the practice depends absolutely on the nurse's ability to deliver the offered services. This ability takes a great deal of planning and networking with other providers of health services. The scope of practice depends on the connections established with other providers and also on the kind of practice envisioned.

For example, if the practice envisioned will deliver services that neither overlap with medical services nor require a collaborative relationship with a physician, the practitioner's connections with physicians will be less important than if the envisioned practice will deliver services that do require collaboration. Referral sources also shape the scope of a practice. If physicians are the referral source, there will be more opportunities for collaborative practice than if referrals are from clients or the community at large. The scope of practice, then, will be shaped as much by pragmatic realities as by philosophical preferences.

However the scope of the practice is shaped, care must be taken to keep it within the legal boundaries of nursing practice as defined by the state's nurse-practice act. The consequences of transgressing the legal limits of nursing practice range from prosecution for practicing medicine or some other discipline without a license[2] to increased exposure to liability for harm resulting from services given.[1(p2)] Prosecution for practicing medicine without a license, if successful, can lead to the suspension or revocation of the nursing license. This dire result can be imposed even though no harm has befallen the client.

Should harm result from illegal practice by a nurse, the nurse may be held personally liable for all the damages suffered by the client. Also, the client is relieved of the ordinary requirement of proving that the harm was caused by the nurse's practice. The client is given the benefit of what is called a "presumption of negligence," which shifts the burden of proof from the client-plaintiff to the nurse-defendant. The nurse might be able to rebut the presumption of negligence by proving that the illegal practice did not cause the injury, but it is difficult to prove a negative premise. The nurse in negligence litigation under these circumstances is at a real disadvantage.

Another serious consequence of illegal practice is that professional malpractice insurance might not cover damages caused by illegal practice. A nurse found liable for damages caused by practice outside the scope of nursing might find that the insurance carrier who pays the cost of the defense will refuse to pay a judgment against the nurse.[3] This refusal would be on the ground that the insurer had not accepted the risk exposure for practice outside the legal bounds of nursing, but only for errors and omissions within the scope of nursing practice.

It is important to understand exactly what coverage a policy provides. The nurse should examine the policy for reference to scope of practice. Even without explicit reference to illegal practice, however, the carrier can take the position that it cannot insure illegal practice. Then the nurse would be under the tremendous burden of having to pay a judgment with his or her own money.

Holistic Health Practices in Nursing

The Chinese proverbial blessing "May you live in interesting times," has come true for nurses today. From a legal point of view, the times present both challenge and risk. It is thrilling to watch the nursing profession gaining autonomy. One feels both pleasure and apprehension about the process, much the same as one would feel witnessing a birth. The desire to assist the process and hurry it along must be tempered with the realization that there are risks attending each struggle to be born.

Among the developments in health care that make the times interesting and risky for nurses in private practice are the growing interest in holistic health practices and the search for alternatives to the medical model of patient care. The difficulty is that holistic practices may or may not fall within the scope of nursing practice as defined by statute.[2] Where they can be demonstrated to be closely related to traditional nursing practice, holistic practices may be incorporated legally into a nursing practice. The problems arise when holistic practices involve applications of modalities that are not within the recognized practice of nursing. When the treatment falls outside the scope of nursing practice as defined by law, the nurse responsible for treatment may be prosecuted for the unauthorized practice of medicine or psychology and runs a greater risk of liability for malpractice.

A related difficulty is that nursing statutes are often couched in language that reflects nursing's long association with the medical model of patient care. Thus, for example, the scope of practice in a given nurse-practice act may be confined to practices based on "scientific knowledge." The bases for holistic practices may not be defined so strictly. Until their bases are more clearly defined, the nurse using holistic practices is operating at risk. The risk may be reduced if the nurse can articulate and document the ways in which the practices used are related to traditional nursing practice. Nursing research is needed to establish (or discredit) as nursing principles those holistic practices that have yielded promising results. The work of Krieger in the "laying on of hands"[4] is an example of what needs to be done to legitimize holistic practices, document their efficacy, and reduce the risk to nurses using them.

Choosing a Legal Structure

After deciding what services to offer, the practitioner must decide on a legal structure for the practice—corporation, sole proprietorship, or partnership. If the practice is incorporated, will it operate on a nonprofit or a profit basis?

Traditionally, businesses have viewed the protection against liability of corporations as a very attractive feature. Corporate businesses have enjoyed limited liability for the debts of the corporation. Shareholders have not been required to make good the debts of a defunct corporation. The fact of incorporation, however, does not protect individuals from tort liability. Professional malpractice is a tort, for which there is no limited liability by virtue of incorporation. Corporations have limited liability for business debts that have been properly capitalized and incorporated. Tort judgments, including professional malpractice judgments, are personal judgments for which the tortfeasors must answer.

Even limited liability for corporate debts may prove to be more illusory than real should the court "pierce the corporate veil" and "disregard the corporate entity." Judges use these phrases when they deny limited liability in a particular case. The courts deny limited liability to ostensible corporations when they find that such corporations are mere conduits or covers for the activities of the controlling shareholders. If it can be shown that a corporation is unable to meet its debts because of serious undercapitalization, and that the corporation is actually the "alter ego" of the person controlling it, the limited liability protection may be withdrawn.

One of the most frequent criticisms of small businesses is that they are undercapitalized. They simply do not reserve enough working capital to meet the overhead of the business during periods of low income when the business is in the process of becoming a "going concern." The economic implications of undercapitalization are clear: without a sufficient cash reserve, the business will fail. Corporations should not rely on limited liability for their obligations if inability to meet them is the result of serious undercapitalization.

The courts have also pierced the corporate veil and tapped the assets of the controlling shareholder when it could be shown that the corporation was organized to avoid liability of the kind for which it should answer. One such case involved a number of separately incorporated operations all owned by the same person. Each so-called corporation was covered by minimal liability insurance. When a person was injured due to the negligence of an agent of the corporation and sought damages in excess of the policy limits, the court refused to recognize each operation as a separate corporation. The court, noting no business reason for the many corporations except to escape liability, allowed the injured party to tap the assets of the owner of all the corporations so that the judgment could be met. Had the owner carried adequate liability insurance, a different result might have been reached.

The best protection against financial liability is careful planning. The nurse should seek help from an accountant skilled in business management and should secure adequate capital. Adequate insurance, both general liability and professional malpractice insurance, is a must. Corporations need malpractice-insurance protection at both the corporate and the individual levels. An attorney's advice during the planning stages is essential. Especially important is knowledge of limits to practice, as defined by law. Consultations with other professionals prior to starting the practice can save money. In business, as in health, preventing problems is always more cost effective than trying to solve them after they develop.

Another standard reason for favoring the corporate form of practice is that fringe benefits, especially medical benefits and retirement plans, can be funded through the corporation. The practitioner is considered an employee of the corporation; in funding benefits, the corporation may receive favorable tax treatment. Needless to say, before fringe benefits can be funded, the business must be making money.

The most attractive retirement programs from a tax-savings vantage are "qualified" plans. The corporation gets an immediate deduction for funding benefits, and corporation employees get deferred taxation of the income of the plan until retirement, when they will probably be in a lower tax bracket. To be qualified, the plan must meet the stringent requirements of the Internal Revenue Service (IRS) and the Labor Code regarding such plan features as participation, vesting, funding, reporting, and disclosure. Getting a plan qualified is an arduous, expensive exercise. It is very doubtful that a beginning nursing practice would be advised to undertake it.

Alternatives to the individually tailored qualified plan are the "standardized" qualified plans available to individuals, whether incorporated or not. One is the IRA, or Individual Retirement Account, through which any individual not covered by another qualified plan may deduct up to $1,500 of annual earnings and deposit it in an account that must be maintained until retirement. Another standardized qualified plan is the Keogh Plan, through which self-employed persons may deduct up to $7500 or 15% of annual income, whichever is less, and contribute it to a plan that must be maintained until retirement. As with the other qualified plans, the income earned by the funds deposited is not taxed until retirement by the participant. These standardized plans probably provide sufficient tax advantages for a beginning business and are not difficult to begin.

A disadvantage of incorporation is the relatively large expense of creating and maintaining a corporation, compared with the expense of setting up a sole proprietorship or partnership. Corporations are defined by law; one must follow the law in drafting and filing the articles of incorporation and by-laws. Corporations also require legal maintenance; the legal services needed to maintain the corporation in good standing will add to start-up costs.

Another disadvantage of the corporation is the potential it creates for double taxation. Both corporation and employees pay income tax. Since the corporation deducts the costs of doing business prior to having income, one way of avoiding double taxation is to make sure that all money is distributed prior to the end of the tax year. Of course, shareholders must pay taxes on the earnings or dividends distributed. A more "automatic" way of avoiding double taxation is for the corporation to take the Subchapter S election. This subchapter allows the corporation of fewer than ten shareholders to be treated as a partnership for tax purposes. Each shareholder pays a proportionate share of the corporate income, whether distributed or not. The Subchapter S election eliminates the need to monitor the books for income to the corporation as the end of the tax year approaches. Subchapter S election may be unavailable to certain professional corporations.

Kinds of Corporations

Within the past ten years, many states have passed professional-corporation acts.[5] Under these laws, professional persons may form corporations for the purpose of providing professional services. Prior to 1969, tax regulations prevented the formation of professional corporations.

Those who practice in states with a professional-corporations act applicable to nurses may be required to organize exclusively under the provisions of the professional-corporation statute. If so, there may be restrictions on stock ownership. (General corporation laws do not restrict stock ownership to any class of persons, although different classes of stock may be sold.) There may be other restrictions on a professional corporation, such as a prohibition against organizing as a nonprofit entity. A lawyer can help investigate these options and restrictions.

California professional-corporation law does not provide for the formation of professional corporations for nurses. Recent changes in the law, however, do apply to nurses. A new provision of the law permits nurses to buy into professional corporations owned by physicians.[6] This law enables nurses who practice jointly with physicians in such corporations to assume a more active role in the ownership and management of joint practices.

Jean Moorhead, the first registered nurse in the California legislature, introduced the legislation, and the California Nurses' Association worked for its passage. It passed in spite of strong opposition from the California Medical Association, which sees it as a threat to its members' preeminence in the health-care delivery system.

The joint-ownership possibilities inherent in California's new law are quite interesting. Nurses are now permitted by law to own up to 49% of the stock in a medical corporation. Stock ownership will give nurses in such corporations a vote in the management of joint practices as well as a share of the profits. Although stock ownership does not guarantee success in collaboration, it is a step toward solutions to some problems of joint practice. It may help to clarify and resolve the problem of defining scope-of-practice issues in joint-practice settings. It can provide valuable experience in owning and managing a business, experience that might otherwise be unavailable to some nurses. Stock ownership probably could be negotiated as part of compensation, so that nurses would not need to invest large amounts of capital to buy into medical corporations. The effect of this new law on the health-care delivery system in California will depend upon the ability of nurses to negotiate favorable terms with the physicians with whom they practice.

Profit or Nonprofit?

Corporations may be organized as profit-making concerns, providing a return on capital invested, or as nonprofit entities. Practitioners looking for a financial return on an investment (beyond the salary received while working for the corporation) may want to organize for profit. The profit corporation has financial appeal for some because one can sell one's shares when one leaves the business. Those who nurture new practices until they are successful may reap double rewards—financial reward from sale of stock plus the personal satisfaction derived from building the business.

A nonprofit corporation, however, may not be sold. The financial rewards of involvement in a nonprofit corporation are limited to reasonable compensation for services. Compensation must be set by the board of directors and depends partly on the success of the business. Nonprofit corporations are regulated strictly so that they do not operate for the private gain of any person. Their assets must be dedicated to the purpose for which they are formed and must be used for similar purposes if the corporation disbands. In spite of the lack of opportunity for profit, nonprofit incorporation is preferred by many.

This preference may be the result of the nonprofit corporation's ability to secure public or quasi-public funding more readily than a profit corporation. Here are some further benefits of nonprofit status:

Nonprofit corporations may be able to provide a valuable public service and at the same time give nurses some exciting management opportunities. Because money is currently both scarce and expensive, public funding may be a desirable alternative to borrowing for a nurse with a commitment to providing a needed service. An added benefit is the tax-exempt status of corporations that satisfy the state and federal requirements for nonprofit organization.

Partnership/Sole Proprietorship

What are the alternatives to incorporation? Those who practice alone have the alternative of sole proprietorship. Those who plan to practice with colleagues may have a partnership. Incorporation can always come later, when the practice has met some initial success. A partnership, because it is less formally regulated than a corporation, may provide more flexibility to the beginning practice.

No matter how the business is organized, an attorney's advice is essential during the planning stages. The National Health Lawyers Association[7] or the American Bar Association Forum Committee on Health Law[8] can refer the beginning practitioner to a lawyer with experience in the health field. Some local bar associations have health-law sections; when appropriate, the practitioner may seek affiliation with one of these groups. Affiliation is especially important if the nurse is to be a consultant on such specialized concerns as scope-of-practice issues or techniques for minimizing malpractice risks.

Apart from advice on legal aspects of health care, the practitioner needs legal advice on business matters—advice any attorney with business experience can give. This attorney can answer basic tax questions, draft and review contracts, help the practitioner comply with local business regulations, review leases, and draft partnership agreements.

The group of nurses that seeks legal services must understand the attorney-client relationship and how it operates. An attorney can draft a partnership agreement by consulting with the partners as a group; however, should conflict about the agreement arise among the partners, that attorney should not attempt to represent any of the disputants. Because the attorney attempted to draft the agreement to meet the needs of the group as a whole, there would be a conflict of interest if the lawyer were an advocate for any single partner. The lawyer may be called on to interpret or explain the provisions of the agreement but should not assume the role of advocate for any partner.

A different situation exists when one partner requests an attorney to draft a partnership agreement. Having undertaken only to represent one partner, the attorney can, without conflict, play the role of advocate for the client. The other partners would be well advised to have the agreement reviewed by their own attorneys.

The partnership creates and defines the legal relationship of the partners. In addition, the common law imposes on partners a fiduciary relationship: the law requires them to use the highest standards of fairness in dealing with each other. Partners may not seek unfair advantages over each other. Nonetheless, disputes may arise even among partners of good will, and the lawyer should avoid conflict of interest. The same caveat applies to groups who have been incorporated by one attorney. That attorney should decline to represent any member of the group in a dispute with another member.

The attorney who drafts the partnership for a group can represent the group in any matter except disputes among partners. Contracts with outsiders, landlord-tenant problems, and the like, ordinarily can be handled for the group by that lawyer.

The partnership agreement should provide flexible guidance concerning the management of the business without including unnecessary detail. It may include such things as the duration of the partnership, capital contribution, duties of the partners, the procedure for dividing profits and losses, buy-sell provisions, death or disability provisions, mechanisms for handling disputes, provision for dissolution of the partnership, and so on. The law requires no specific provisions in partnership agreements and will recognize the existence of a partnership even when there is no formal, written agreement. Partnerships are regulated less formally than corporations because of differences in the ownership-management relationship. In corporations, ownership and management are separate functions.

In partnerships, these functions usually are combined. The law seeks to protect the interests of owners, especially those of minority shareholders, by imposing on management rather formal rules for the conduct of the corporation. In the partnership, this safeguard is unnecessary because the owners can manage the partnership to any extent they agree. Partners have only themselves to blame if disputes arise because they have neglected to define their relationship with sufficient explicitness (and if, as a result, management suffers).

Unlike corporations, partnerships are not legal entities. They are associations of persons doing business together for profit. Partners are taxed as individuals, but the partnership must submit tax returns to the state tax board and the IRS. However, a corporation may act as one of the partners in a partnership. Each partner is answerable for the debts of the partnership.

Minimizing the Risks of Malpractice

Malpractice is an issue of great concern to nurses considering private practice. Malpractice may be defined as harm inflicted on another by the failure of a professional to meet the applicable standard of care.

In some states, nursing negligence that results in harm to a patient is not considered malpractice. Instead, it is considered ordinary negligence, to which the usual statute of limitations applies.[9(p855)] Some states, including California, have a special statute of limitations applicable to suits against health professionals, including nurses. This statute limits the time during which an action may be brought for malpractice.[9(p855)] Such a statute benefits the health professional because, under ordinary negligence law, the injured patient has a longer time to seek redress of an injury.

Nurses should know the statute of limitations applicable to them and should be familiar with the procedure for bringing a malpractice or negligence suit to trial. Familiarity with some of the legal issues and procedures can demystify this rather threatening subject and give the nurse more sophistication in dealing with legal issues. Continuing-education programs dealing with risk management, malpractice litigation, and the like, should be a part of every professional's preparation.

Two particularly important issues for the nurse in private practice are expanded liability for the independent contractor and the applicability of standards of care.[9(p871)]

Until rather recently, malpractice was not a major concern of nurses. Nurses have felt secure against the charge of malpractice because, by tradition, they have been employees of hospitals, doctors, and others. This security is unwarranted because any person who commits a tort, defined as a civil wrong (including malpractice) against another, is personally liable for the resulting harm. Despite their legal accountability, nurses have in the past often escaped involvement in malpractice cases because of the policy of *respondeat superior*. This legal doctrine makes the employer responsible for acts performed by employees in the course of employment. Because hospitals are more likely to have money than individual nurses, hospitals frequently have been the preferred targets in malpractice actions, even though an employee caused the injury.

In such cases, hospitals may protect themselves, depending on the jurisdiction, by one of several methods. Some states permit the hospital to be indemnified by the employee in a separate action.[10(p25)] For example, if a registered nurse gives a patient the wrong medication and serious harm results, the patient may invoke the principle of *respondeat superior*, sue the hospital, prove the cause of injury, and collect a judgment against the hospital. The hospital in turn may sue the nurse for indemnification. The hospital would not be permitted to be indemnified if it, too, were negligent—for example, if it failed to exercise due care in hiring the nurse. Then the hospital would bear the cost of the judgment in whole or in part.

Some jurisdictions, including California, do not permit indemnification. The hospital may, however, file a cross-complaint against the nurse in the same action as that brought by the plaintiff-patient. The doctrine of *respondeat superior* assures the plaintiff recovery, yet the hospital can protect its interests as well.

When the nurse steps out of the employer-employee relationship, the doctrine of *respondeat superior* becomes inapplicable. However, it is not entirely clear how the law will perceive the relationship between nurses and physicians who are not in an employer-employee relationship, but are engaged in collaborative practice.[11(p384)]

In general, the more the nurse exercises independent judgment in the practice, the more closely he or she approaches independent-contractor status. Independent contractors are liable for their own malpractice. The doctrine of *respondeat superior* traditionally has not been applied to independent contractors. As nurses assume new, independent roles, they need adequate malpractice insurance more than ever.

At the present time, it seems unlikely that nurses who are in joint or collaborative practice with physicians will be recognized as completely independent contractors for all purposes. Because physicians in such settings usually are required to supervise nurses at least to some extent, there appears to be a relationship between the physician and nurse that keeps the nurse from being an independent contractor and creates vicarious liability in the physician.

Nurses whose practices do not overlap into areas of medical practice are more likely to be recognized as independent contractors. Their liability is greater for two reasons: such nurses will be solely accountable for any malpractice they may commit, and there are more areas of potential liability.

Informed consent is a new concern for the nurse in independent practice.[10(p97),11(p394)] In the traditional hospital or office setting, nurses are not responsible for informing patients about the various risks of medical interventions. This responsibility lies with the physician. Neither does present practice require that nurses obtain formal consent for their nursing interventions. The general consent obtained by the hospital from the patient at the time of admission is considered adequate. However, if nursing care is given in a nurse's office, there is no general consent form unless the nurse requires one. It may not be necessary or desirable to formalize consent with a written document in every case, but in all cases in which there is potential risk to the client, consent should be written.

Related to the issue of informed consent is the question of whether or not the client understands the status or role of the nurse.[11(p396)] Nurses, particularly those in an independent practice, should clarify their roles so that the client does not mistake a nurse for a physician or some other health professional. Patients who are not informed of the caregiver's identity and role as a nurse might claim that they would have withheld consent had they known a nurse was treating them and that therefore they gave no consent. Medical treatment without the patient's consent can be interpreted by the law as battery and may lead to litigation.

The standard of care to which a nurse in the private practice of nursing should be held depends on the kinds of nursing interventions he or she performs. Nurses who practice skills and make judgments that require a high degree of expertise will be held to a high standard of care.[9(p876)] A question arises about nurses who perform functions ordinarily performed by physicians—should their standard of care be that of nurses or that of physicians? A nurse practicing collaboratively with physicians and performing duties ordinarily performed by physicians could be held to the same standard of care as a physician. If a question arises about the adequacy of the nurse's performance, the nurse could be judged by reference to the skill and learning possessed by the physician who performs the same skills. This approach is unsatisfactory, not because the nurse is unable to meet that standard, but because the standard of care should be set by the profession to which it applies.

In the past ten years, advanced clinical training programs for nurses have proliferated, and there is no scarcity of experts within the profession to define the applicable standards of care. It appears that there is a lag in the law concerning this issue, due largely to unfamiliarity with the various levels of education and skill that exist within the nursing profession.[9(p871)] In many respects, the nurse in private practice may require more education and skill than one who practices under the supervision of an institution. The former will be called upon to exercise independent judgment in matters such as when to refer a client to another health professional. Although it is true that a nurse practicing in a hospital also has a duty to call to the attention of other health professionals changes in the condition of patients, the hospital setting dilutes the risk for a nurse who fails to do so. In private practice, the nurse who fails to refer the client to an appropriate health professional might be held solely liable for the result. A more important reason why the nurse in private practice should have more education and skill than the hospital nurse is the protection of the public from unsafe practice.

The importance of a standard of care defined by the nursing profession is underscored by the issue of expert witnesses in professional-malpractice cases involving nurses. Unless the law recognizes different levels of nursing practice and education, there is a danger that the wrong standard of care will be applied. This misapplication could happen either because the law permits the wrong expert to testify about the applicable standard—as when a physician testifies about nursing standards—or because it permits a nurse without the requisite education or experience to testify as an expert witness. In either case, the application of the incorrect standard might lead to injustice. The nursing profession should press for the right to define its own standards of care and must insist that these standards be properly applied. The result will benefit both the nursing profession and the general public.[9(p878)] The development of standards will clarify exactly which services are available to the public and who can safely deliver them. When roles are delineated, there will be greater professional opportunities for nurses. The public, too, may benefit by reduced costs of health care. When nurses are directly accessible to the general public, consumers can seek a service from the most appropriate practitioner and bypass the inappropriate practitioner, thereby eliminating waste and saving health-care dollars.

Organizing for Political Support

The new independent roles of nurses demand a new attitude toward politics. Nurses must organize politically to achieve nursing goals. A first step is to show that these goals are worthy of support because they will enhance the well-being of the whole community. The organizing task is tremendous; it involves nurses in changing relationships with other health professionals, consumers, legislators, and insurance companies or others who reimburse for services.

Although the battle ahead will be taxing, it can be rewarding. Already some victories are in view. California nurses have won the opportunity to purchase stock in medical corporations. The California Medical Association opposed this right, even though it promises mutual benefits to both professions and to health consumers as well. This victory is significant because it signals a changing relationship between the nursing profession and the California legislature. Instead of merely reacting to legislation, nurses have begun to plan for legislation. Nurses are now planning for legislation that will create economic structures to accommodate their expanded roles. This change is due to many factors. Not the least among them is a growing sophistication among nurses about the political process.

On another front, simultaneously with their recent victory, California nurses are still very much on the defensive regarding proposed legislation known as Senate Bill 666. This legislation would undo the whole structure of nursing law in California and undermine attempts to improve the quality of education for nurses by substituting institutional licensure for the current regulatory system. Senate Bill 666 is viewed by many as an unforeseen attack, to be resisted at all costs, on the profession. This legislation may be an experiment by persons outside nursing who wish to launch an attack on the whole system of licensure of professionals. Nursing is the target because of its apparent vulnerability to such an effort. The California Nurses' Association has been mobilizing resistance to this bill and at the same time addressing needed changes in current law. This bill could turn out to be the stimulus that the California Nurses' Association needs to create a tough, cohesive organization capable of preventing any such further attacks.

Other organizing efforts are under way to place the private practice of nursing on a sound economic basis. One such effort is a project known as "The Next Step," the drive by psychiatric/mental-health nurse specialists to obtain reimbursement through third-party payment. California nurses are actively organizing to lobby each member of the legislature to create the proper environment for the introduction of reimbursement legislation. The effort is being orchestrated by several work groups throughout the state and is gaining momentum. As the organizing effort grows, the nurses involved will certainly gain a sense of their own effectiveness and the respect of the legislators, for these nurses are a powerful and articulate group.

Nurses who are interested in creating their own practices are well qualified to carry forward the process of political organizing. They must have the appropriate mix of independence and cooperative skill necessary for mobilizing for political goals, or they would not be planning private practices. The political process accommodates many different personal styles. Some will work best behind the scenes, while others will enjoy the challenge of keeping the public informed about what it can expect from nursing.

Successful political organizing requires a constituency. Memberships in the American Nurses' Association and its local affiliates are absolutely necessary for developing political power. There are a number of other groups that can be very helpful in developing support for the goals of nursing, not because they give any direct support, but because nurses on the boards of such organizations can meet and influence people informally. Health-systems agencies, the regional planning vehicles for health care, are one such group that can help nurses meet people they should know. Involvement in one group tends to open doors to other voluntary agencies. This process can keep nurses in touch with the health needs of the community and at the same time expand nurses' opportunities for influencing political decisions regarding health care.

Author's Biography

Phyllis Gallagher graduated from St. Mary's School of Nursing, Milwaukee, Wisconsin in 1958. She received her B.A. in English and graduated with honors in 1973 from California State University, Fullerton. In 1977 she received a law degree from Loyola Law School, Los Angeles.

Besides maintaining her private law practice, Phyllis writes a monthly column on the legal aspects of nursing for a widely circulated nursing magazine. She has just served a two-year term as president, Region 1, California Nurses Association. She is actively involved in the following: Orange County Bar Association, member of Health Law Section and Human Rights Section; American Bar Association, Forum Committee on Health Law; National Health Lawyers' Association; and the California Society for Health Care Attorneys. In addition, she serves on the Orange County Health Planning Council, Health Promotion Task Force; and is a member of the Advisory Committee, Nursing Home Advocacy Program.

REFERENCES

1. Anderson RD: *Legal Boundaries of California Nursing Practice.* Sacramento, Calif, Robert D Anderson, 1978.
2. Warren DM: Legal considerations in the search for holistic health. *Holistic Health* **3**:106, 1978.
3. *California Jurisprudence,* ed 3. San Francisco, Bancroft-Whitney, 1973, p 705, §415.
4. Krieger D: Therapeutic touch: The imprimatur of nursing. *Am J Nurs* **75**:784, 1975.
5. Jorrie R, Wolf, RW: Selected practical problems with professional associations and professional corporations. *St. Mary's Law J* **10**:248, 1978.
6. California Assembly Bill No. 1112, which amends §2500 of the California Business and Professions Code and §13401 of the California Corporations Code.
7. National Health Lawyers Association, 522 21st St, NW, Suite 708, Washington, DC 20006.
8. American Bar Association Forum Committee on Health Law, American Bar Association, 1155 East 60th St, Chicago Ill 60637.
9. Eccard WT: Revolution in white—New approaches in treating nurses as professionals. *Vanderbilt Law Review* **30**:855, 1977.
10. Hemelt MD, Mackert ME: *Dynamics of Law in Nursing and Health Care.* Reston, Va, Reston Publishing Co, 1978.
11. Bell WS: Medico-legal implications of recent legislation concerning allied health practitioners. *Loyola of Los Angeles Law Rev* **11**:379–398, 1978.

FURTHER READING

Bullough B (ed): *The Law and the Expanding Nursing Role.* New York, Appleton-Century-Crofts, 1975.

———, Bullough V (eds): *New Directions for Nurses.* New York, Springer, 1971.

———: *Expanding Horizons for Nurses.* New York, Springer, 1977.
Jacox AK, Norris C (eds): *Organizing for Independent Nursing Practice.* New York, Appleton-Century-Crofts, 1977.

CHAPTER

4

Research Aspects of Independent Practice

Marilyn T. Oberst, RN, EDD

The nurse in independent practice today stands at the frontier of new knowledge about people and health. Paradoxically this "new" frontier may be very old indeed, a reversion to attitudes held before nursing's almost total immersion in and subjugation by the organized health-care system. It is not easy for the nurse to function independently in an agency setting, though some practitioners manage to do so remarkably well. Those who do are characterized by the ability to articulate clearly the nursing rationale for their nursing behavior; that is, they are knowledgeable in their field (nursing), have a clearly defined philosophy regarding the nature of man, and are guided by this philosophical/knowledge base in predictable ways.

This chapter examines some aspects of nursing's knowledge base for practice and explores a variety of ways in which the independent practitioner can participate in and use research. The purpose is to stimulate increased awareness of the research potential inherent in independent practice, rather than to serve as a methodological primer.

The Practice Base: Intuition vs. Science

Nursing practice and nursing research are interrelated activities that cannot profitably exist in isolation from one another. Research relevant to practice, whether it involves simple testing of an isolated nursing measure or highly sophisticated theory development, must arise from and be grounded in the realities of practice. Conversely, nursing behaviors require a scientific basis if they are to be more than intuitive reactions or traditional responses to each situation encountered.

Unfortunately, nursing actions often appear to be entirely intuitive, as the following examples illustrate.

Situation 1. A nursing student, assigned to care for a patient during the immediate postoperative period in a recovery room, was informed by the charge nurse that she was going to call the patient's physician because ". . . this patient will be in trouble soon." "Trouble? Now what does that mean?" wondered the student. She had just completed a routine check of all the physiological indicators being monitored and had observed no changes in these or in any other aspect of the patient's behavior. Concerned that she had missed something, the student pressed for an explanation. "Look," said the nurse, "this patient is going into shock—don't ask me how I know, it's just something I feel. When you've been in this business as long as I have, you'll get to know these things too." Within 15 minutes the patient was exhibiting the classic signs of surgical shock and was returned to the operating room for religation of a major vessel in the abdomen.

Situation 2. Meredith New, a recent nursing graduate, was being oriented to the chronic care unit by Henry Old, clinical specialist on the unit. As Ms. New and Mr. Old made their rounds together, he introduced her to various patients. The two stopped briefly to discuss the previous night's baseball scores with one patient whom they found engrossed in the sports page. In the middle of this conversation, Mr. Old said abruptly, "Excuse me, I'll be right back," and left the room. He returned a few minutes later carrying a filled syringe. Seeing him, the patient said, "Oh, is that my pain medicine? I was just going to ask for it."

As soon as they reached the hall Meredith said, "Hey, Henry, that was really great. He looked perfectly comfortable to me. What did you see that I didn't see?" Henry chuckled a bit and replied, "Well, it's hard to say exactly. I guess it comes with time; you'll pick it up after you've been here awhile."

In both situations described, the clinicians correctly assessed the state of the patient and initiated appropriate actions. Was their behavior motivated by instinct? Or were these apparently intuitive responses the product of a system of cue utilization, originally reasoned from empirical observation? McLachlan argues that reasoned inference is often operative in such situations and cites the following example from her own experience in pain management.

> Over the years I had been seeing clues to pain, storing them in my memory, and when all the clues presented themselves at the same time I recognized pain almost instantly, so quickly that it appeared to prompt action without reasoning. But that learning had taken 25 years' observation of behavior . . . this was accidental, random learning and too slow.[1]

Unlike the two clinicians, who were unable to describe their own thinking, McLachlan was able to cite the specific cues she used to determine the presence of pain. Then, by testing a series of assumptions, she was able to conceptualize some manifestations of pain. This conceptualization enabled her to sharpen her practice skills and to teach others. How long will it be before the student in situation 1 learns the subtle, almost imperceptible signs of impending shock? How many patients will Ms. New judge to be "not in pain" before she realizes that the acute pain model she learned in school is inadequate for assessing long-term chronic pain? Once she comes to that realization, will she settle back, a replica of Mr. Old, content to appear magically, syringe in hand, at precisely the right moment? Will she instead begin to organize systematically her thinking about pain and gradually build a coherent framework for assessment and intervention?

It should be noted that, because they are the result of experience and observation, seemingly intuitive behaviors are very often correct. For the professional, however, being right is not necessarily sufficient. In its entirety, the nursing process requires thinking and acting within a framework. Assessment implies that one knows what to look for, and diagnosis implies that one can recognize it when one sees it. Before intervening, one must have reason to believe in the efficacy of the action taken, and evaluation requires clearly delineated goals and the means to measure them. Implementation of the nursing process at a professional level requires the use of theoretical or conceptual models as a means of structuring experience.

Theories and Models

A theory is a set of statements or propositions that provide a systematic explanation of the relationship between concepts. Thus, a theory is really a succinct way of characterizing real phenomena in terms of the essential or critical factors believed to be involved. Some theories, often referred to as macrotheories, are broadly conceived attempts to describe whole segments of human experience; sociological theories of social functioning and some current theories of holistic health are examples of this type of theoretical system. Other theories, called microtheories, focus on a more narrowly defined range of behaviors, such as decision making or grief. Stevens notes that theory of some type is inherent in the writings of many nurse authors, though it may not necessarily be so labeled.[2] Stevens offers guidelines and useful suggestions for extracting, analyzing, and evaluating such inherent theoretical linkages.

Dickoff and James[3] delineate four distinct levels of theory:

1. *Factor-isolating theories.* This level of theory involves the identification, definition, and naming of variables.

2. *Factor-relating theories.* At this level, previously identified factors are conceptualized in relation to each other. As the interrelationships of the variables become clear, a listing of factors can be produced and a schematic picture of a given situation begins to emerge; such theories are sometimes called situation-depicting theories.

3. *Situation-relating theories.* Theory at this level examines causal relationships among variables and takes into account time factors and directionality as well. Because this third level allows us to say that factor B always occurs in the presence of factor A or that factor A will increase when factor B is increased, theories of this type permit us to make predictions.

4. *Situation-producing theories.* Also called prescriptive theory, this is the fourth and highest level. It allows us to state unequivocally that A is a sufficient cause of B, not merely that they occur together. Thus, if B is the desired goal, than A is prescriptive for its production.

The nature of nursing practice requires that actions be taken to achieve particular goals or outcomes: we act to reduce anxiety, increase mobility, foster independence, maintain vital functions, and so on. Ideally, therefore, prescriptive nursing theory will provide essential direction for nursing action.

Theory construction is a relatively new endeavor for nursing; the interested reader should explore Newman's *Theory Development in Nursing*.[4] At the present time, few truly prescriptive theories exist, although there are a number of conceptual models or frameworks containing prescriptive elements. With adequate testing and refinement, high-level theory may eventually evolve from these frameworks. A conceptual framework is an organization of ideas, a way of thinking about things one perceives and experiences. The term *conceptual model* usually denotes an abstract version of a framework, often presented in schematic form, which symbolically represents the significant variables and their interrelationships.

Selecting a Practice Model

Since the object of nursing is the individual person, sick or well, nursing theory should focus on the client. Useful theories for nursing will be those that expand knowledge and understanding of clients and their responses and will be relevant to the nurse's function as helper/caretaker.[5]

Ellis notes that theories significant for nursing have specific characteristics.[5] These may serve as guidelines in evaluating the potential utility of a theory for a particular practice:

1. *Scope.* The theory encompasses the interrelationship of numerous biological and behavioral concepts and provides a framework for ordering observations about phenomena.
2. *Complexity.* The theory goes beyond readily apparent, simple ideas and treats individuals or multiple variables in depth.
3. *Testability.* The theory should be capable of being tested in practice and have sufficient flexibility to allow change in its constructs when indicated by practice realities.
4. *Usefulness.* The theory must be applicable to practice and provide guidance for action.
5. *Information generation.* New hypotheses for testing and, subsequently, new information should be generated by the theory.
6. *Terminology.* The theory should utilize language that is meaningful in nursing.

No existing practice model possesses all of these characteristics to equal degrees in all situations. The primary consideration in selecting a model must therefore be its potential utility in the particular practice in which it is to be applied. A framework useful in an obstetrical/well-child practice may not encompass all variables appropriate to a geriatric or chronic-care practice. Usefulness in practice will also be determined by the practitioner's response to the model. The model must be understandable; it must use ideas and constructs with which the practitioner feels comfortable. Since theories arise from a basic philosophical viewpoint about the nature of man, this base must be in accord with the practitioner's own philosophy.

Some practitioners may find that no single model entirely meets their needs. They may wish to augment a broad theory, developed specifically for nursing, with narrower and more specific constructs from other fields.

If such a theory is adopted, it is important that the theory be tested carefully for the appropriateness of its language and constructs as well as for the accuracy with which it represents reality. Uncritical use of apparently clear theories can sometimes lead to serious errors in practice. An example is the way in which Kübler-Ross's theory of the stages of dying has been utilized. From her observations and interviews with hospitalized, dying patients, Kübler-Ross formulated a model that describes the dying process as a series of stages: denial and isolation, anger, bargaining, depression, and finally, acceptance.[6] Her conceptualization came at a time when little other information on the subject was available; it filled a disturbing knowledge gap and was eagerly seized upon by a number of professions.

Although Kübler-Ross had presented the stages as typical, and not ineluctable, the users began to treat the schema as fact. The dying individual was expected to conform to the theory. Deviations from the behavioral stages were considered pathological, and in some cases practitioners made attempts to force people into the "proper" sequence. In reality, the various emotional responses to impending death are not necessarily chronological; there is often movement back and forth. In addition, the stages are not necessarily mutually exclusive; overlap and intermingling occur frequently.

A number of factors contributed to misuse of this particular theory. First, it was assumed by its users to be at a higher theoretical level than was, in fact, the case; at best, it is situation-depicting theory, but it was used as though it were predictive and/or prescriptive. Second, it had been tested inadequately. Although Kübler-Ross's constructs may have had some validity in the highly structured hospital situation from which they originally arose, there had been no attempt to verify the constructs in other settings. A final factor has to do with the basic philosophical issue of behavioral norms and their application. By treating the described stages as "ideal," one establishes normative behavioral expectations for perhaps the most personal and individual human experience. Despite "ideal" stages, there is no single right way to die.

It is imperative, then, that the models or frameworks selected for use in practice be continually subjected to scrutiny and evaluation in the context of practice. As noted earlier, theory and practice are interrelated activities; neither can be adequately developed in isolation from the other. Figure 1 (see page 80) schematically illustrates the developmental continuum of the theory-practice link.

Independent nursing practice may well be the ideal setting for testing, revising, and expanding current practice models and for developing new models. Freedom from organizational constraints should facilitate creativity of approach and enhance control of the research situation.

Research and Independent Practice

Using Existing Research

Nursing research and nursing practice have traditionally been thought of as separate activities—researchers performed studies and developed theories, while practitioners gave care to patients. Until recently, few practitioners had the requisite knowledge to design and carry out nursing studies, while those who were academically prepared to do so were, for the most part, employed in university settings with limited access to patient populations. As the number of practitioners with advanced degrees increases, however, a combined research-practice role may emerge.

The nurse in independent practice must, of necessity, enlarge and make visible the research component of that practice. Every practitioner need not personally undertake research, although many will probably do so. However, the practitioner must, at the very least, be an intelligent consumer and critic of research in order to assure that the nursing actions taken have a base in nursing science. The fact that the practice is independent precludes reliance on traditional agency policies and procedures as a rationale for behavior. Neither

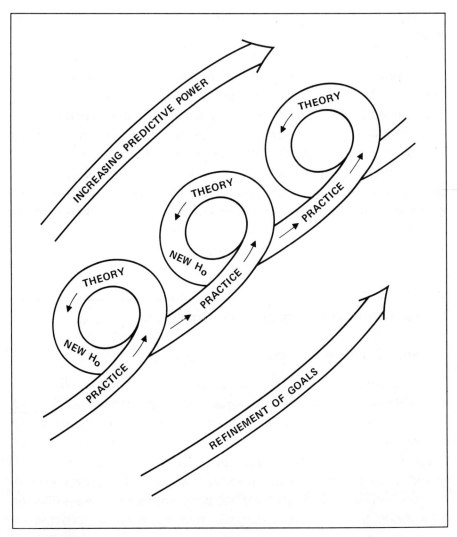

Figure 1. Theory-Practice Continuum

good intentions or intuitive actions are adequate defenses in a court of law. Accountability, a characteristic of any profession, is much more likely to become a reality for the independent practitioner. If one elects to fly solo, the technical skills of flying, learned by rote, are insufficient. Also needed are an understanding of aerodynamics to enable accurate response to varying conditions of weather and wind, as well as knowledge of the current air-traffic regulations that govern flight. Finally, assurance that the plane is structurally sound and has been built in accordance with the latest and best knowledge available is essential in order to assure maximum safety, efficiency, and efficacy.

In solo practice, it is equally important to remain cognizant of the latest knowledge in the field. The most important rule, then, is *read*. The reading program should regularly include at least one general nursing-practice journal (more, for a general practice), several journals in an appropriate specialty, several nursing-research and theory journals, and material from related fields pertinent to the practice. Reading privileges can probably be arranged at the library of a local nursing school, hospital, medical school, or university. Journals and other materials might also be shared among colleagues informally or through a journal club. The second rule is *think critically*. Publication does not automatically confer authority. Articles and research reports should be examined for theoretical soundness, logical consistency of the arguments presented, and accuracy of clinical content. Discussion with colleagues is often helpful in bringing new insights. The third rule is *apply with care*. There has been a tendency in nursing to pick up seemingly good ideas and apply them on a large scale without adequate attention to whether or not they actually work. Interventions fully tested for effectiveness in one setting may not necessarily produce the same result in another. Application of new knowledge should therefore be undertaken carefully and should include evaluation of outcomes. The fourth rule is *establish a dialogue*. Questions or comments about the material read should be communicated to the journal or the author. Sharing one's own ideas and experiences with others at professional meetings and through publication can be an effective method of clarifying one's thinking and often leads to productive professional relationships.

Establishing a Climate for Research

In one sense, all clinical practice is experimentation. Whether one uses a new approach or applies a proven method, the client's individuality creates a unique situation on each occasion. As Feinstein notes, "A clinician performs an experiment every time he treats a patient."[7(p14)] He indicates that "every aspect of clinical management can be designed, executed, and appraised with intellectual procedures identical to those used in any experimental situation."[7(p27)] The nursing process and the research process are similar in several respects. The assessment and diagnosis phases, for example, are analogous to formulation of the research question and development of specific hypotheses for testing. Also, both nursing and research require planned implementation and the collection of data relevant to outcomes, followed by final evaluation of results. The major difference between the two is in their primary goal. Normally, the goal of practice is a specific patient-care outcome for an individual. In research, the primary goal is increasing knowledge that will be applicable in many situations. (In some types of clinical studies, individual outcomes may also be achieved.) Given this similarity of sequence and process, good practice can easily become the beginnings of good research.

When the nursing process is fully implemented by a master clinician, vast amounts of data are produced relevant to every stage of a client's illness and recovery. Much of these data are presently lost simply because they are not fully recorded in the individual record. Even more potentially useful information about commonalities among groups of clients is virtually inaccessible if there is no systematic approach to recording data so they are retrievable. Nursing diagnoses can be utilized as a focal point for the organization of empirical observations in such a way as to provide data for varied research studies.

The diagnostic process involves organizing data about the state of the individual into logical categories and assigning a label that provides a meaningful representation of that state. A diagnostic label is therefore operationally defined in terms of subjective and objective signs and symptoms. To qualify as a nursing diagnosis, a label must represent a client state that is appropriately acted upon by a nurse. A number of listings of nursing diagnoses are currently available in the literature. They vary considerably in style, organization, and underlying framework for classification. The most complete list is

Campbell's,[8] which is taxonomized according to Maslow's hierarchy of needs and specifies interventions for each diagnosis. Several diagnostic systems have been developed from other nursing frameworks. The practitioner may elect to work with an existing diagnostic list in its entirety, select portions from several lists, or add additional labels; the diagnoses selected, however, must fit logically within the chosen theoretical practice model and must be consistently applied.

The diagnostic file card shown in Figure 2 illustrates one possible approach to organizing information about nursing diagnoses for later use. A separate card is used for each diagnosis for each client. Because the cards are an adjunct to, and not a replacement for, the regular client record, considerable latitude and flexibility in use of abbreviations and symbols is possible. This shortened annotation

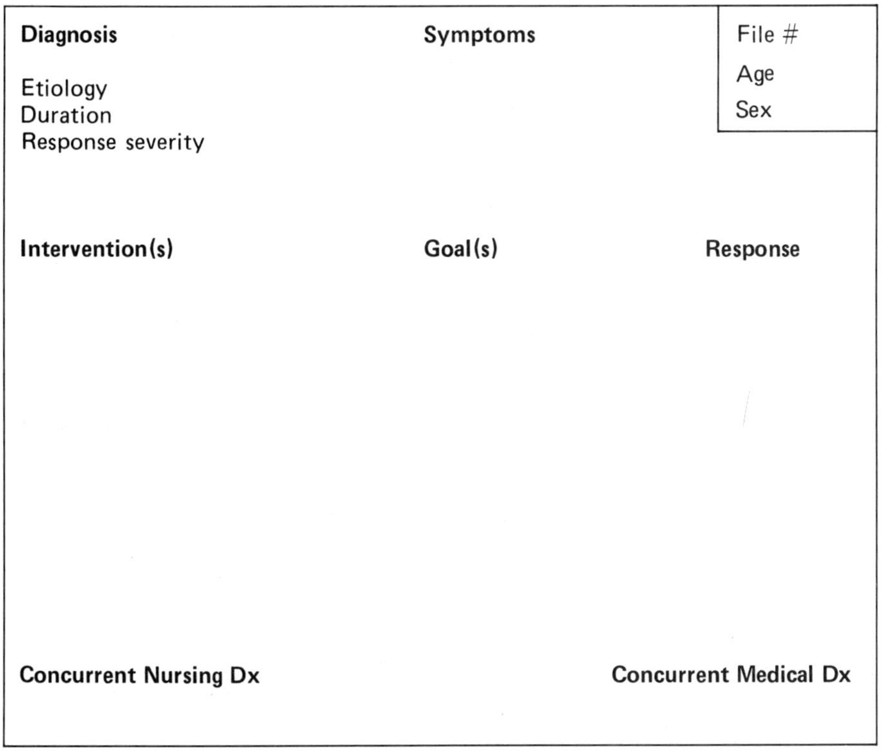

Diagnosis	Symptoms	File #
Etiology		Age
Duration		Sex
Response severity		
Intervention(s)	Goal(s)	Response
Concurrent Nursing Dx		**Concurrent Medical Dx**

Figure 2. Diagnostic File Card

allows large amounts of information to be recorded in relatively compact form.

The information required under each suggested heading is discussed below:

Diagnosis. Diagnosis should be a single label, representative of a particular state, based on the assessment. Labels should be consistently applied.

Etiology. Etiology is the probable cause or source of the diagnosis. Gebbie and Lavin[9(p59)] suggest four etiological categories: anatomical, physiological, psychological, and environmental. Other, more detailed breakdowns may also be utilized and would be especially appropriate when most of the diagnoses encountered fall primarily into one or two of the etiologic categories, as might be the case in a mental-health practice.

Duration. Gebbie and Lavin[9(pp58-59)] also suggest four divisions for this temporal description: intermittent, chronic, acute or situational, and potential. Again, other categories may be utilized as necessary.

Response severity. This category describes the distress level of the client in response to the diagnosis. Five levels are suggested:[9(p50)]

1. None—shows no evidence of a problem, able to handle needs.
2. Minimal—shows some evidence of a problem, primarily related to knowledge or anticipation of a problem.
3. Moderate—can handle problems with supervision or minimal assistance.
4. Severe—client is dependent; problem is acute and probably reversible.
5. Very severe—client is dependent; problem is long term and probably not reversible.

Symptoms. Included here are all objective and subjective signs and symptoms from which the diagnosis was inferred; symptoms of other problems should be recorded separately. Uncertainty about the relationship of a particular finding to the diagnosis may be recorded in parentheses or with a question mark.

Intervention(s). Included here is a brief list of diagnosis-specific nursing actions undertaken to treat the client. Dating each entry will distinguish simultaneous from sequential interventions.

Goal(s). Goals are the specific measurable client outcomes desired as a result of each intervention. Inclusion of a time frame for goal accomplishment may be useful.

Response. Listed here is an indication of the degree to which the intervention was effective in achieving the desired goal.

Concurrent diagnoses (nursing and medical). This brief list of other known diagnoses and major treatment modalities is the last entry on the card.

Figure 3 shows a partially completed diagnostic file card on a client referred for dietary and behavioral management of obesity. The nurse practitioner formulated the diagnoses of unresolved grief on the basis of symptoms noted on initial assessment; the question mark indicates uncertainty about the source of this symptom, later found to be more highly related to limited movement in the dominant hand. Note that the second intervention listed is "switch from individual to group" obesity treatment with the goal of increasing interaction with others. Obviously, the primary goal of the obesity treatment remains weight reduction; the change in mode of treatment is the diagnosis-specific intervention for this client's unresolved grief. The final intervention / goal set listed might also be appropriate for treatment of the client's obesity problem. Multiple-problem situations such as this one often necessitate a somewhat different intervention approach than might be useful in treating any of the single diagnoses

Diagnosis Grief, unresolved (limited insight) **Etiology** Psychol **Duration** Chronic (14 mos.) **Response severity** mod → sev	**Symptoms** Preoccupation with thoughts of deceased spouse Fréquently tearful Slumped posture Disheveled appearance (?) Slow verbal response	**File #** 0093 **Age** 62 **Sex** F
Intervention(s) 6/2 Encourage verbalization, listen, give feedback, request consensual validation of impressions	**Goal(s)** a. Open acknowledgment of problem b. Insight	**Response** a. Good after several sessions b. Partial
6/9 Switch from individual to group for obesity treatment	↑ Interaction with others	Excellent after initial hesitation
6/9 Set limits on crying while conveying empathy; expand discussion topics	Refocus attention	Fair (some hostility to limit setting)
6/16 Assist in identification of positives: strengths, experiences, expectations	↑ Feelings of self-worth, ↑ Interest in life	Initially negative
Concurrent Nursing Dx Obesity, chronic (primary presenting problem) Limited ROM hand/wrist (4 mos. post wrist fracture)		**Concurrent Medical Dx** Hypertension

Figure 3. Active Diagnostic File Card

in isolation; careful review of the diagnostic file may facilitate identification and clarification of these differences.

This format was developed for use with diagnoses of primarily anatomical and physiological etiology. When the etiology of diagnoses encountered in a particular practice is more likely to be psychosocial or environmental, additional contextual variables should be included. Such factors as family structure, personal and extrapersonal resources, social-support system, and health beliefs may markedly influence the extent to which a given intervention is effective. When such factors are significant, nursing actions must be situation specific as well as diagnosis specific.

The cards may be cross-indexed by using color-coded tabs for each diagnosis and for major intervention modes. Because it allows rapid hand sorting, a coded punch-card system, such as McBee Keysort®, may be useful when much data and multiple variables are involved.

The diagnostic file may be used in a variety of ways. Its primary usefulness to the practitioner will be as a tool for evaluating practice effectiveness. Although the effectiveness of specific interventions with individual clients is evaluated as matter of routine, effectiveness in the aggregate is examined rarely. The file facilitates access to all occurrences of a particular diagnosis, whether primary or not, for review. At this point, the practitioner/researcher can use the data to answer these questions:

- What is the overall effectiveness of therapy for clients with a particular diagnosis?
- Were the presenting symptoms similar enough across cases to warrant this diagnostic label in each instance?
- What between-client differences might account for differential response to various intervention modes?

The file may also be utilized to provide data for descriptive research, for validating current diagnostic labels and categories, and for generating hypotheses for subsequent testing. These questions might be explored:

- Does the diagnostic nomenclature clearly delineate mutually exclusive conditions?
- What behaviors are characteristic of persons with a particular diagnosis?
- Do certain diagnoses occur in groups? Does this diagnostic cluster represent a syndrome different from the individual diagnoses of which it is composed, and does it therefore require a different intervention approach?
- Are the delineated outcomes reasonable and logical expectations of interventions? Are the outcome measurements utilized adequately?
- What unexpected outcomes occur as a result of specific interventions? How might these be used productively in other situations?

Information from the diagnostic file might also be made available to other practitioners and researchers who lack access to a sufficiently large data pool. If the practice is small, a cooperative data-gathering effort among several nurses would be mutually advantageous.

Research Possibilities in Independent Practice

The possibilities for research by independent practitioners are as varied as the nature of the practice, the practitioners themselves, and the client group they serve. There is need for extensive descriptive research of the health and illness phenomena and the nursing actions they elicit, as well as experimental studies that test those actions. The American Nurses Association Commission on Nursing Research, in an effort to assist the profession to focus its research on development of the knowledge necessary to improve practice, has identified a number of broad priority areas.[10] These identified practice priorities include studies that will:

- reduce the complications of hospitalization and surgery;
- improve the outlook for high-risk parents and high-risk infants;
- improve the health care of the elderly;
- increase knowledge of life-threatening situations, anxiety, pain, and stress;
- explore adaptation to chronic illness;
- develop self-care systems and group-care systems;
- facilitate successful utilization of new technology;
- identify nursing interventions that promote health;
- facilitate successful application of new knowledge to patient care;
- define and delineate healthy states;
- increase understanding of addictive and adherence behaviors, and of under- and over-nutrition;
- evaluate outcomes of different patterns of delivery of nursing services.

Many of these priority areas are pertinent to the work of the nurse in independent practice and are problems to which research could be addressed in this setting. The priorities fall into two broad categories: those relating to illness care and those relating to health (or wellness) care. The following discussion explores some additional research questions and practice problems that might be examined in each of these categories.

Health-Care Research

The terms *health care* and *health-care system,* as they are presently used, are misnomers for what is, in fact, illness care. Illness and its effects remain the focal point of most activities subsumed under these labels. Even preventive health programs focus, for the most part, on case-finding and early diagnosis of illness. The focus is not on taking a more positive approach to the creation and maintenance of a healthy state. Although nursing has traditionally had greater involvement in providing care to essentially well populations than have other health-care disciplines, we have generally failed to operationalize our definitions of such terms as *health, maximum well-being, optimal functioning,* and the like.

Nurses must ask relevant questions about health and wellness. What are the indicators of a healthy state, and what behaviors on the part of individuals produce that state? What characteristics distinguish the well individual from the ill individual? What are the norms for health, and how do these vary across ethnic and sociocultural groups? What factors prevent optimal functioning or enhance health? What factors influence the individual's health-belief system and health-seeking behaviors? What actions by nurses will foster health and facilitate wellness? How do health and developmental norms differ cross-culturally? What is healthy aging, and what behaviors promote health in the aged? What adaptations to aging are personally satisfying to the elderly, and to what extent can such adaptation be guided? What is the effect of prenatal teaching on subsequent mother-child relations and the growth, development, and health of the child? What nursing strategies are useful in increasing parenting skills?

These are but a few of the questions that might be raised about health. Answers to such questions would provide a foundation for the development of health teaching and health-maintenance strategies appropriate for nursing action and would begin to define measurable outcome goals for such action. Descriptive and correlational studies addressing many of these basic questions about health could be undertaken by nurses in independent practice.

Illness-Care Research

Many nursing studies have explored aspects of response to illness and examined the effectiveness of various nursing interventions. Most of these have examined illness phenomena in the hospital or institutional setting and should now be replicated in other settings. The questions raised below have been organized according to the type of practice in which they might be studied, but many are relevant to several practice specialties.

There are many pertinent questions for the nurse with a gerontologic practice, in addition to the issues related to healthy aging already mentioned. What assessment tools are useful in differentiating between illness and normal age-related changes? What are realistic outcome goals for treatment of the elderly? What are the effects of family structure, personal loss, and social-support systems on development of illness and recovery from illness in the aged? How are functional capacity, cognitive capacity, and coping style best measured? Which nursing interventions are most effective in improving memory, orientation, and morale; how can these outcomes be measured? What specific recommendations will be helpful to families of aged, ill individuals in managing problems of incontinence, physical disability, and mental deterioration?

In a psychiatric or mental-health practice, the major challenge may well be identification of outcome measures that define clearly the abstract goals of therapy. How is success or failure measured? How do the characteristics of individual clients—sociological, psychological, demographic—influence treatment goals for them? Which interventive and affectual styles are most productive for individuals and groups? What are the predictors of good adjustment or adaptation after hospitalization? What factors influence patients' decisions to seek and continue treatment?

For the nurse in general practice, the questions are far ranging and diverse. What factors influence assumption of a sick role versus a well role in individuals with minimal disease or disability? What is the differential effect of illness and disability at various stages of the life cycle? What nursing behaviors help patients adjust to chronic illness? What is the relationship between teaching style, client knowledge, and subsequent compliance with treatment; what variables affect this relationship? What nursing measures are most effective in reducing or relieving the stress of terminal illness for client and family? Which measures of complex, multifaceted outcome variables such as coping, recovery, psychological distress, and adjustment can be utilized to evaluate care effectively?

Ethical Considerations

Issues of research ethics and informed consent should be considered carefully before any study. If one accepts the basic tenet that the individual has the right to be involved in decisions that affect his or her life, it follows that all participation in research must be voluntary and fully informed.

In a practice discipline, the line between research and practice is a fine one, at times barely perceptible. Testing an intervention with a client in order to find an effective treatment for that individual is not research; the only goal is a specified treatment outcome for a single individual. Systematically testing the effectiveness of the same intervention in a group of clients in order to make generalizations applicable to other situations is research, even when treatment goals for individuals are accomplished in the process. In general, informed consent is required if any aspect of the research involves activities or actions to which the patient would not otherwise be exposed in the course of treatment. Experimental studies in which randomization or other nonbiased techniques determine subject assignment to treatment group cannot be performed without subject consent. Consent is required even if all the interventions being used for comparison (experimental and control treatments) have been tested for safety and are in common use. The issue here is that the treatment decision, normally a matter of clinical judgment, is being left to chance; this fact must be made known to the subject.

Usually data normally available from records may be used without consent to generate hypotheses or control quality (as previously described for the diagnostic file) even if the findings from the grouped data are published. If, however, individual case studies are shared with colleagues or published, clearance should be obtained from the client. The client's right to privacy may be threatened by publication. Even when names are disguised, other aspects of the case may make possible identification of the client by others. Confidentiality is a particularly important consideration when sensitive material such as drug addiction, alcohol dependence, or abortion are involved, but applies whenever information supplied by the client for the express purpose of expediting treatment is used in ways that may violate privacy.

When more formal research is planned, the study must be designed to pose minimal risk to the research subjects. The concept of risk involves not only physical harm but also psychological distress, which may be induced by certain psychological tests, probing questions about highly personal material, or extended use of subject time for nontherapeutic purposes. In most agency settings, clinical research proposals are subject to rigorous review prior to approval. The scientific reviewers examine the study design, underlying theory, and all methods and tools of measurement to determine their safety, clinical and research soundness, and probable efficacy in meeting the stated objectives. An ethical review board, usually a formally constituted human-subjects review board, examines carefully the relative risks and benefits for the individual subject and for society; the board pays particular attention to measures planned to protect human subjects from threats to safety, confidentiality, and privacy. For the nurse researcher in independent practice, rigorous review of the proposal, though not required by law unless federal funds are being utilized, is imperative. Consultation with a qualified researcher during proposal development is recommended. Once the proposal is ready, the researcher should seek formal written review from a number of experts who have no vested interest in the study. Reviewers should include a researcher, an expert in the clinical area, and legal counsel; involvement of a lay reviewer should also be considered. Although the review does not relieve the researcher of responsibility or accountability for harm that may result, review substantially reduces the risks for subjects and is evidence that due care was exercised.

A final ethical consideration lies in the possible conflict of interest when the practitioner assumes a dual role—researcher and clinician. Clients may feel compelled to participate in research proposed by a professional upon whom they rely for care, thus diluting the voluntariness of their consent. Care must be taken at the outset to assure clients that they are not required to participate and that they may withdraw at any time without prejudicing the relationship with the practitioner or jeopardizing treatment. The clinician who contracts to provide certain services must be vigilant that research procedures do not compromise those services. The responsibility as clinician must remain primary and take precedence over the research commitment, even at the expense of lost data.

Some Difficulties and Solutions

Two difficulties frequently encountered by clinical researchers are shared by the independent practitioner who undertakes research. The difficulties are sample size and investigator bias.

It may be difficult for the nurse with a small or part-time practice to get a subject population large enough to assure meaningful results within a reasonable time frame. The most productive approach to this problem is to collaborate with one or more colleagues with similar clients. Care must be taken, of course, to assure that all observations and measurements are standardized and that experimental manipulations are identical in all settings.

Observer bias can be a threat to the validity of findings whenever the individual applying an experimental manipulation is also responsible for measuring the results. In any type of study, data may be missed, enhanced, or misinterpreted by an observer interested in demonstrating a particular point. Bias can also exist when the researcher inadvertently influences the subject to respond in particular ways; this influence can be especially troublesome when the outcome variable is the client's subjective response. In the latter case, an independent observer is necessary. When the measurement tools are fairly objective, periodic checks on research/observer reliability by a second observer may suffice. The practitioner in solo practice must attempt to control such sources of bias by designing the study carefully and obtaining outside assistance when collecting data.

The clinician who wants to do research has many potential sources of assistance often no more than a phone call or letter away. The most important first step, thinking through and clarifying the nature of the problem and identifying prior research and knowledge relevant to it, should be taken independently. Next, it is often useful to check the validity of clinical ideas and observations with other practitioners and to recruit one or more interested colleagues as co-investigators or consultants. The neophyte researcher may find a number of recently published books useful during the early stages of proposal development,[11,12] but also should consider taking graduate courses in research methodology.

Experts in research methodology and nursing theory may be called upon for advice and consultation during proposal development. Individuals who have done prior research in the problem area, identified in the literature review, may be willing to assist in developing a project that tests their ideas in another setting. These experienced researchers may be able to suggest other resource people. Graduate faculty from a local college or university nursing program are another potential source of help.

Summary

This chapter has reviewed some of the ways in which the independent practitioner might view the clinical experience from a research perspective. Suggestions for evaluating the quality of practice, developing a data base, verifying knowledge, testing theoretical assumptions, and developing new knowledge have been given. Although it is unlikely that most nurses in independent practice will have the resources, time, or interest to undertake highly sophisticaled research, small-scale exploratory and descriptive studies are well within the realm of possibility. Research originating or carried out in the autonomous-practice setting could well map the future by describing health and illness phenomena in ways that will clarify consumer needs for care and delineate a logical nursing-practice base for meeting those needs.

Author's Biography

Marilyn T. Oberst has been a research assistant and instructor of nursing research at Teachers College, Columbia University, New York City. In addition, she was project director for a nursing research facilitation grant from Department of Health, Education and Welfare. Ms. Oberst is associate editor of *Cancer Nursing* and is a frequent contributor to nursing journals. Presently Ms. Oberst is director of nursing research at Memorial Sloan-Kettering Cancer Center, New York City.

REFERENCES

1. McLachlan E: Recognizing pain. *Am J Nurs* **74**:496, 1974.
2. Stevens BJ: *Nursing Theory: Analysis, Application, Evaluation*. Boston, Little, Brown, 1979.
3. Dickoff J, James P, Weidenbach E: Theory in a practice discipline. *Nurs Res* **17**:415–435, 1968.
4. Newman MA: *Theory Development in Nursing*. Philadelphia, FA Davis, 1979.
5. Ellis R: Characteristics of significant theories. *Nurs Res* **17**:217–222, 1968.
6. Kübler-Ross E: *On Death and Dying*. New York, Macmillan, 1969.
7. Feinstein AR: *Clinical Judgment*. Huntington, NY, Robert E Krieger, 1967.
8. Campbell, C: *Nursing Diagnoses and Intervention in Nursing Practice*. New York, John Wiley & Sons, 1978.
9. Gebbie KM, Lavin MA: *Classification of Nursing Diagnoses*. St. Louis, C V Mosby, 1975.

10. ANA Commission on Nursing Research: Nursing research priorities. *Nurs Res* **25**:357, 1976.
11. Diers D: *Research in Nursing Practice.* Philadelphia, J B Lippincott Co, 1979.
12. Brink PJ, Wood MJ: *Basic Steps in Planning Nursing Research: From Question to Proposal.* North Scituate, Mass, Duxbury Press, 1978.

FURTHER READINGS

Research Methodology
Abdellah FG, Levine E: *Better Patient Care through Nursing Research*, ed 2. New York, MacMillan, 1979. A comprehensive basic text.
Babbie ER: *Survey Research Methods.* Belmont, Calif, Wadsworth Publishing Co, 1973. Offers good suggestions on questionnaire construction.
Campbell DT, Stanley JC: *Experimental and Quasi-Experimental Designs for Research.* Chicago, Rand McNally, 1963. A classic focusing on design and issues of internal and external validity.
Cohen J: *Statistical Power Analysis for the Behavioral Sciences*, rev ed. New York, Academic Press, 1977. Useful in determining sample-size requirements for differing levels of measurement; assumes a basic knowledge of statistics.
Daniel WW: *Biostatistics: A Foundation for Analysis in the Health Sciences.* New York, John Wiley & Sons, 1974. An introductory text that utilizes examples from the health-care field.
Kerlinger FN: *Foundations of Behavioral Research*, ed. 2. New York, Holt, Rinehart & Winston, 1973. Good methodological source book for studies involving psychosocial variables.
Polit D, Hungler B: *Nursing Research: Principles and Methods.* Philadelphia, J B Lippincott, 1978. A comprehensive basic text recommended for its readability and completeness.
Selltiz C, Wrightsman LS, Cook SW: *Research Methods in Social Relations.* New York, Holt, Rinehart & Winston, 1976. A basic text with a sociological perspective that gives detailed information about a variety of data-collection methods.
Wooldridge PJ, Leonard RC, Skipper, JK: *Methods of Clinical Experimentation to Improve Care.* St. Louis, C V Mosby, 1978. Focuses on experimental design and the testing of nursing interventions.

Nursing Theory
Hardy ME (ed): *Theoretical Foundations for Nursing.* New York, MSS Information Corp, 1973.
Murphy JF (ed): *Theoretical Issues in Professional Nursing.* New York, Appleton-Century-Crofts, 1971.
Newman MA: *Theory Development in Nursing.* Philadelphia, F A Davis, 1979.
Nursing Development Conference Group: *Concept Formalization in Nursing: Process and Product.* Boston, Little, Brown, 1973.
Riehl, JP, Roy, C: *Conceptual Models for Nursing Practice.* New York, Appleton-Century-Crofts, 1974.

Rogers ME: *An Introduction to the Theoretical Basis of Nursing.* Philadelphia, F A Davis, 1970.
Stevens BJ: *Nursing Theory: Analysis, Application, Evaluation.* Boston, Little, Brown, 1979.

The Diagnostic Process
Campbell C: *Nursing Diagnosis and Intervention in Nursing Practice.* New York, John Wiley & Sons, 1978.
Feinstein AR: *Clinical Judgment.* Huntington, New York, Robert E Krieger, 1967.
Gebbie KM, Lavin, MA: *Classification of Nursing Diagnoses.* St. Louis, C V Mosby, 1975.

Evaluation of Research Reports
Downs FS: Elements of a research critique, in Downs FS, Newman MA (eds): *A Sourcebook of Nursing Research,* ed 2. Philadelphia, F A Davis, 1977, pp 1–12.
Fleming JW, Hayter J: Reading research reports critically. Nurs Outlook **22**:172–175, 1974.
Leininger MM: The research critique: Nature, function and art. Nurs Res **17**:444–449, 1968.

Research Ethics
Beecher HK: *Research and the Individual: Human Studies.* Boston, Little, Brown, 1970.
Freund PA (ed): *Experimentation with Human Subjects.* New York, George Braziller, 1969.
Hershey H, Miller RD: *Human Experimentation and the Law.* Germantown, Md, Aspen Systems Corp, 1976.

CHAPTER
5

Obtaining Hospital Visiting Privileges

Donna Lee Wong, RN, MN, PNP

Beginning a practice is a great challenge for nurses. Before planning begins, the nurse must give some general thought to his or her professional practice. Two terms commonly used to denote the expanded roles of nurses outside traditional settings are *independent practitioner* and *private practitioner*. Generally, both titles imply professional responsibilities and privileges that arise from self-employment obligations associated with an agency. For this reason, *independent* is descriptive. However, in a health-related discipline, in which comprehensive care involves a multidisciplinary approach, it is almost impossible to define one's own role as independent. The need for collaboration and consultation with other professionals and for direct patient care outside the usual "office" boundaries necessitates an interdependent practice. As a result, sanction to continue one's care of clients in other settings, such as hospitals, clinics, or community nursing services, is a necessity.

This chapter deals specifically with gaining visiting or staff privileges from a health agency. The author's discussion stems from her experience in obtaining privileges at two community hospitals. Since neither institution had previously dealt with such a request and since the literature could provide little direction or guidelines, the process was a mutual learning experience. Although this chapter discusses both the practitioner's request for and the agency's granting of visiting privileges, the focus is primarily on the request for privileges. Perhaps if these experiences and insights are shared, other practitioners will benefit from the precedent and the process will be expedited. Because of the hospitals' very cautious processing of the author's request, permission was granted approximately 20 months after the initial request to the first agency and 8 months after request to the second. The much shorter wait after the second request, which was initiated after the first but concurrent to its review, was in part due to the fact that a precedent had been partially established.

The Beginning

The author's need for visiting privileges did not arise during the early phase of her practice. Rather, the need came because the hospital placed restrictions on the visiting of her patients and their families and because of the perceived disruption caused by her involvement with the staff nurses. The following is a brief history of the long and sometimes frustrating struggle for obtaining staff privileges.

The author's clinical expertise is in counseling families who have a child with a potentially terminal illness, usually cancer. Initially her referrals came from a group of hematologists and oncologists who were associated with a voluntary organization, the Community Leukemia Fund. The fund has three goals, one of which is to provide counseling and support to families of cancer patients. Such services are provided by two clinical nurse specialists with master's degrees.

All fees for counseling service are paid directly to the nurses by the fund. Since that time, word-of-mouth has made the practice well known. In addition to families referred by the fund, the author now counsels any family who seeks help in a crisis, usually death of, or a serious health problem of, a child.

The first child referred to the author was a 6-year-old male with aplastic anemia. Her initial two visits with the family took place in the hematologist's office. However, the third visit was arranged at the hospital because the child had been admitted for more intensive therapy. Since this hospital was unfamiliar to the author, she arrived at the pediatric unit, introduced herself to the head nurse and nursing staff, and briefly explained her role as "a nurse who has been working with the parents and child for the past two months, helping them adjust to the treatments, the possibility of failure, and perhaps the child's death." The author believed this introduction and the fact that the physician had informed the hospital of her visit was sufficient. Indeed, at that visit the head nurse was very cordial and the staff nurses were eager to learn how the author became involved with the family and how the staff could better support the family. Because there was no area of the chart on which the author could record her findings, she gave the nurses a verbal report and wrote a separate summary to the referring physician.

On her return visit to the hospital, the author's reception was quite different. The nursing supervisor immediately approached the author and requested that she and the head nurse meet. The supervisor made it very clear that the author could see the family only as a visitor, and claimed that the author upset the hospital routine. These rules were set forth: the author could visit the family only during scheduled visiting hours, could not read or write in the patient's chart, and was not allowed to talk to the nurses. The author agreed to abide by the first two rules but strenuously objected to the last one, arguing that even "visitors can talk to the nurses." Fortunately, the conference ended with a compromise in the author's favor.

However, this temporary arrangement was far from satisfactory. The author needed 24-hour visiting privileges and some means of permanently communicating her assessment and interventions with the family. In addition, she could not meet the family's needs during weekly one- or two-hour sessions. To be effective, she needed to plan care jointly with the nursing staff, using staff findings and her own input about the child, siblings, and parents. The day after this meeting, the author began the process of applying for visiting privileges.

Applying for Visiting Privileges

Preliminaries

Before applying for staff privileges, one should clarify one's professional role, specify one's reasons for requesting privileges, and define those privileges necessary to fulfill that role. If possible, it is also advisable to delineate those reciprocal responsibilities assumed once visiting privileges have been granted. Although such reciprocal obligations are usually mandated by the institution's rules and regulations, emphasizing one's own recognition of them prior to acceptance by the institution reinforces one's professional accountability and ethics.

The following are hypothetical examples and excerpts from the author's letters to the hospitals regarding the above:

> A. *Clarification of a professional role and reason for requesting privileges:* I am a clinical nurse with a master's degree. I practice independently and specialize in counseling families who have a child with a life-threatening illness. Several of these families reside in the _____ and periodically use the inpatient/outpatient services of _____ (institution). Because I work with these families on a long-term basis (usually from the time of diagnosis to cessation of therapy or the terminal phase and postdeath), I will have occasion to visit them while the child is hospitalized. I realize that this role is a new one for nurses, but I feel that the patient's physical and

emotional welfare can be enhanced by my working jointly with the nursing staff.
B. *Definition of privileges necessary to fulfill professional role:*
   1. Twenty-four hour visiting privileges to see those families currently referred to me.
   2. Access to these patients' hospital records.
   3. The right to communicate both in writing, via the client's chart, and verbally with the hospital staff to facilitate continuity of care.

   (The privileges needed will depend on the nurse's role. Practitioners involved in physical assessment, diagnosis, and treatment of health problems may require a much more extensive list of hospital privileges—for instance, obtaining and recording a health history, performing a comprehensive physical examination, using appropriate instruments for assessment, requesting necessary diagnostic tests, and identifying and managing specific minor common illnesses.)
C. *Reciprocal responsibilities and obligations:*
   1. Assurance of confidentiality pertaining to the client, hospital records, and all interactions with health-team members.
   2. Compliance with the present administrative policies of _____ (institution).
   3. Promotion of continuing education among the staff by participation in conferences, seminars, or workshops on mutually determined topics.

Approaching the Institution

The next step is directly approaching the agency with one's request. Because the request is for visiting privileges as part of the nursing staff, it might be appropriate to write or call the director of nursing rather than the hospital administrator. This approach serves two purposes: First, it supports one's role as a nurse (not as a physician's assistant, psychologist, and so on); second, it initiates a rapport with a person who can be a very valuable advocate. Before making the written request, it may be wise to make an appointment to meet the director of nursing. This face-to-face meeting not only serves as an introduction but may also dispel or minimize any anxiety the hospital administration may have. Also, if the request has been preceded by any negative events, a personal interview allows the nurse to clarify those events from his or her perspective. Frequently, the administration learns only one aspect of the incident; incomplete understanding by the administration can later become an obstacle to favorable response.

The applicant should present a positive attitude about receiving hospital privileges. The argument "this has never been done before" is weak; it should be countered with the fact that nurse staff privileges are being granted in various institutions throughout the country. In addition, with nurses' expanded roles and their greater visibility, it is almost inevitable that a private practitioner request hospital privileges, especially from institutions dedicated to innovative, high-quality patient care. (Precedents have been set by hospitals regarding the procedure for granting of nurse visiting privileges, as outlined in this chapter and in the literature.[1] An informative booklet, "Guidelines for Appointment of Nurses for Individual Practice Privileges in Health Care Organizations," is available from the American Nurses Association, 2420 Pershing Road, Kansas City, Missouri, 64108.)

Another argument, not always expressed directly but frequently perceived as a reason for denying nurses such privileges, is fear that the nurse in an independent role will create conflicts with other professionals. In the author's opinion, supported by various publications, the basis of this concern is economics.[2,3] For example, during an initial interview the associate administrator asked the author if she expected to establish a full-time psychiatric practice once she received visiting privileges. The author made it very clear that hers was a nursing role. If a family required psychiatric care, she would refer them for specialized assistance. It was obvious that the administrator was concerned about the author's "taking over business" from attending physicians.

Nurses must be ready to answer these concerns with data supporting the need for their services. Obviously, local physicians will not be pleased to see another health-care provider set up a practice and compete for the same limited population that formerly was served by them alone.[2] However, nurses should be able to defend their position by presenting data regarding why clients choose the additional service.[4]

Once the nurse makes a formal request for privileges, the institution usually asks for a professional resume, evidence of current licensure as a registered nurse, and proof of professional liability insurance coverage. The importance of a resume cannot be overemphasized. The resume should be attractive and professionally printed. It should list the usual biographical data plus those activities that enhance one's professional education and experience, such as speeches given, awards received, publications, and professionally related voluntary community work. The resume is an advertisement of the applicant's professional achievements and may constitute the most important evidence used by various credentialing committees to decide whether to grant the requested privileges.

Professional liability insurance coverage is a very significant issue in health institutions. Because lawsuits are an ever-present threat, institutions require maximum insurance coverage. Although one hospital accepted the author's insurance coverage, the other stipulated that liability insurance must have limits of $1,000,000 each claim and $3,000,000 each aggregate. The nurse must be prepared to increase professional liability insurance coverage to comply with the demands of hospitals.

At some point, the institution generally requests personal and professional references. Like the resume, these are an extremely important reflection of one's qualifications. One's references should reflect forethought to their appropriateness as judges of the professional nursing role the nurse will assume, their credibility, and the political atmosphere of the agency. The author chose as references (1) a nurse colleague involved in similar counseling work, (2) the dean of nursing at the university where she was then teaching, (3) parents of a child whom she had worked with for some time, and (4) the hematologist who referred families to her. The last reference was chosen specifically in light of the political nature of the hospitals, where physician input was highly influential. The selection of a family was particularly advantageous because the parents clarified the author's role and emphasized the value of her support in very straightforward language. The author requested that each of these individuals send her a copy of the letter of reference in case some areas could be misinterpreted by those reading it.

The importance of selecting appropriate references and submitting a well-written resume is emphasized clearly by the following experience. In one of the hospitals, the director of nursing discussed with the author each of the hospital members who would review her request. The nursing director expressed concern about the chief of the medical staff, who had a rather skeptical and narrow view of nurses' abilities. The author knew nothing of this physician except his name and area of specialty. Several months after the author's application, this same physician referred to the author an adolescent patient of his who had undergone a hysterectomy for ovarian cancer. The physician explained who he was and how impressed he had been by the author's professional background and letters of recommendation, especially from the counseled family. Ironically, formal granting of privileges by the hospital had not yet been received. Later, he was delighted to inform the author of her acceptance. Still later, when the director of nursing learned of this incident, she agreed that the physician had been favorably impressed and had influenced the committee's decision to grant nurse staff privileges.

Granting of Privileges

From the nurse's point of view one of the difficulties of the review by a hospital is that the nurse cannot be present during the review. When

applicant cannot speak for himself or herself. To minimize this drawback, the nurse should seek an advocate among those involved in the credentialing process. Generally, the best advocate is the director of nursing, who, because of his or her rank, holds more authority than others in nursing administration. However, roles vary with institutions and should be carefully evaluated. For example, a clinical specialist or nurse practitioner who has close knowledge of the applicant's role may be as effective an advocate as the nursing director, provided this specialist is involved in some capacity in the review process.

The advocate's role is to support and clarify the applicant's qualifications and communicate to the applicant the progress of the granting procedure. Because the credentialing process can be quite prolonged, it is helpful to have a periodic update of the status of the proceedings. An application for visiting privileges may not be viewed as a priority by the administration. Unless there is some follow-up procedure, the request may be tabled for an indefinite period of time. To avoid unnecessary delays, the nurse should request a tentative time schedule for ruling on the request. One approach is to request a list of the committees reviewing the application and the dates of their scheduled future meetings.

The review process differed in the two hospitals. In both settings, the author was interviewed by the directors of nursing. In one hospital, the director of nursing advised the author to write the hospital director directly and state her request. Subsequently a meeting was arranged among the three. In the other hospital, the director of nursing advised the author to write directly to her and send a copy to the hospital administrator. The director of nursing and the associate administrator in turn discussed the request and arranged for a joint interview. In both cases, the review was closed after the initial interviews.

The review process of the second hospital was as follows: (1) after the joint meeting and following receipt of requested references and professional credentials, the director of nursing sent a letter to the president of the medical staff regarding the request for visiting privileges; (2) following the medical staff's confirmation of the request, the director of nursing drafted a job description entitled Pediatric Clinical Specialist (Private Practitioner), which was then reviewed by a specially appointed joint committee of representatives from medicine, nursing, and administration; (3) following the joint committee's recommendations regarding the job description, it was revised, reviewed, and finally accepted, and; (4) visiting privileges were formally granted by the hospital administrator.

Unfortunately, the specific procedure followed by the first hospital is not known. During the 20-month review process, a new director of nursing was appointed. She was unwilling to disclose the internal events leading to granting of privileges. However, the following procedure has been established for future applicants: (1) the nurse applies to the director of the nursing service and submits at least four letters of recommendation and also letters of introduction from one member of the professional nursing staff at the hospital, who serves as a sponsor; (2) the director of the nursing service submits the application to the Executive Committee of the Nursing Service, which acts as a credentialing committee; (3) this committee recommends to accept, defer, or reject the request and forwards its findings to the Medical-Surgical Committee (composed of members of the board of governors) through the director of the hospital, and; (4) this committee may interview the applicant and submits recommendations regarding the request to the board of governors, which makes the final decision.

Granting of visiting or staff privileges is not a new concept. Hospitals and other institutions have subjected physicians to review for many years. In most institutions there is a well-established review process. Unlike the review procedure for nurses, the physician review process has the physician apply directly to a credentialing committee composed of medical staff members. Once approved by this committee, the application is sent to the board of governors (trustees), which makes the final decision. In the two hospitals mentioned, nurses had no voting power in either of the medical credentialing committees, yet in granting of nurse visiting privileges, medical approval contributed to the decision to approve or reject. Review of nurses by both nursing and medical staff, however, seems to be the trend.[1] Also, only one hospital required the approval of nurse privileges by the board of governors. In the other institution, only medical staff required approval by the board. Nurse staff privileges were granted by the hospital administration, which was under the authority of the hospital board of trustees.

Because the author's request was unprecedented, neither hospital had a job description befitting her role. Therefore, a major portion of the time spent in reviewing the acceptability of this new role was used to formulate an appropriate job description. Each job description was well detailed and specified the practitioner's qualifications, function, classification of privileges, and the reciprocal responsibilities to the institution. Basically each topic elaborated on those included in the initial letter of request. For this reason, it proved valuable to have included in that letter a clear and succint statement of role, privileges sought, and responsibilities to be assumed.

In one hospital, the director of nursing shared the first draft of the job description with the author prior to submitting it for approval from the appointed committee. This courtesy was appreciated because it eliminated the possibility of misunderstanding or misrepresentation of the request before the committee ruling. Fortunately, this problem did not occur in either situation, but if it had, it would have meant more delay and possibly some disquietude among the review members. It is probably advisable for applicants to request copies of job descriptions, if available, as soon as they apply for visiting privileges. If the description is not acceptable to the applicants, changes can be made more easily during the negotiating stages than when a final decision is pending. If a new job description must be drafted, the applicant should ask to review the first draft. If there is resistance to this request, one can offer the reasoning that a job description is in effect a written contract between an individual and an organization. As such, the job description should be mutually agreeable, protective, and advantageous.

The formal granting of visiting privileges is only one part of the acceptance process. A private nurse practitioner who extends practice to more traditional health-care settings sometimes meets resistance from the nursing staff. The author's experience was fortunate: The staff felt that assessment of the family's adjustment to the child's illness was valuable, necessary information. Reciprocally, the nurses' daily involvement with the family afforded the author insight into intrafamilial relationships that were extremely difficult for the author to analyze in brief, hour-long sessions. Therefore, the two roles complemented each other and enhanced the quality of care offered to the family.

A relationship of mutual respect is not always so easily formed. Staff nurses may view the independent practitioner as an outsider who comes to dictate how they should administer care. The practitioner's greater knowledge of his or her practice area may pose a threat to nurses who do not possess these skills. Family nurse practitioners, pediatric nurse practitioners, and nurse midwives often write orders that must be honored by other nurses and considered by physicians in the patient's overall treatment plan. Therefore, it is essential that a workable professional relationship exist.

The atmosphere of receptivity varies in each institution. Before beginning to practice there, the applicant is wise to gain some insight about the institution. Sometimes it is helfpul to meet with the director of nursing and ask advice on how best to introduce the practitioner to the staff. Other personnel who usually can evaluate the atmosphere of the unit include clinical nurse specialists, head nurse, supervisor, and patients. Patients develop a keen awareness of the personality of the institution. Touring the hospital and spending time in the inpatient/outpatient service is a valuable way of developing opinions about the practice setting and beginning to develop a rapport with the staff.

Once the nurse is granted visiting privileges, a formal introduction to the staff is appropriate. Key nursing staff—such as nursing administrators, clinical specialist, supervisor, head nurse, team leaders, and/or primary nurses—should be invited to meet the practitioner. If the practitioner's practice involves other disciplines, such as social service, medicine, pharmacy, and so on, individuals from these areas should also be invited. Although the meeting is usually arranged by the nursing service, the practitioner should discuss the importance of this event with the director and submit a list of people he or she would like to meet.

An advantageous time to schedule the event is near the afternoon change of shift, when it is convenient to both day and evening staff. A day should be chosen when a full complement of nursing staff is present—not a holiday or vacation period. Sometimes two meetings may be necessary, but spending the time to establish a preliminary relationship based on mutual understanding of each person's role pays great dividends in the long run.

A major purpose of the introduction is to clarify the practitioner's role. Although one's role can be described in general, theoretical terms, it is advisable to include some practical examples of how the practitioner functions. Relevent biographical data might also be included. A case history works very well and can be used effectively to stress the need for collaboration and joint planning of care among the disciplines. The meeting should end with an opportunity for people to ask the practitioner questions.

Summary. These are the guidelines for obtaining visiting privileges:

1. Clearly define the *nursing* role. Be prepared to clarify and support its unique contributions as compared to other related professions.
2. Delineate those visiting privileges necessary to implement one's role and the reciprocal responsibilities one is willing to assume.
3. Approach nursing administration regarding the need for visiting privileges as soon as possible, preferably before actual visiting begins.
   a. Discuss credentialing procedure.
   b. Request tentative schedule for completing the granting procedure.
   c. Request periodic confirmation of progress.
   d. Choose personal references and advocates judiciously; submit well-prepared resume.
   e. Review job description while in development stage.
4. Prior to receiving privileges, obtain preliminary sanction to visit the family, including some method of written communication and form of personal identification for security purposes.
5. If visiting begins prior to granting of privileges, introduce yourself to the nursing staff, especially key members (supervisor, clinical specialist, head nurse, primary nurse, team leader, and so on); explain nursing role, emphasizing need to collaborate and plan care jointly.
6. Upon granting of privileges, plan formal introduction with

attention to invited members, time and date of meeting, clarification of role, and opportunity for discussion.

Advantages of Obtaining Visiting Privileges

Applying for visiting privileges may take a considerable amount of time, particularly if the applicant and agency are embarking on a new project. It is often feasible to practice in an institution before one receives formal visiting privileges. For example, nursing educators often have a working arrangement with an insitution that allows them to see their private patients. However, there are several significant advantages to formal visiting privileges. They facilitate one's own practice through an established and recognized staff position. The scope of practice is clearly defined, setting forth safeguards for both the institution and the nurse. In the event of grievances from either party, there is an approved course of action to be followed. In essence, formal privileges allow for much greater security in one's role than if only a verbal contract exists.

Visiting privileges allow for multidisciplinary communication and coordination of patient and family care. All professionals, but especially those in private practice, sometimes need the advice of another professional who has a special area of expertise.[5] Association with an institution provides for a network of consultants in a variety of disciplines and facilitates the referral of certain patients for specific services. The private nurse practitioner who extends the boundaries of practice becomes a liaison between the family and the physician, hospital, clinic, and providers of care in the home. Families frequently comment that the author's flexibility to practice in a number of settings bridges the gap they experience as care is begun in one facility and terminated in another.

Visiting privileges provide a means of emotional support and continuing education for the nursing staff. They are able to discuss their impressions with another nurse who is knowledgeable about the patient. In turn, the nurse practitioner can validate his or her assessments with the staff. This dialogue often provides an opportunity for mutual support and learning. While the unstructured-practice area provides practitioners with independence, freedom, and flexibility, it also deprives them of reinforcement, feedback, and evaluation from others.[6] Through association with an institution, practitioners are afforded peer review, which enhances their objectivity about and accountability to the client.

Last, and probably most significantly, staff privileges extend the realm of nursing beyond the traditional and stereotyped roles. This new role establishes a precedent for nurse colleagues to follow and facilitates practicing within expanded roles. Already, the issue of nurses practicing independently in rural health-care settings has changed the reimbursement process; this trend will probably continue as nurses increase their visibility and impact upon health care.[7] Recognition of the new role through staff privileges promotes consumer education regarding nursing and the available alternatives to care. It is a process by which nurses can demonstrate their particular area of expertise, gain recognition in their own and other professions, and make an impact upon the health-care delivery system.

Author's Biography

Donna Lee Wong received her BSN degree from Rutgers and her MN from the University of California, Los Angeles. She received her pediatric nurse-practitioner certificate from Seton Hall University, South Orange, New Jersey in 1974. She has held various clinical positions in pediatrics and has also taught at Seton Hall University. She has written extensively in the area of pediatrics. Since 1975 she has maintained a private practice as nurse counselor to families whose child has a life-threatening illness. Currently, she is a pediatric nurse consultant at Hillcrest Medical Center in Tulsa, Oklahoma.

REFERENCES

1. Bergeson P, Melvin N: Granting hospital privileges to nurse practitioners. *Hospitals* **49**:99–101, 1975.
2. Wershing S: Nurse practitioners vs. doctors: A battle is on. *Med Economics* **55**:73, 1978.

3. William MK: Reimbursement: An issue of conflicts. *Ped Nurs* **4**:61–62, 1978.
4. Brown BS: Reimbursement woes. *Ped Nurs* **4**:5, 1978.
5. Kohnke MF, Zimmern A, Greenidge JA: *Independent Nurse Practitioner*, Garden Grove, Calif, Trainex, 1974, p 77.
6. Wong DL: Private practice—at a price. *Nurs Outlook* **25**:258–259, 1977.
7. McAtee P: Rural health care. *Ped Nurs* **4**:14–15, 1978.

FURTHER READINGS

Allen P: A joint practice in the hospital. *AORN* **30**:150–164, 1979.
Kimbro C, Gifford A: The nursing staff organization: A needed development. *Nurs Outlook* **28**:610–616, 1980.
Melvin N: Developing guidelines for clinical privileges for nurse practitioners. *ANA Publication* **(G-135)**:62–67, 1979.
Rowan F: The privileged nurse. *ANA Publication* **(G-135)**:68–73, 1979.

SECTION

II

Clinical Practice

CHAPTER

6

Hanging Out a Shingle: The Joys and Sorrows of Private Practice

Karlene Kerfoot, MA, RN

The step to private practice is one that should be considered at length and taken very seriously. It is now possible for the professional nurse to practice in a variety of settings, and an increasing number of nurses are choosing independent practice. The responsibilities of private practice are enormous, but so are the rewards. This chapter gives a set of guidelines to the practitioner contemplating a private practice and is food for thought for the already-practicing independent nurse. The insights of this chapter are the result of the author's private practice in psychiatric mental-health nursing over several years.

Nurses choose to enter private practice for a variety of reasons. Their motivations can be positive and helpful to initiating independent practice or negative and a hindrance to successful practice. Positive motivations are founded in the practitioner's belief that he or she can, in private practice, deliver a high quality of patient care. The belief in oneself as a competent, concerned clinician who has more to offer the patient than he or she is presently being allowed to deliver can be the source of energy behind a successful private practice. Nurses have entered private practice out of frustration with the disease orientation of the traditional model of institutional health care. Some feel this model does not allow them to offer patients the health teaching and intervention services that are essential to nursing care. If the nurse believes strongly in his or her ability as a clinician and has an equally strong commitment to a new way to deliver nursing services, the energy and motivation to struggle through the initiation and maintenance of a private practice follow easily.

Some nurses choose private practice as a result of their failure in an institutional setting. Many excellent nurses have experienced this kind of failure for a variety of reasons and have gone on to become successful in private-practice settings. Institutions often cannot tolerate practitioners with a sense of direction, purpose, or vision that is incongruent with the institution's goals of patient care. Consequently a parting of the ways is inevitable.

Termination from an institution because of clinical and interpersonal incompetence is a different kind of problem. Nurses who choose private practice after a history of failure run a high risk of further failure. A great deal of maturity and clinical experience are required to evaluate one's practice accurately and to deal with both positive and negative aspects. Nurses who evaluate their clinical competence inappropriately and enter private practice out of anger and for revenge usually fail. The competition in private practice is keen, and only the most competent and energetic nurses will succeed. Nurses who go into private practice are pioneers; they must be the best nurses the profession has to offer.

Defining Private Practice

For many years, nursing has been concerned with self-definition. Nursing has borrowed theoretical frameworks from other disciplines and used them to develop practice; it has also generated theoretical frameworks of its own. In spite of all the theorizing, nursing is still defined in many ways. Probably the most important quality that the practitioner in private practice can have, however, is a sense of what nursing practice is all about—a definition of practice. The definition of practice should guide all decisions the practitioner must make, such as whom to admit into practice, what nursing interventions to choose, and how to evaluate nursing care.

Some nurses have built practices around one theoretical frame of reference. Levine's[1] wholistic nursing, Roy's[2] adaptation model, and Berne's[3] transactional analysis are examples of theoretical frameworks on which private practices have been based. An independent practitioner, however, need not necessarily subscribe to only one theoretical frame of reference. An eclectic model can be useful, but this eclectic model must be well defined. The term *eclectic* can be used to disguise a lack of in-depth knowledge of any theoretical framework. Amorphous theoretical paradigms are not conducive to productive outcomes. It is especially critical for the nurse in private practice to evaluate the soundness of the theory upon which he or she bases nursing care. If a theoretical model is chosen that cannot be supported by research, then the nurse runs the risk of jeopardizing the health of clients. For example, a practitioner might choose stress as a theoretical model upon which to base nursing care. The nurse then adopts the model that stress is the result of inadequate diet and builds a practice around that principle. This model is fine as long as the practitioner's reccommendations are based on sound research and not on some vague feeling or belief that some types of food have curative properties. Professionals have in the past based their advice to clients on inadequate and poorly researched assumptions. This practice is dangerous in the current legal climates. A theoretical framework should have the potential for sustaining scientific inquiry.

A theoretical framework limits, guides, and defines practice and provides guidelines from which to make clinical decisions. For example, the practitioner who adopts a systems theory as a framework must take into account many more people and factors than just the patient. The practitioner, therefore, admits into practice people who understand this concept and support the practitioner's involvement with the family, if necessary. If the client cannot agree with this approach, then a contract is not possible, and the client is not admitted to the practice.

Theoretical frameworks can be based on single theoretical models or be synthesized from the practitioner's educational and clinical experiences. The adoption of a framework adds consistency, clarity, and rationality to one's clinical practice. This framework is valuable when one explains one's practice to clients and professionals. The practitioner who has a good sense of what his or her clinical practice is about is likely to project a sense of confidence to clients, colleagues, and self.

Preparing for Private Practice

An important issue for the practitioner contemplating private practice is the educational and clinical preparation and qualifications necessary for the task. Some nurses with only a minimum of formal preparation and clinical experience succeed in private practices. However, they are the exception. A nurse cannot be overprepared for private practice; in order to succeed, one should be at least as well prepared as the competition. The nurse in private practice frequently competes with physicians, psychologists, and other professionals. In other situations, the nurse might work collaboratively with these professionals. The practitioner's skills and preparations should make possible both competition and collaboration.

Extensive preparation will protect the practitioner in several ways. Prior clinical experience protects the practitioner professionally. If a clinician has an extensive clinical background, he or she should make more-intelligent decisions and work more effectively with clients.

Extensive preparation also protects the clinician from damaging malpractice suits. Private practitioners making difficult clinical decisions should be able to show these decisions were taken from a background of sound academic and clinical preparation. The better the practitioner's preparation and ability to apply academic and clinical knowledge to practice, the safer it will be for the clinician to practice privately.

The two most common forms of private practice are solo and group practice. Both forms of practice result in a high degree of professional isolation. Few neophyte clinicians are prepared for the professional isolation of solo practice. While in training, nurses are socialized to work in groups. Hospital nursing provides many peer experiences that allow nurses to consult and get feedback in unfamiliar situations. Community-health nursing also provides nurses a reference group with which to discuss patient problems. Whenever one sees nurses practicing, one usually sees nurses and other health-team members working in groups. In solo practice, one works alone. When a problem with a patient arises, one must rely on one's own resources to determine the solution. There is no built-in feedback mechanism in one's own practice. Most nurses work under supervisors and with peers who offer suggestions about areas needing improvement. In private practice, no one monitors the nurse's practice but one's self. The opportunity to learn from the feedback of others is absent. The experienced, clinically competent practitioner can exist successfully in this environment. The neophyte clinician probably cannot.

Solo practice obviously affords the highest degree of professional isolation, but group practice also can be an isolating experience. In a busy practice, a partner does not have the time to respond adequately to the clinician who is feeling the effects of isolation. And a business partner is not necessarily the best supervisor and teacher. The nurse must often look elsewhere to have specific needs met.

In addition to the usual undergraduate and graduate preparation, there are several ways to prepare oneself clinically for private practice. The first way is to get experience in a health-care institution that offers the opportunity of primary nursing. The second way is to enter group or partnership private practice before solo private practice.

The nurse who works in an institution that structures the delivery of nursing services within the framework of primary nursing learns the diagnostic and nursing skills needed in private practice. Primary nursing is a system in which one nurse is accountable for a caseload of several patients, a work environment similar to private practice. Primary nursing, as opposed to team nursing, prepares nurses for the responsibility of 24-hour patient care. Theoretically, primary nursing can teach many of the organizational skills that are necessary to plan individualized care for a client. These skills can be easily transferred to private practice. Many health-care facilities, such as hospitals, public-health-nursing agencies, and mental health centers, are now instituting primary nursing. If a clinician can prove he or she is a clinically competent primary nurse and feels good about the responsibility, the step to private practice is an easier one.

Inexperienced practitioners may be wise to consider group or partnership practices, as opposed to solo practices, as a first step to independence. In group or partnership practice, the clinician has the advantage of working with colleagues who can provide some feedback and consultation about patients when needed. Also, in group practice, mundane aspects of running a business are shared. Many experienced practitioners prefer group or partnership practice to solo practice. In addition to the isolation of solo practice, it is difficult if not impossible to plan time away from the practice without covering for evenings, weekends, and vacations. Group practitioners can eliminate this problem with a system of off-hours coverage.

Visibility and Credibility

One cause of failure in many private practices is insufficient lead time. The person considering opening a practice should allow a lead time of six months to one year. The practitioner needs this time to develop his or her visibility, credibility, and referral sources in the community.

Many excellent practitioners have made the mistake of thinking that clinical skills are all that is necessary to open a successful private practice. Clinical skills are very important. However, visibility and exposure in the community are also important to the success of a private practice. Just as a politician spends a great deal of time and money on publicity, so the practitioner initiating a private practice needs to consider methods of making his or her name familiar to people in the community. Voters often choose a politician because they know one name better than another; clients often choose clinicians this same way.

Becoming visible among professionals is one way to gain exposure. Professionals are excellent referral sources. A nurse considering private practice should become known in nursing groups. Exposure comes by joining nursing organizations and by initiating contact with nurses who work in institutions. Visibility among other health professionals is also important. Professionals are usually active in service organizations, such as mental-retardation organizations or the Heart Association. A nurse can join these organizations or offer services as an unpaid or paid consultant, as a board or committee member, or as a workshop leader.

Clinicians frequently overlook the importance of becoming visible among lay groups. Word-of-mouth referrals are among the best referrals. Nurses can join lay groups, especially those appropriate to the nurses' practices. For example, nurses who work with mastectomy patients can join Reach for Recovery. If the nurse works with diabetics, the Diabetic Association is an excellent patient group to join. Psychiatric nurses can join mental-health associations. In addition to becoming members of groups, nurses can volunteer to serve on boards of these organizations. Board membership by nurses is not only an avenue to visibility, it also gives nurses the opportunity to educate the public about health practices. Also, organizations often welcome volunteer speakers and consultants; nurses can fill these roles.

Media presentations increase visibility in the community. Most television and radio stations have talk shows that frequently treat health issues. The nurse who participates in these programs enhances his or her visibility. The practitioner can write opinions to the station, call the moderator to air opinions, and extend an offer to be interviewed on a particular topic of interest to the general audience. For example, a practitioner working with cancer patients offered to participate in a program during National Cancer Week. Her appearance consequently precipitated many calls from potential clients. Newspapers, especially those in smaller towns, welcome articles written by health-care practitioners on aspects of health or illness. In addition to articles written by the practitioner, newspapers often print feature stories on nurses in the still-unusual role of private practitioner. A practitioner conducting a workshop can solicit newspaper exposure.

Another way to become known in the community is face-to-face discussion of your private practice with potential sources of referral in the community. Physical therapists, ministers, school psychologists, physicians, and other resource people might be interested in your practice. The nurse should approach these discussions as an opportunity to learn about the other person's practice as well as to give information about his or her own. Professionally printed brochures describing the new practice are effective promotion pieces to bring to these discussions.

Many practitioners report that an advertisement in the Yellow Pages often is effective. Clients may chose a practitioner by scanning a list of available professionals in the telephone book. The advertisement should appear professional—not showy—yet include enough information to convey to the reader a sense of the practitioner's practice. Potential clients are often familiar with other disciplines but are not aware of the ways a nurse can help them. Subtle, understated professional advertisement can attract clients.

A successful private practice is no accident. Success is the product of long-term planning, and the key to success is developing visibility within the community. Becoming visible in the community, however, is not enough. It is also important to attain a reputation as a credible practitioner.

Credibility can be developed in several ways. Education is the first and probably most common way of attaining credibility in our society. In the past, the public has assumed that the practitioner with the most degrees is the most credible, although this attitude might be changing. In some communities, clinicians seem to need an MS or PhD degree to compete with other practitioners for clients. In other communities, the degrees seem less necessary.

Certification is as important as educational credentials. Of course, the nurse should seek certification by a nursing organization. The American Nurses Association has worked long and hard and effectively to establish a reputable certification process for various types of nursing practice. In some situations, it is desirable to seek other certification. In highly specialized practices, such as sex therapy or marital therapy, nurses can seek further, more specialized, certification in order to compete with other clinicians and to obtain information in highly specialized areas. The trend toward more—and more-specialized—certification certainly will continue. Certification probably will be necessary to receive third-party payments. The nurse considering private practice would be wise to become credentialed as soon as possible in nursing, and as necessary in other areas that are related to his or her private practice.

In addition to education and certification, practitioners need to be viewed as experts in the field. Positions and titles are one way of obtaining this credibility as an expert. For example, a professorship at a university or positions of clinical specialist or supervisor in a respectable hospital imply expertise. The reputation as an expert will facilitate the move to private practice.

Another advantage of working in an institution is that one develops contacts with a group of potential clients in many health-care situations. For example, a nurse who works in an outpatient diabetic clinic becomes known to diabetic patients and their families. When the nurse leaves the outpatient clinic for private practice, many patients will follow him or her into private practice. Even if they do not become clients, they can recommend the nurse as a credible practitioner.

Clinicians occasionally move to new communities. Such clinicians face a great challenge, but the method remains the same: consider the potential sources of referral, become visible among these groups, establish credibility among professional and lay groups, and become well known in the community. Until they become better known, many nurses at first work part time in institutional settings and part time as private practitioners. Eventually, they are able to become full-time private practitioners.

The nurse considering private practice should anticipate that not everyone will support the decision to become an independent practitioner. Some nurses meet a great deal of hostility from other nurses. If one encounters this hostility, it is perhaps natural—but unproductive—to brood about the lack of support from one's peers. A better approach is to encourage and support the people who are in turn supportive. As one's practice and reputation grow, people who once denied support may have a change of heart. It is not unusual for innovators to be criticized. As nurse colleagues become more familiar with the model of private practice, their support will grow.

Other health professionals have attacked the right of nurses to enter private practice. The basis for this attack is usually a well-disguised fear of competition for a limited number of clients. It is no coincidence that these attacks usually occur in areas where health professionals are heavily concentrated. These attacks take many forms: Nurses have been accused of practicing medicine without a license, breaking of zoning laws, and not complying with the state's nurse-practice acts. The wise clinician will anticipate attacks and prepare answers to likely charges. The practitioner who has defined his or her practice carefully is at an advantage when responding to attacks.

Specific Skills for Private Practice

Before beginning practice, private practitioners should assess what skills they will need. The skills a practitioner will need depend to some degree on the type of client he or she anticipates. A practitioner who treats diabetic clients will need different skills than a practitioner whose clients have marriage problems. However, both hypothetical practitioners share a common need for certain physical and psychological assessment skills. As mentioned before, it is a mistake to separate the physical and psychological aspects of client care. The clinician in private practice must be able to assess all phases of the client's life. This ability to assess is crucial both to the client's welfare and the legal safety of the clinician. Each new patient admitted to the practice should be given a thorough diagnostic assessment interview on the first encounter. Before intervening, the practitioner must be absolutely sure about the nature of the client's problem.

At one time it was taboo for nurses to *diagnose;* a more neutral term, *assessment,* was used. Regardless of the terminology, the practitioner must possess the assessment skills to differentiate between physical and mental problems. This ability is especially critical in private practice because there is no one around to correct diagnostic errors. For example, a psychiatric mental-health practitioner admitted to her practice a client who complained of depression. She failed to do any diagnostic screening and automatically assumed the patient's problems were psychological. After six months of therapy, the patient failed to respond and on his own obtained a medical examination. The patient was diagnosed as diabetic; the depression and mood swings were the result of the disease. A thorough assessment would have enabled the nurse to relate the symptoms to untreated diabetes. This practitioner essentially wasted the patient's time and money by not diagnosing the problem earlier. This patient did not take legal recourse; however, he made the clinician very uncomfortable until a compromise was negotiated. Most lay people assume that nurses are experts in physical as well as mental problems, and they should be. It makes no sense to treat physical or mental problems without considering the possibility of an interplay between the two.

If a nurse considering private practice is unprepared to do a thorough assessment, further training should be undertaken. Many schools of nursing are now offering excellent workshops and short courses on nursing assessment or diagnosis. The nurse might take specialized workshops that deal only with areas such as mental status exams or some particular aspect of physical assessment. If these workshops are not available, many continuing-education departments will respond to requests by clinicians and offer requested courses.

The most reliable and valid type of patient assessment is structured, standardized assessment that integrates physical, emotional, and social aspects. The clinician need not conduct the assessment interview mechanically, checklist in hand. The nurse can use a variety of communication techniques to elicit the needed information in a style that is conversational and acceptable to the client. Some note taking is mandatory, since most people cannot remember all of the information, but notes can be taken unobtrusively.

A structured assessment format has several advantages: It ensures that the nurse has elicited all the information needed for a thorough assessment. If, however, the interview is conducted haphazardly, the information elicited probably will be disorganized and inadequate. The structured format serves to keep standardized information readily available in a client's record. This format facilitates compiling statistics from the practice. Also, clients usually respond favorably to a thorough assessment interview. Thoroughness serves to increase the client's confidence in the clinician as a professional.

Many schools of nursing have developed health-assessment tools for students to use. These can be adapted easily to the needs of a private practice. The focus of a practice will determine what format will be most useful for the initial assessment. A synthesis of several types of health-assessment tools might be needed.

Specific rating scales are available for specialized areas of practice. The fields of psychiatry and psychology have many rating scales designed to screen for specific problems. For example, some private practitioners in mental health request every new client to fill out a Minnesota Multi-Phasic Personality Inventory (MMPI). The Beck Depression Inventory is a self-administered questionnaire that assesses the client's level of depression, and the Hamilton Depression Scale is a clinician-administered scale that also indicates the level of depression. Scales can be used in a variety of ways in practice: For example, the two scales rating depression can be used initially to assess the level of depression and potential for suicide and can be used later to monitor progress and assess therapy. Goal-attainment scaling has been developed as another way of objectifying the outcome of therapy. The therapist and patient mutually decide particular attainable goals, and periodic assessments are made that indicate movement toward these goals. Goal-attainment scaling is easily adapted to clinical practices other than mental health and can be a useful way to document progress. There are rating scales available for other areas of a specialty practice, such as family therapy and work with children. The Denver Developmental Screening Tool has been widely used by nurses. Other similar rating scales are now available.

The clinician should scour the literature to uncover specific rating scales that might be incorporated into the practice. Rating scales are objective, facilitate patient involvement in therapy, and offer the clinician a wealth of data to evaluate.

Some nurse practitioners require clients to undergo physical examinations before admitting them to the practice. Some nurses perform these physical exams themselves, whereas others require that the physical be obtained elsewhere. Many physical problems can masquerade as mental-health problems and vice versa. It is essential that the nurse identify the problem accurately before initiating treatment. Not only does the client benefit, the nurse is on much sounder legal footing if he or she can document that a thorough assessment is an integral part of the service.

Even nurses who choose not to work in the area of illness and instead choose a practice of health teaching or prevention must constantly keep alert to the possibility of an abnormal process. For example, the bulk of the author's private practice is giving sex therapy to women without partners and to couples. Although sex therapy is considered a practice with an essentially healthy clientele, many people cannot be admitted for sex therapy because some other problem is causing the sexual dysfunction. By far the largest number of clients who cannot be admitted for sex therapy have unrecognized depression. Physical problems, such as diabetes, can be the cause of the sexual problem. If a practitioner in this field cannot effectively screen out inappropriate clients from the practice, the sex therapy would probably fail. The practitioner could be held liable for a misdiagnosis, and the patient would be exploited financially. In any type of practice, the importance of assessing the physical and mental functioning of the client cannot be underestimated.

In addition to recognizing the symptoms of physical and medical illnesses, nurses need to be alert to the potential effects on clients of prescribed drugs, over-the-counter medications, and street drugs. Part of the assessment interview should be a thorough drug history, including drug education included when indicated. Nurses also need to know when to refer patients for drug therapy. Part of knowing when to refer is (1) being able to diagnose a condition and (2) knowing when drugs can help. To neglect to refer a patient for medication out of ignorance is a serious omission.

Unfortunately, some people in private practice have allowed themselves to become specialized at the expense of not keeping up to date on developments in general nursing care. The good practitioner maintains a sound foundation of general nursing and superimposes a specialty upon it.

Selection, Transfer, and Termination of Clients

Because they need clients, nurses starting out in private practice are often tempted to admit any client who applies for services. This decision can lead to problems. The nurse needs to establish beforehand what cases will and will not be accepted into the practice. The nurse should not be swayed from this decision by a need for more patients. It is much easier to explain to the client initially that one is not equipped to handle the particular problem than it is to terminate the patient after several sessions.

One decision a nurse must make is whether to treat clients on medication. If a medicated client is accepted, the nurse, of course, must become an expert on how the medication works and what its side effects are. In order to discuss the side effects and therapeutic effects of the medication, the nurse needs to develop a good working relationship with whoever prescribes medications. The responsibility for clients on medication is an issue that the nurse will have to settle before accepting these clients in the practice. Some nurses choose to not admit clients who are on medication. In the author's view, this decision is unfortunate because nurses have essential services to offer medicated as well as nonmedicated clients. Although there is an extra responsibility involved in treating medicated clients, many nurses are able to develop excellent collaborative relationships with physicians. Nurse and doctor seek each other's advice about the medication and illness. This trend is very encouraging.

A second decision the nurse must make is whether to treat severely ill patients. Some nurses limit their practice strictly to health teaching and health maintenance and therefore do not admit clients with severe physical and emotional problems. If one chooses to work with more impaired clients, then certain support systems must be built into the practice, such as a 24-hour answering service and a system of backup coverage if the clinician is not available.

Some practitioners state initially that they will exclude seriously ill patients from the practice. That is, the suicidal client would not be admitted to a psychiatric/mental-health practice, nor would the client with poorly controlled diabetes be admitted to a health-teaching practice. However, in spite of effective screening procedures, there is no way to make certain that the client will remain in a steady state of health. Many clients experience a variable course of illness and health. The clinician must be prepared to handle clients who become worse and are suddenly beyond the scope of his or her practice. At the point the client becomes worse, the nurse has two choices: The nurse can keep the patient and seek outside consultation or transfer the patient to another clinician with more expertise and resources to treat this particular condition. Neither solution is as simple as it sounds. For example, a suicidal client can interpret transfer as rejection. The clinician who chooses to keep the patient, however, might not be able to supply services the patient needs, such as 24-hour coverage and backup. Abandoning a patient without referral is a serious breach of professional ethics—a breach for which the nurse can be held legally liable.

The easiest way to facilitate transfer of a client in the middle of therapy is to ensure that potential termination is addressed at the very beginning of the relationship. The first session with the client should structure the relationship. The client needs to know from the very beginning what the limits of practice are. The client needs to know, for example, if the nurse is available 24 hours a day, accepts phone calls after hours, or makes house calls in crises. Clients also need to know what the nurse will and will not treat. Clients feel much more comfortable if they know in advance the structure of the relationship. Many clinicians state at the beginning of a relationship that they will follow a patient as long as a particular condition is stable and that some conditions are beyond their scopes of practice. For example, the clinician can indicate that severe depressions or particular medical conditions are beyond the clinician's ability to treat and still protect the patient's best interests. If the client understands the contract from the beginning, there is less likelihood of angry feelings if the client is transferred or terminated.

Within the course of the relationship, clients occasionally view the clinician as omnipotent and able to handle any situation. For example, chronic illness often brings on marital difficulties. The nurse who treats the chronic illness may be asked for marital therapy also. Most nurses understand some counseling techniques and can help families with first-level interactional problems. However, it is imperative that the nurse be able to diagnose the severity of these peripheral problems and make appropriate referrals if the situation is beyond his or her level of expertise.

Nurses also need to discover their own limits and value systems. A nurse's ability to give the best care might be prejudiced by attitudes held toward the client. For example, some nurses are very uncomfortable with certain lifestyles or sexual practices, such as community living or homosexuality. The nurse should recognize this bias and not give care if there is a possibility his or her work will suffer as a result. It is impossible for clinicians to work well with all types of people. Clinicians need to understand their limitations and do the work they do best. Much harm has been done to clients' self-esteem by clinicians who project their own value system onto their patients. Ideally, clinicians are accepting of peoples' differences, but reality does not cleave to the ideal.

Maintaining a Practice

Professional Matters

Visibility. Continued visibility is essential for the successful continuation of private practice. Satisfied clients are the best source of referrals. However, to ensure continued success, the practitioner needs to maintain the referral sources already developed in the community. It takes a great deal of time and energy to work with clients and then spend one's free time in community organizations. Some community work can be discontinued, but it is essential to remain fairly active in the community.

Continuing Education. Private practitioners can become isolated and out of touch with their professional peers. This isolation can lead to professional stagnation. Private practice imposes a hectic time schedule. Potentially this schedule can prohibit contacts with peers and colleagues, making it difficult to stay abreast of new professional developments. Continuing-education workshops are no longer viewed as glamorous trips to exotic parts of the country. Workshops are now seen as financial losses because the practitioner must abandon practice for the duration; lose that income; and, in addition, pay for transportation, tuition, and living expenses. Because workshops are so expensive, some private practitioners meet only the minimum continuing-education requirements. Over a period of years, their practices suffer because the practitioners do not keep abreast of the latest innovations in their discipline. The interchange with peers and colleagues at conventions and workshops is probably as important as their content. Private practitioners must balance the inconvenience and expense of workshops against their benefits and select workshops judiciously.

Referrals. A good way to maintain good relations in the professional community is to follow up on referrals. If a client is referred to a nurse by another professional, it is courteous to give feedback about the client's progress to the person who first referred the client.

Many clinicians complain that they receive no feedback after they refer a client to another clinician. The referring clinician usually appreciates a letter describing the assessment visit and future therapeutic plans. The patient must give permission, however, for this exchange of information. Release-of-information forms should be used.

It is important to state explicitly the purpose of the referral—consultation or continued treatment—at the very beginning. A clinician should not assume that a patient has been referred for treatment; sometimes the referring clinician wants only a consultation. If the purpose of the referral is not clear, the nurse should write the referring clinician.

Fees. Many nurses starting out in private practice are reluctant to discuss fees with clients. Because they usually work for others, nurses are not accustomed to considering the economics of practice. The private practitioner must overcome this reluctance. The best time to discuss fees is the first meeting. Misunderstandings can be avoided if the nurse and client can agree on a financial contract as well as a therapeutic contract. Clients may be embarrassed to discuss fees. If so, the nurse should discuss rates, methods of payment, and so on.

Satisfied clients usually pay. Dissatisfied clients may express anger through late or voided payments. The nurse should pay special attention to patients who are not happy with services. Most lawsuits occur, not just because the client is dissatisfied, but also because the clinician does not acknowledge the dissatisfaction. Turning over delinquent accounts to a bill collector might further anger already alienated patients. It is much more effective for the clinician to handle this problem personally and strive for a satisfactory compromise. The clinician who is unwilling to negotiate invites expensive, time-consuming litigation.

Missed Appointments. Missed appointments tax the practitioner's ability to maintain a caseload. Clients miss appointments for a variety of reasons. The majority of clients miss because they forget.

Sometimes clients feel guilty about missing appointments and hesitate to call for another one. Other clients miss appointments because they are dissatisfied with the service. In either case, a phone call or letter to the client is a good idea. This letter or call assures the client that the clinician is interested and is concerned about the missed appointment. The client is reassured and flattered that the clinician has taken the time to write or call. If there is an indication that the client is dissatisfied, the letter should elicit feedback from the patient. Some client/clinician relationships cannot be salvaged. However, if the client and clinician can discuss their differences openly, a real understanding can take place, which can lead either to continued treatment or to referral. Practitioners who methodically follow up on missed appointments tend to keep patients because misunderstandings are solved.

Innovations. After the practice is established, the clinician can consider other methods of delivering services. Patients are beginning to show interest in many forms of health-care services, partly because of the publicity new modes of treatment have received. For example, self-help groups have engaged the public interest because they promise eventual independence. Practitioners can offer many kinds of self-help groups—teaching groups, problem-solving groups, groups with families, and so on. An advantage of such groups is that they allow the clinician to see many patients in a short period of time and at a low cost to the patient. Groups are just one example of new ways private practitioners can deliver services. There are many others.

Seasonal Variations. Variability of income can be a problem for private practitioners. Even successful practitioners experience peaks in demand for services. For example, mental-health practitioners can predict that need for their services will be low in the summer and will peak in the spring and fall. Some medical conditions also display seasonal patterns. Clinicians who work with children and families may be busier in summer than in winter. Slack periods are excellent times to attend workshops and seminars, catch up on reading, and start new forms of treatment—such as parent teaching groups and therapeutic children's play groups.

Business Matters

The successful practitioner must also have a good head for business—that is, pay attention to detail, have management skills, and understand investments. Good client care accounts for only half of a practitioner's success. A practice is also a business; it must be managed well. Nurses who take courses in accounting and in small-business management have valuable skills. A nurse who does not have these skills should seek advice from someone who does. Starting a business is no simple matter. The business needs as much attention as the patients. A well-managed practice inspires confidence in the practitioner. The best patient care may go unrecognized in an inefficiently run practice. Therefore attention to good business practice is essential.

Employees and Colleagues. The practitioner must be aware of the effect on patients of others associated with the practice, such as receptionists, partners, and operators of answering services. It is not enough to employ an answering service. Operators need clear instructions about how to treat patients who call. A gruff answer from a receptionist or a partner can alienate clients and harm the practice. For these reasons, the ability to manage people is essential to the practitioner.

Records. Well-maintained clinical records are an important, but often neglected, aspect of private practice. Everything of significance should be recorded. Even though it might never happen, the nurse should assume he or she will have to defend every professional decision in court. Clinical records should reflect the nurse's awareness of that eventuality. In many states, patients have the legal right to their records. Nothing should appear in the record that is opinion or that belittles or slanders the client, family members, or other professionals. One should keep the record as if the patient will look at it or as if it will be subpoenaed in a court of law.

Careful record keeping is not merely a legal precaution; it is an aid to practice. Beginning practitioners might not see the necessity of extensive record keeping. When the caseload becomes large and after several years of practice, the necessity becomes apparent. A careful, written record is especially important when the practitioner terminates private practice or chooses to transfer clients to another practitioner. It makes much more sense to keep good records from the beginning rather than to face the embarrassment of inadequate records during litigation or at the time a patient is transferred.

The nurse can choose from among several systems of record keeping. The advantage of any system is that it gives structure to the details the nurse records. Without such structure, a nurse might forget to record important information. Whatever system is chosen, the best time to record information is as soon as possible after the nurse sees the patient because one's memory for detail is greatest then. Nurses who have trouble handwriting notes should invest in a dictaphone and pay the fee for having the notes typed; the convenience is well worth the price.

Supervision. Practitioners can choose to contract for ongoing supervision from another professional. This supervision can be mutual peer review; this arrangement of mutual supervision costs nothing. The clinicial might prefer to pay a fixed fee to have another nurse supervise. Some practitioners chose to contract for the services of several professionals in addition to nurses. Supervision can be requested from mental-health professionals, from clinical specialists working in hospitals, and from other professionals.

Probably it is best for the practitioner to contract for supervision. Peer supervision from a partner in the practice is rarely sufficiently objective to ensure high-quality supervision. As the practitioner gains experience, he or she can offer supervisory services to other clinicians. Unfortunately, many clinicians in private practice ignore the need for supervision and consequently do not continue to grow in their practice. Ongoing supervision is an excellent insurance policy against the professional isolation of private practice.

Difficulties and Rewards

Private practice can be a very glamorous and rewarding experience. Some nurses respond very well to the challenge; others may find private practice oppressive. Those who can overcome the difficulties will find limitless opportunities for developing creative approaches to patient care. The experience of being responsible and accountable directly to clients can be very exciting.

An important problem private clinicians face is the inability to get away completely from the responsibility of client care. Nurses who work in institutions know that someone is there to take over their duties when they leave. Although group practice can diffuse responsibility by allowing a rotating call schedule, private practitioners have far greater responsibility. The concern about a client who is responding poorly can be very weighty at times. This concern is present even during vacations and leisure activities. A clinician must respond to this responsibility, but must also develop a system to relieve the pressure of work. A distinct separation between work and leisure is important to avoid "burn out" in private practice.

Private practice is a new field for nurses. The clinicians who are striking out on their own are a remarkable group of people willing to take risks, face criticism, and risk failure because they believe that a new delivery system will allow them to give better patient care. Because the field is new, there are many unanswered questions, but there is also the exciting opportunity to develop new answers. Private practice is never boring—it may be exhilarating or maddening, sad or joyful—but always rewarding to those who meet its challenges.

Author's Biography

Karlene Kerfoot has maintained a part-time psychiatric/mental-health private practice since 1973 in Iowa City, Iowa. Her practice has consisted mainly of clients with sexual dysfunction problems but also includes clients with problems of depression and other mental-health problems.

She is currently a doctoral candidate in nursing and commutes to the College of Nursing at the Medical Center of the University of Illinois in Chicago, Illinois. In addition to various staff and supervisory positions in psychiatric nursing, she has taught at the University of Iowa and has been a chief of the Aftercare and Community Services Division of a mental-health center in Cedar Rapids, Iowa. She received her BSN in 1965 and her MA in 1970 from the University of Iowa. Her research interests are involved with the psychobiology of depression in women.

REFERENCES

1. Levine M: Wholistic nursing. *Nurs Clinics N Am* **6**:253–264, 1971.
2. Roy C: *Introduction to Nursing: An Adaptation Model.* Englewood, NJ, Prentice-Hall, 1976.
3. Berne E: *Transactional Analysis in Psychotherapy.* New York, Grove Press, 1961.

CHAPTER
7

A Joint Psychiatric Practice

Margaret T. Campbell, RN, MSN,
Owen E. Clark, MD

The West Seattle Psychiatric Service is a joint venture by a psychiatrist and two psychiatric nurses who, in a general psychiatric practice, treat severely disturbed adults in a community setting. The professional climate of this collaboration cannot be likened to other circumstances of practice—a nurse practicing alone, a physician employing nurses, or individuals sharing office space. We would like to describe our practice arrangement—how it got started and how it is working out—and its pleasures and problems, both expected and unexpected. Because we know some of the possibilities and limitations of this kind of practice, we offer our reflections on the prerequisites for such a collaboration.

The idea for our joint practice first surfaced in the summer of 1975. A half-dozen people (doctors, nurses, psychologists, and social workers) were involved in the initial deliberations. Casual mention of the idea sparked interest in many persons whose interest slackened with the need for commitment and work. The group narrowed to three, who agreed to a six-month trial beginning that fall. On January 1, 1976, the operation was formalized as a private nonprofit corporation, and we have continued to practice with the same personnel and format.

The practice evolved in three distinct phases. The first phase ran from July to October 1975. The planning, exploration of feasibility and possible directions, and the opening of the office took place then. The second phase occurred during the next three months. By this time we knew we had a potentially successful enterprise and could envision a pleasurable working relationship. The third phase was the period from January 1976, when we made the commitment and officially incorporated the organization, to the present time.

Background

Earlier in 1975 we were working together in a community mental-health center that was cutting back its program and staff severely (not an infrequent happening in such agencies). It was a time of turmoil and many rumors. It was then that the possibility of starting an independent collaborative practice was first mentioned. When talk became action, the ranks of the interested dwindled quickly from ten people to three. The number of practitioners was a real concern. Too few might mean insufficient diversity and too great a work load; too many might not be supported by the projected client load.

A year of working together had given us an awareness of each other's theoretical biases and ways of approaching work. We had practiced in several settings and knew the usual bureaucratic pitfalls. Although working together in our own clinic seemed highly desirable, giving the fantasy shape alternately overwhelmed us, excited us, and made us anxious.

Our first consideration was feasibility. The most crucial factors considered were anticipated focus and design of the association, promotion and start-up, and projection of income and expenses.

Initial Planning

We spent several hours talking about our previous work, our philosophies of mental health and mental illness, our feelings about mentally ill people, our thoughts about treatment, and our thoughts about what services we could offer realistically. In retrospect, it is clear we were building the foundation of a mutually agreeable partnership.

Previously, we had experienced the confusion of "role-blurring," with its inherent lack of professional accountability. We were eager to practice medicine and nursing respectively, unfettered by an organization's expectations that we perform in other professional roles. Organizations had defined for us rules, tasks, and functions in which we were not expert. We wanted to expand *service* to clients without confusing them or frustrating ourselves.

As psychiatric/mental-health nurses moved into private offices, their practices appeared, to us at least, to duplicate the roles of other mental-health professionals, rather than to present a discrete alternative. The solo practice of a psychiatrist often precludes treatment of the seriously ill. One person alone is unable to provide the optimal social support and to be as available as needed. We believe the treatment of the serious emotional illnesses requires an approach quite different from that of the traditional private psychiatrist or that of most "medication/evaluation meetings" in public centers. We agreed that our foremost purpose was to establish a viable alternative to existing mental-health services, but this alternative had to allow the fulfillment of our professional aspirations.

We enjoy working with people labeled "seriously disturbed." This patient group needs mental-health services most critically but is least able to get these services. We believe these people have major psychophysiological disturbances (that is, "mental illness"). Furthermore, we believe doctors and nurses working together are best prepared to confront the complex issues of the serious emotional illnesses.

These patients have recurrent episodes with frequent and expensive hospitalizations. Often they cannot get quality, continuing care. Even when quality and continuity are available, they are often not available from the same person or groups of persons in one fixed location. Patients are offered a "program," not the help of people. Clinics often open and close with indifference to the clients' need for consistency.

Like patients with any chronic illness, the mentally ill, even at their best, do not function as well as healthy people. The goal of our treatment is to restore patients' level of functioning to the pre-illness level and, after that, to maintain that level. When patients become worse, our goal is to forestall the process and stabilize the condition as quickly as possible. Treatment outside the hospital is usually most desirable. However, at times we lack the support system to accomplish this goal. At those times, hospitalization, either full or part time, is indicated. When we planned the practice, we decided to assume responsibility for hospitalizing patients.

Our conceptual model emerged as predominantly biological (pejoratively referred to as the "medical model"[1]). We have no quarrel with the idea that schizophrenia, manic depression, and psychotic depression are psychophysiological problems with concurrent interpersonal components. They can be treated psychophysiologically or interpersonally. The evidence points to a combination of both approaches as the most effective treatment.[2] Although chemotherapy is vital, the therapeutic relationship with the patient is the cement that binds the curative process. If we were to offer "comprehensive" service, we would have to provide expensive and varied programs (and thereby become another mini-mental-health center). We decided to do what we knew how to do, do it well, and utilize *existing* community services, rather than attempt duplication.

As these thoughts and biases crystallized, our philosophy was affirmed. The philosophy has not altered significantly during three years of partnership. This philosophy was the foundation for our planning. We committed it to paper, a process that clarified our thinking and gave our planning direction. The final document was a three page prospectus. This prospectus later proved an invaluable introduction, explanation, and springboard for discussion.*

Defining Focus and Scope

The intent of our association was to have mixed disciplines provide services in one setting. Emphasis was upon matching the skills of the personnel to the perceived psychiatric needs of the community. The target population selected were persons with severe mental illness, those with psychotic and potentially psychotic conditions. We planned to provide diagnostic evaluations, regulation of psychotropic medication, and psychotherapy of a supportive nature. We planned outpatient services primarily, but we were willing to treat patients in hospitals whenever necessary and allowed. We were willing to make home visits and would arrange for partial or full hospitalization. We did not foresee extensive exploratory therapy, forensic psychiatry, or treatment of children or adolescents. We preferred to avoid treating people with sociopathic, developmental, or addiction disorders, although we decided to deal with these problems when they were coincidental to a primary illness. We arrived at these limitations to practice after candid appraisal of our clinical interests and skills. These primary objectives emerged: having readily available services and providing continuity of care directed at the presenting problems. After we agreed on the focus of our work, we began to consider community supports and promotion. At this time, we also made a conditional commitment to each other to operate for

---

*See Appendix I on page 193.

six months and reevaluate at the end of that time whether we could or should continue.

After the philosophizing came the time for practical decisions. We decided to (1) locate in a geographic area with which we were familiar and where we had patient and professional ties, (2) locate an office with reasonable rent and accessibility to public transportation, (3) rent equipment and modest office furniture to avoid large cash outlays, and (4) hire a secretary who had experience with mental-health clinics. These decisions were crucial to our success.

The name of our group was selected: The West Seattle Psychiatric Services. "West Seattle" designated the community we wished to be associated with; the "Psychiatric" denoted the focus of our practice.

Setting the Stage

The next step was the office location. We approached the administrator of a small general hospital, a person with whom two of us had worked before. The hospital was in the process of expanding services to the community and was most interested in psychiatric services. They still hoped to reopen their small psychiatric unit, but saw an outpatient practice as a highly desirable alternative. A small frame house adjacent to the hospital was offered at a most modest rental, which included heat and janitorial services. With great delight we accepted the offer. After some cleaning and painting, it became a pleasant, attractive office with a large reception (living) room, two interviewing rooms, and a kitchen.

Furnishings were brought from our homes or rented ($31.59 a month). We rented comfortable chairs, small sofas, table, and lamps for the interviewing and waiting rooms. Renting office furniture was more expensive; we supplied a desk, a locking file cabinet, and an FM radio. We also rented an office typewriter for $17 per month.

At that time we decided to create a cash reserve of about $1,500, to which we contributed equally. This money was to be the capital outlay with which to open the office—that is, pay the rent deposit and advance, install the phones, buy some paint, get the necessary office and paper supplies, rent the furniture, and so on. In addition, we acknowledged our willingness to receive no salaries for three to six months. We agreed to allocate initial profits to operating expenses.

We were fortunate to hire a secretary who had worked with us all in the mental-health center. She was familiar with the billing procedure to insurance companies and public assistance, was a skilled bookkeeper, and knew how to handle stressful situations with patients. She took over her job expertly.

We ordered a small supply of inexpensive stationery and business and appointment cards. (We noted that prices varied greatly among vendors.) An appointment book and office supplies were bought at a small office-supply house.

The clinical record is by design brief. None of us believes in keeping voluminous records. We use a modified "problem-oriented" record. It lists the client's problem, our observations and thoughts about that problem, and our treatment plan. The record is purposely concise and without detail. The record is color coded. The four sheets are: information form (pink); history and overall treatment plan (blue); ongoing treatment report and supplemental plans (green); billing information (orange). These sheets were bought from a local printer for $12 per 500. A simple manila folder holds the record together.

Telephones are important but expensive equipment. Our system has two lines and two phones. The second line rings automatically if the first is in use. Both phones are in the reception area, but one has a long cord to provide privacy. We give the number for only the first line to the public, which keeps the second line usually free for outgoing calls. This simple system has proved to be most adequate for us. The psychiatrist already had an answering service; that number was given for after-hours calls. If clients call after hours (there have been remarkably few after-hours calls), their calls are returned as soon as possible by the primary therapist.

A guiding principle at the beginning was to keep expenses to a minimum while giving the sense of an established, dependable operation. In any business it is important to keep the consumer, the service, and the cost congruent. Ours is a warm, pleasant office, yet the emphasis is on service, not on trappings.

Community Liaison

Beginning and expanding depends on delivering services and maintaining relationships with people who refer patients or are potential sources of assistance. Because the community itself is the referral source, we spent three or four weeks contacting potential referral sources. It's very important to identify the community agencies that are apt to refer patients. Even more important is making personal contact whenever possible with the people who are apt to make referrals. We telephoned various agencies and found out who might make referrals; then we asked for appointments to discuss our plans with them. We met with people at the state hospital, the local and nearby community-health centers, the in-patient units of local hospitals, the local department of public assistance, the Family and Childrens' Service, and the two community-health nursing agencies.

We used the county medical-society directory and made brief visits to physicians who had their offices in our general geographic area. We introduced ourselves, described our plans, and left business cards. One of the community-health nursing agencies gave us some staff-meeting time to discuss our plans. The public assistance office in our area asked us to participate in an in-service training course, which we did without fee. We also visited a newly established home nursing agency, where we had the opportunity to discuss and clarify our common goals. This agency also provides services to the severely disabled in the community, particularly those being relocated from the state hospital to group homes in the city.

We realized that we would need to know the community resources available for our clients. Severely disturbed adults are frequently ostracized and experience problems finding suitable living arrangements and employment. We were already aware of many community agencies serving this population. Starting with those agencies, we made personal visits or telephone calls to tell them about our plans and to learn about their facilities, their services, and their admission criteria. (Whenever possible we designate one individual in that agency as our contact; a single contact facilitates referral immensely.) As we talked with people, we learned of other programs and contacted them. We started a file of individuals and agencies; as programs open, close, or merge, we update the file.

In preparation for opening the practice, we read the state nursing and medicine practice acts. We also talked with other professionals about our plans, people such as other nurses in private practice, the Washington State Joint Practice Commission, and our respective professional organizations. Although we sought direction, guidelines, and assistance, we were given instead expressions of interest and encouragement. In retrospect, we realize it is inappropriate to look for direction and guidelines from these sources. We were assured that the Nurse Practice Act in the state of Washington would permit the practice we intended.* (The most concrete and beneficial advice about how to set up the business came from relatives who were in business.)

We also contacted the public, official mental-health agencies. Although we would not be under their jurisdiction in any way, it seemed wise to inform them of our intent. These agencies included the county mental-health board, the local and state offices of the Department of Social and Health Services, and Region X of the then Department of Health, Education and Welfare.

Our sources of referral and the cooperative relationships we enjoy with many agencies to this day makes us believe this time in preparation was wisely spent.† Of particular note is our collaboration with the mental-health center, with which we work closely to plan treatment for several patients we have in common.

---

*Since that time, the Nurse Practice Act has been expanded. The State Board of Nursing has revised it to include the "Certified Registered Nurse." In the case of psychiatric/mental-health nursing, certification by the American Nurses Association is required for state of Washington certification. Rules and regulations are now written for prescriptive authority for duly certified registered nurses. Just how these rules will affect our practice is uncertain, but we will begin the certifying procedure this fall.

†As of 1979, 30% of the patients were referred by community physicians, 13% by public assistance counselors, and 8% by the local mental-health clinic.

We are ever mindful that personal relationships are the bond between us and other health/social/medical services. A frequent complaint registered by community care givers is that communication with a mental-health clinician ceases after referral is made. We stay in contact with our referral source through written reports and phone consultations.

The Treatment Program

At the initial interview, patients are apprised of the group nature of our clinic. It is not unusual for two of us to participate in that interview. We explain that we share information and give the reason why. We believe forthrightness wins the patients' cooperation and participation in planning their treatment program. After the initial interview, the majority of patients are seen by one therapist. This assignment is made by mutual consent and depends on current caseload, open appointment times, and the therapist's inclination to work with particular problems or people. Sometimes a patient begins with one therapist on a temporary basis and then switches to another.

The therapeutic relationship is developed and sustained by using supportive and problem-solving techniques.[3] Except in times of personal crisis when the client is seen more frequently, weekly, biweekly, or monthly appointments are the norm. Some clients are in contact only once or twice a year but they are nevertheless technically in treatment and perceive the therapist as "my therapist."

Keeping in mind that the patient is part of a family or social system that may be supportive of our therapy goals, we wanted to be available to these other people, work together with them, and enlist their assistance with the treatment planned. Family conferences and joint sessions are expected and encouraged. Management of medication during an acute phase usually requires monitoring by a third party. Careful monitoring can mean the success or failure of an attempt to prevent hospitalization. Making alliances with the positively supportive persons in the patient's environment is imperative. We offer families of patients counseling and teaching as well as encouragement, support, and reassurance.

Client teaching and informed consent are an important component in our treatment planning. We believe people should know what we think the problem is, how we would prefer to handle it, and what the treatment entails. We try to enlist their cooperation at the outset. If that cooperation is minimal or lacking, we adjust our goals and change the treatment strategy accordingly. Explanations are given about the choice of medicines, their effects, side-effects, and interactions with other drugs.

As any mental-health professional knows, determining the client's aspirations for treatment is most essential. Negotiation skills come into play when the client's expectations, or purpose for seeking help, are quite different from our range of skill or stated purpose. Often we have to set aside our goal-directedness and focus on the interpersonal relationship. Other times we cannot establish a dynamic relationship but can offer medication and consistent, long-term contact. Flexibility is a hallmark of our approach to our work.

An example might be a psychotic person who says, "The only thing I need is a job." Allaying the psychotic symptoms is necessary before the client enters the competitive labor market. We negotiate our goals and time frame, treat the thinking or mood disorder, and then refer to vocational-training or job-placement agencies. In this case, designing a complex therapy program (introspective analysis of psychological conflicts or family restructuring) is contraindicated.

Financial Considerations

At the outset we agreed that no one held the goal of becoming wealthy through private practice. We did not envision making a profit beyond reasonable salaries. Earning a reasonable income while pursuing professionally worthwhile and pleasant work was our financial goal.

We moved from a projected to a concrete budget as we learned what to expect. The most productive approach to projection was to estimate how many patients could be seen monthly and how many weekly. We then determined what fees could be charged and how much would probably be collected. Balanced against that figure were fixed expenses—such as rent, secretarial salary, telephone, initial supplies, and professional-liability insurance—and estimated ongoing overhead expenses, such as postage, stationery, and syringes for fluphenazine injections. After we estimated both costs and incomes, we could begin to project salaries.

Our fees for service were set at the same level as those comparable professionals and clinics in this area—$40 per hour for the psychiatrist and $25 per hour for the nurses. (These have since been raised to $50 and $30 respectively.) However, these figures are simply guidelines from which we negotiate downward. Ninety percent of our clients pay less.

There are several methods by which clients pay for services. We have used direct payment at the time of the appointment; deferred direct payment over time; direct billing to the insurance company with payment coming directly to us; consignment of the insurance payment by the patient to us; Medicaid; and Medicare. As one goes through that list, the billing procedure becomes more complex. There is a direct ratio between the complexity of the payment process and the chances of being paid in full for an hour of professional service. There is also a direct ratio between the complexity of the process and the time lag before payment is received. The rate at which Medicaid pays for service is a set amount, and it is fairly simple to estimate that income. Medicare, however, is a complex system that changes frequently. The ratio of Medicare patients to other patients is low in our practice. Consequently, it has not been a determining factor in our budget.

In late 1975, we hired an accountant, who discussed with us the merits of cash versus accrual accounting systems. The cash system was adopted because, with this system, tax is paid on collected fees, not on earned fees. In other words, our business and operations tax is paid on collected income, not on service that has been billed but not yet paid for. Likewise, our individual income tax is due on salary paid, not on salary earned but not yet paid. A considerable amount of money is always in arrears (our paychecks are often late). Also, we have no inventory; consequently, the cash accounting system is the best for us. The tax laws have a significant role here—once the system (cash or accrual) is set up, it cannot be changed without Internal Revenue Service approval. The accountant set up our bookkeeping system, taught our secretary how to do the payroll and quarterly taxes, and audits our books at year's end.

Working hours were tentative at the outset. As the client load increased, we have stabilized our hours: the doctor works 20 hours per week (4 hours, 5 days a week); one nurse works 20 hours per week (2½ days); and the other nurse works 27 hours each week (3½ days). These arrangements allow us to work part time yet give clients full-time coverage. Our salaries are based on hours worked. In early 1976 we worked more hours than we were paid for. As the patient load increased we were able to increase the paid hours so that our salaries are now a true reflection of hours worked.

Determining the hourly rate was perhaps our most difficult interpersonal decision. The decision meant estimating the dollar worth of one staff member relative to another. It is easy enough to talk of education, experience, and relative contributions in general terms. When exact dollar figures are assigned, the truth emerges. Education and experience notwithstanding, the questions really are: Is one person risking more than the others? Is one more vital to the operation than the others?

As a personal note, each of us tended to underestimate our contributions; perhaps this attitude was part of the glue that melded our practice. After we reached a decision, we deliberated again to make sure there were no second thoughts. We have not changed the ratio of reimbursement, which fact speaks well for our decision.

Nonprofit Status

The first few patients were seen about September 1. Again, timing was in our favor. The mental-health center was without a psychiatrist for an interim period, and several former patients were referred to us. They in turn spread the word. The emergency room and crisis team referred people for us to evaluate. The public-assistance case workers also referred clients. That first month we saw 38 patients, many of whom are still with us.

On October 1, 1976 we held an open house. In mid-September, we mailed announcements about our practice and included invitations to an open house.* Announcements were sent to all the agency representatives we had spoken with earlier, the physicians in the area, mental-health professionals we knew, personal friends, and patients. It was a sunny fall day; the festive occasion gave us an opportunity to thank people for their encouragement and support. That day marked the end of "shall we do it?" and the beginning of "what do we have here?"

What we had, when all was said and done, was essentially two registered nurses working in a psychiatrist's office. "Ethics" issues have been raised in psychiatric circles about fee splitting, and we wanted to avoid any suggestion of fee splitting. (The fees that were coming in were used to meet expenses, particularly the secretary's salary. None of us earned anything for four months, September through December, 1975). The patients were coming in; we could see the potential that was there. Still, we did not want an employer/employee relationship between doctor and nurses. What kind of business relationship would allow the service we had in mind yet be professionally equitable to us all? After two or three months, the nonprofit corporation emerged as the solution.

We wanted an organization that would allow the greatest flexibility in achieving our goals. We considered a partnership practice between the two nurses, but business acquaintances uniformly and unequivocably discouraged a partnership. (Business partnerships have been aptly compared to marriage—very simple to get into but extremely difficult to get out of.) Another possibility was to become solo practitioners who shared office space and office expenses. By contracting with the psychiatrist for medical consultation, the two nurses could have worked out a method whereby psychotropic medications could have been prescribed. However, in this structure accountability and responsibility are diffused and the patients may

---

*See Appendix II on page 196.

be confused about roles. For this reason, we decided against shared solo practices. A professional service corporation could not be considered because the principals of such a corporation must have the same professional status.

We spent some time in the public library learning about business organizations and charitable foundations. In order to apply for and accept funding from charitable sources, the tax laws require not-for-profit status. We considered support by foundations as one way to augment our income. Because we did not envision making a profit beyond payment for our salaries and because as board members we would have no interest in financial gain, the nonprofit corporation appeared to be the best alternative for us. A corporate attorney was retained. After reading the prospectus and discussing it with us, he clarified the legalities and drew up the articles and bylaws of our corporation.

Corporation bylaws require a statement of purpose. We wanted a statement sufficiently general not to be challenged by the Internal Revenue Service, yet specific enough to define what we wanted to do.

The articles of incorporation state that the purposes of the West Seattle Psychiatric Services are to (1) provide services to emotionally distressed adults, (2) be a resource for the training of professionals, and (3) study the effectiveness a service of this kind can have in this community. All of these purposes are addressed primarily to the severely disturbed population.

Some of the important features of a nonprofit organization should be mentioned here. We are all salaried staff of the organization. As such, our remuneration has to be within the community's prevailing range for comparable positions in similar agencies. This point is critical because the Internal Revenue Service is suspicious of any charitable organization that pays exorbitant salaries. Any furnishings and equipment purchased by the corporation must be, at its dissolution, donated to a charitable agency. (In view of this ruling our earlier decision to rent these articles was affirmed.) The most positive feature of the nonprofit organization is its ability to accept grants from private foundations and charitable donations from individuals. In two or three weeks the attorney had completed his work, and the corporation was registered at the state capitol.

We then had to request nonprofit status from the Internal Revenue Service. This recognition as a nonprofit organization is necessary before applying for funding from charitable organizations. This process, however, is one that for us, at least, took 13 months and involved many pieces of correspondence. The IRS granted their approval in February 1977, retroactive to January 1976. Though the bureaucratic machinery runs slowly, the IRS representatives were patient with our questions and explained the procedures adequately. We suspect our low level of sophistication with tax forms was detrimental to speedy resolution.

At the beginning of December, the hospital learned that the building we occupied was to be removed for fire safety reasons. By this time, the hospital administration had decided that the former inpatient unit would not be reopened and offered a portion of it to us as office space. We had two months to alert our patients. The move was made quickly, without incident.

With careful decorating we have been able to disguise for the most part the "hospital" appearance of the quarters. We now have four interviewing rooms, all good sized, a locked storage room for records and supplies, and an open clerical-reception area.

How It All Worked Out

After two and one-half years, we can report separately on each of the three purposes stated in the bylaws. The first—service for emotionally disturbed adults—is the primary concern. At first, the patient load was small, with a core of about 25 people whom we knew from the past. In the two and one-half years since then, 347 patients have been in treatment. (For statistical purposes, we consider any patient seen two or more times as having been "in treatment.") One hundred and fifty-six, or 45% had psychotic diagnoses according to the *Diagnostic Standard Manual II*. The duration of therapy varied from two sessions, to several visits for several weeks, to monthly visits over the two and one-half years. Fifty-two percent of the 347 had the onset of symptoms more than three years prior to seeing us. Forty-seven percent of them had also been hospitalized for psychiatric reasons at some time in their past. Nearly one-third had a positive family history of psychiatric illness. Fifty-seven percent were prescribed neuroleptics, including lithium carbonate but excluding antianxiety medication. Eighteen patients had been hospitalized during the course of treament, two of them on two occasions. At any given time we have had an average of 110 patients in active treatment. Sixty-one percent of those terminated from treatment were referred to other social/vocational/mental-health agencies.

To evaluate the severity of the illness we use indicators of daily functioning, such as psychiatric condition, social adjustments, marital stability, and economic productivity. Using these criteria, we can describe at least 65% of our caseload as having severe difficulty psychologically, socially, interpersonally and/or vocationally. These figures confirm that our service is accomplishing what it set out to do. We believe it vitally important that a new venture establish for itself what it will and cannot do, announce that to the community, and then consistently do it.

The second stated purpose of the corporation is to be a resource for the training of professionals. Before the practice began, one author had a clinical faculty appointment, which has been continued, at the University of Washington School of Nursing (without salary). Over the three years several nursing students have utilized the practice to meet various course objectives: eight basic generic and RN baccalaureate students taking a psychosocial-clinical course, a graduate psychosocial-nursing student in systems-oriented community mental-health nursing, two students gathering data for their theses, two who worked with a family in therapy, and two who conducted a group for the management of stress. The relationship with the faculty has always been one of congenial support, interest, and encouragement.

The third purpose of the corporation is to study the effectiveness a service such as ours can have in the community. One way to test others' opinion of effectiveness is to apply for grants. In the spring of 1977 we submitted proposals to three charitable foundations in Seattle and were turned down. There are several explanations for that.

Our success seemed apparent after one and one-half years. We wanted funding to conduct demonstration projects. We requested additional funds to conduct home visits and family therapy with the seriously disturbed. Third-party payers do not, almost without exception, allow payment for psychotherapy conducted in the home or for psychotherapy for persons other than the identified patient. Yet these two forms of intervention are highly desirable if one intends to (1) be available to extremely emotionally distressed people (which means going to them when necessary) and (2) intervene in the social system of these people (which means involving people with whom the patient has intense relationships). These two services are also expensive. Home visits involve travel time, which at an hourly rate of $30 to $50, is generally prohibitive for all but the affluent. Family therapy is most effective in 90-minute sessions rather than 50- or 60-minute sessions. If the family is dysfunctional, the combined talents of two therapists may be required. Additionally, the most usual time to conduct family sessions is in the evening, when the family is most apt to be free but when the therapists are at the end of the work day. All of these factors increase the cost of this service, which cannot be met by the patient. It was for these reasons that we approached the foundations.

Writing grant proposals is an art, as is selecting the foundation most likely to accept one's proposals. We had much to learn. Also, there seem to be vogues in grant proposals. In 1977 in Seattle, foundations were interested mostly in vocational rehabilitation, sheltered workshops, and job-training programs.

A graduate student provided great assistance in the proposal writing by gathering data about our program. Without that help we could have mustered neither the enthusiasm nor the time to write the lengthy proposals. Our energy had been nearly expended searching out information about the many existing foundations—who they are, what they are or are not interested in funding, and what their procedure is for reviewing applications.

A second way to study the effectiveness of a service is to conduct a cost analysis, which we have not seriously undertaken. Superficially, at least, it appears that by keeping the overhead expenses to a minimum and maintaining a comfortably full caseload, quality service can be provided at a most reasonable cost. We intend to examine this thought more closely.

General Discussion and Considerations

Complementarity. A joint practice presupposes individuals who complement each other's clinical skills and personalities. Collaborators need not be alike; in fact, extensive similarity in skills and personal attributes might detract from effective collaboration. Rather, complementarity resides in a felicitous blend of personal and professional styles. Mutual respect for each other as professionals and people gives the group a flexibility it can capitalize on. Responsibility for one's actions is always retained, but tasks can be shared and authority delegated.

An advantage of joint practice is that it has none of the isolation of solo practice. In our case, the strain of caring for this difficult group of patients is greatly reduced by sharing. At the same time, a truly generalized practice is possible since we are able to work with patients through the various stages of their illnesses, from the acute to the rehabilitative. We have the time and capability to see the patients individually, with family members, or in groups. Time is available for collaborating with other agencies on the patient's behalf (and with their approval and participation). This collaboration is a predominant aspect of our treatment plans. Vacations and time away are easier to arrange and are less disruptive to the therapeutic process because our caseloads are not exclusive.

Another advantage is readily available consultation. If a plan of treatment needs immediate institution or alteration, it can be accomplished with ease. Often we need validation and clarification of the treatment plan and encouragement that the progress (or lack of it) is a reasonable expectation.

Supervision. Clinical supervision is achieved informally when we discuss therapy during the course of the work day and formally when the three of us meet once a week. This time is set aside to discuss clinical and business issues and to raise problems and concerns to the group.

The nurse author has also contracted with a psychosocial faculty member for clinical consultation on a regular, ongoing basis. At those times, the more intense subjective reactions to the nurse-patient relationship are discussed in greater depth. These consultations also provide the opportunity to discuss current psychosocial nursing literature.

Respective Roles. Notwithstanding all that has been said about joint practice, there are two primary reasons why the psychiatrist is the sine qua non of our practice. They are (1) economic realities and (2) medical-management issues. The economic realities will be addressed first.

The patients upon whom we have focused our attention are the seriously mentally ill. They are often only marginally employed and unable to finance their treatment. More often than not they are on public assistance. The regulations are unbending in regard to payment for medical services. There is a set amount that is paid to any psychiatrist for one hour of therapy per month. Because the two nurses work "under the direction and supervision of the physician" they can be paid for time spent with these patients. This is the only feasible way that a fee-for-service arrangement is possible. (A possibility would have been to work on a contractual basis with the Department of Social and Health Services, but that relationship would have meant bureaucratic constraints we purposely wanted to avoid.) Without the association of a psychiatrist, a psychiatric nurse could not support herself or himself if the caseload were primarily of indigent patients.

If mental-health services are covered in a particular insurance policy, the private insurance carriers in the state of Washington will reimburse policy holders for services rendered by a registered nurse. Blue Cross and Blue Shield, however, are exempted from that ruling. Many employers in the Seattle area use these two particular programs. Thus, if the seriously emotionally ill person is employed, the chances are very good that the insurer will be Blue Cross or Blue Shield. Again, the services of the registered nurse are not reimburseable in the area of mental health.

Second, the psychiatrist is essential because of medical management issues. These are essentially (1) hospital-admission privileges and (2) authority to prescribe medicines. A few hospitals are beginning to establish policies under which registered nurses can admit patients for treatment. At the time when we were designing our practice, these events were only trends. Because we were more interested in getting to work than in being innovators, we opted for traditional roles. The psychiatrist writes the initial prescription and has the sole responsibility for hospitalization.

The nurses can and do on occasion follow a patient during hospitalization, but it is the exception rather than the rule. The reason lies with time and travel constraints. In a busy schedule it is often unfeasible for the nurses to find time to drive to and from the hospital. The psychiatrist already has an office practice adjacent to one of the hospitals and a part-time hospital practice. Consequently, the psychiatrist usually hospitalizes patients, providing continuity of care. The fees for hospital service are billed separately from our group-practice fees. Upon discharge, the patient frequently returns to the clinic for continuing therapy and follow-up care.

It would be feasible in some cases to have the entire team continue the treatment program during hospitalization. Our particular inclinations and the geography of the city dictated our decision at this time. A great deal of energy is required to obtain admitting privileges—making applications, attending meetings, and so forth. We have not seen any real need, not had any strong desire, to expend that energy. However, we will watch the activity closely and be prepared to move in new directions.

Prescribing medicines traditionally has been the prerogative of the doctor, but the practice of nursing is being defined in various new and exciting ways. As mentioned before, the state of Washington has expanded its definition of nursing practice to include a "certified registered nurse" with the authority to prescribe medicine. Considering the fact that 59% of our patients are getting medication, there has been remarkably little problem in this area. (We believe that neuroleptics should be used, when indicated, in effective dosages for as long as necessary. After patients stabilize, the dosage should be as low as possible yet still remain effective).

As time goes on we will of necessity adjust our role and functions to reflect the current statutes. For now we are content to work as we are.

We have said we adhere to the medical model. One negative aspect of the medical model as it usually exists is an authoritarian attitude—perhaps of doctors toward nurses, professionals towards other workers, or staff toward patients. Maintaining the balance between democracy and autocracy and between directing and enabling is seldom easy. Again, in our previous work we had experienced irresponsibility and lack of accountability camouflaged as "role blurring." As nurses and physicians we are all accountable for our therapy decisions, but the reality is that the physician has the greater responsibility under the law. Thus, when a difficult problem arises, the final decision must be in keeping with the medical standards of the community. This fact does not mean that there are no psychiatric-nursing standards or that they are in controversy with medical standards. Still, when difficult decisions have to be made (for instance, involuntary admission), we find the "customary standards of medical practice" to be the best rule of thumb. Our primary purpose is to provide service, not to challenge tradition and statutes.

We reach decisions about the office and its operation by consensus. When we cannot agree on problematic issues, we agree to the solution of the "leader." At various times, over various issues, each of us has assumed the role of leader. Each will accept responsibility for tasks according to his or her felt competence in certain areas. When no one is particularly competent to do a task, we are willing to learn.

(An example is keeping account books.) We take formal votes very infrequently—when we elected the officers of our corporation and as dictated by parliamentary procedure during our annual meetings (required by our bylaws.)

The medical model can be problematical when the nurse or doctor "directs patients" or "enables the healing process." We try to approach clients as if we were in a teaching situation, assuming the clients know more about themselves than we do but want to learn more. Our position is that we possess skills that might help clients learn more about themselves. We convey that we do not have a secret, magic potion or procedure. What we have is time, training, concern, hope, and a willingness to participate in a therapeutic process as long as the relationship is mutually beneficial.

Acceptance. A realistic appraisal of acceptance by the local medical community should be given. Their reluctance to recognize new roles for nurses is a definite difficulty. It is rare for a client to complain to the referring physician about not seeing the psychiatrist enough; most clients express satisfaction with their therapeutic relationship. It is not so rare for referring physicians to demand direct (and exclusive) contact with the psychiatrist. Public agencies, however, will consult freely with the nurses.

Physicians are a significant referral source (30% of our caseload). We share clients with many physicians in the community. By and large doctors are conservative, suspicious of the role of the nurse, and reluctant to deal with nurses. There are a few overt rebuffs, many inquiries about the role of the nurses, and little genuinely enthusiastic encouragement. The burden is on the nurses to prove to each doctor that psychiatric nurses produce results as good or better than that obtainable by someone seen by a psychiatrist.

In all cases, patients referred by a physician are evaluated by the psychiatrist, who gives the referring doctor feedback. The psychiatrist always indicates when, and in what role, a nurse will be involved. Subsequently the nurse may deal directly with the physician, especially if the referring physician is accepting of the role of the nurse.

We are always straightforward about our mode of operation and spell out who has what responsibilities. We attempt as far as possible to have the staff person most involved with the client correspond with outside agencies. Practically, this policy cannot always be put into effect. We test each physician's tolerance for direct contact with the nurse, draw back when we encounter emotional resistance, but encourage the professional relationship as far as the physician will allow.

We believe nurses will find acceptance in fact several years before they are accepted in theory. There is increasing acceptance by the more flexible and often younger physicians. We anticipate that the process will take years to decades; older physicians with rigid role expectations may retire rather than change their opinions.

Finances. Finances were and remain a concern. The population we are treating is by definition disenfranchised—usually economically and often socially. More and more insurance companies will reimburse outpatient psychiatric treatment, but payment is often predicated on a recent hospitalization. The coverage seldom extends to nursing therapy. Public assistance has a modest payment schedule and is only for a physician's services.

Fees are arrived at after open discussion of our sliding scale. The fees range from $5 to $50, depending upon the patient's ability to pay. The average collected fee per session is $22. We encourage patients to pay on a cash basis and not incur large balances. Through vigilance and active pursuit of accounts receivable by our secretary, we have brought our uncollectables down to 5% from an approximate 10%.

We are now treating as many patients as we can comfortably handle. To be able to increase our caseload, we would probably have to seek a new kind of client—children or alcoholics, perhaps—and abandon our specialization in severe emotional disturbances. Another way to expand the caseload might be to focus on other forms of therapy, such as marriage counseling or stress management. A third approach would be to refer patients to other services—services further away than those we now recommend. At some point over the next year we will have to address objectively the issue of where to go from here, if indeed we want to go anywhere. The rising cost of living makes it imperative to expand our receivables or lower our expenses (although our expenses are already near the bone) if our incomes are to keep up with the increased cost of living.

Diversity. In independent practice, especially at the beginning, there is a tendency (and need) to put one's total effort into the enterprise. There is much to do: learning about business, building a caseload, developing referral sources, and acquainting oneself with a whole range of things, such as bookkeeping and tax forms. The ultimate purpose is the joy of working directly with patients. It is ironic that, even with so much to do and learn, the private practitioner often sacrifices professional diversity for the sake of the practice.

When one works in agencies, one often hears others say how rewarding an independent practice would be. Without agency constraints, one would be able to write, teach, read, be a volunteer consultant—be a "whole" professional. What happens instead is that, to make a comparable income, one immerses oneself in clinical work that consumes most of one's time and energy. (Perhaps the stringency of practice is the reason so many professionals have half-time practices and work half of their time in agencies. They have secure income, the joy of doing one's own thing, and built-in diversity.) The private practitioner has to make time for professional diversity (just as for personal life) if he or she is to keep alert and informed. However, getting time away for professional meetings, conferences, and faculty activities is difficult for the practitioner with commitments to patients. Also, the expense of such activities can be prohibitive, especially since vacations and holidays are "without pay."

Being totally accountable for one's work (no agency policies or governing body to use as excuses for not going the extra step) is comforting, exhilarating, anxiety provoking—and tiring. The novice independent practitioner must work harder than ever before!

Solitude. Private practitioners often cannot escape self absorption and loneliness. Although there are three people in our practice, there are still many occasions when we function quite apart from each other. The separation from one's peers is at times a relief, just as going on a well-earned vacation away from friends and responsibilities is fun and refreshing. If the isolation lasts too long, however, it quickly evolves into separation anxiety. This phenomenon is well known among physicians. Some traditional ways of dealing with it are staff meetings, grand rounds, professional society meetings for continuing education, lunch gatherings, and so on. In Seattle, in 1975, there were no comparable activities for psychosocial nurses in independent practice.

We began to hear of other psychosocial nurses who were in independent practice. We made contact, and 5 of us met for dinner in the fall of 1976. Since then the group has grown to about 12; we have decided to formalize our meetings. In the fall of 1978 the group began to meet monthly for three hours for the purpose of satisfying some of our felt need for collegiality, mutual support, and continuing education. There has been an expression of interest for using this group as a forum for peer review.

Summary

The West Seattle Psychiatric Services is a nonprofit corporation formed by a psychiatrist and two psychiatric nurses. Its primary purpose is to treat, on an outpatient basis, people who are severely disturbed. The treatment approach used for this particular population is flexibility, accessibility, and long-term problem-solving offered on an individualized basis.

This organizational design is one in which members of two professional disciplines can work together as colleagues. Through collaboration their mutual goal of enhanced health service is attained, while cost to the patient is held down. The professionals' self-direction and personal accountability are maintained.

This type of joint practice can be an adjunct to the traditional public and private mental-health services. In some instances it can be an alternative to them.

Authors' Biographies

Margaret Thomas Campbell is a graduate of Hartwick College, Oneonta, New York, and Boston University. Before joining the West Seattle Psychiatric Services with Dr. Clark, she was program coordinator of the inpatient unit at Highline–West Seattle Mental Health Center in Seattle, Washington. She was director of nursing at the Capital District Psychiatric Center in Albany, New York. She has been on the faculty of Syracuse University School of Nursing and The Home Treatment Service of Boston State Hospital. Currently she is in private practice and serves as an auxiliary clinical instructor at the University of Washington School of Nursing.

Owen E. Clark is a graduate of Harvard College in History and Science. He attended medical school at the University of Washington, Seattle, and did postgraduate work in psychiatry at Yale University. He has been director of a military mental-hygiene clinic and associate clinical director of a community mental-health center. He now has a part-time private practice in Seattle.

REFERENCES

1. Thomas M: The medical model versus psychiatric nursing. *J NY State Nurses Assoc* **2**:30–38, 1971.
2. May P: *Treatment of Schizophrenia.* New York, Science House, 1968.
3. Lamb HR: *Community Survival for Long-Term Patients.* San Francisco, Jossey-Bass, 1976.

FURTHER READINGS

Bellack L, Loeb L: *The Schizophrenic Syndrome.* New York, Grune & Stratton, 1969.
DeYoung C, Tower M: *Out of Uniform and into Trouble.* St. Louis, C V Mosby, 1971.
Greenblatt M, Solomon MH, Evans AS, et al (eds): *Drug and Social Therapy in Chronic Schizophrenia.* Springfield, Ill, C C Thomas, 1965.
Klein DF, and Davis, JM: *Diagnosis and Drug Treatment of Psychiatric Disorders.* Baltimore, Williams & Wilkins, 1969.

CHAPTER
8

A Multidisciplinary Group Practice

Keville Frederickson, RN, EDD

Nursing journals often feature articles that attempt to define and somehow limit the role of nurses. Many times these articles elicit a flurry of letters charging that such definitions are counterproductive, restrictive, time consuming, and in general not necessary.

Such definitions, however, have their uses. For instance, the psychiatric nurse who is a private practitioner in an interdisciplinary group needs such a definition. Nurses who collaborate with practitioners from other disciplines and share a common clinical base must internalize and project a strong professional identity if they are to practice independently yet avoid isolation. This attitude is particularly important in the psychiatric/mental-health nurse because many of the techniques and skills for the practice of psychotherapy are common to many professions. The unique contribution that can be made by nursing to the mental-health field must be emphasized.

Initiating Practice

The preceding two paragraphs are part of a philosophy that developed over four years as a result of my experiences in a private interdisciplinary psychotherapy and counseling practice. I became involved with private practice after I completed a master's in psychiatric/mental-health nursing education and a doctorate in nursing education. Polishing my skills in psychiatric/mental health was part of my postdoctoral plan; however, I was unsure of how to begin a private practice. This problem solved itself during a workshop I presented to professionals in the community. A psychologist approached me with the idea of an interdisciplinary group practice. At the time, the partially formed group consisted of two psychologists, one of whom was also a licensed marriage counselor and a social worker. The emphasis was on individual adult and child therapy, group therapy, and marriage and family counseling.

At the same time that I was approached, another candidate, who would eventually join us, was being considered. He was a licensed marriage counselor with a background in psychology.

Shortly after our initial meeting at the workshop, I met with the psychologist at the group office to discuss the details of the practice. I found the proposal attractive not only because it offered me a chance to begin practice but also because I welcomed the opportunity to interact with professionals from other disciplines in an atmosphere of mutual respect.

Licenses

During our interview, the first issues we explored were the problems of licensure, legitimacy, and professional jealousy. The state of New Jersey seems prone to inter- and intradisciplinary professional struggles. For example, marriage counselors are licensed by the state after they have practiced psychotherapy for five years (of which at least two must be counseling under the supervision of a licensed marriage counselor) and after they pass an examination. The educational requirements are a master's degree in social work or a doctorate in a human-relations field such as psychology, social work, or nursing.[1]

I decided to seek a license as a marriage counselor. According to New Jersey law, a licensed marriage counselor may advertise as a "counselor" or "therapist." This process involved investing approximately two and one-half years and two hundred dollars. The investment was timely: shortly after my decision to get the marriage counselor's license, a nurse in New Jersey encountered the legal and professional tangle I had been concerned about avoiding.

In October of 1977 the right of a nurse to practice with a master's degree in psychiatric/mental-health nursing was challenged by psychiatrists and psychologists (through their respective state boards) who wished to protect the practice of psychotherapy as their exclusive domain. According to the nurse's legal counsel, the two professions pursued their investigation in order "to define 'turf' without regard for professional parity." The New Jersey Board of Nursing responded that the practice of psychotherapy by a psychiatric/mental-health clinical specialist who holds a master's degree in psychiatric/mental-health nursing and is eligible for certification by the New Jersey Nurses' Association is within the scope of the New Jersey Nurse Practice Act. This statement by the nursing association did little to deter the prolonged legal maneuvers of the psychology and medical boards.

Although the eventual outcome of this case favored the nurse and upheld her right to practice privately, the cost in time, money, and energy was great. My decision to get a marriage counselor's license was, in a way, insurance against such legal harassment. I mention this case not so much for its outcome but to apprise nurses considering private practice of the legal problems they may encounter.

Supervision

To meet with the requirements of the marriage counseling license, I had to comply with certain requirements for supervision. These requirements created a philosophical conflict for me. I believe psychiatric nurses should be supervised by other psychiatric nurses even though many of the therapeutic techniques are held in common by professionals from psychosocial disciplines. Socialization occurs in subtle ways during supervision, and I believe nurses should be socialized by nurses. This opinion has led to many heated discussions with my colleagues.

In order to comply with both the law and personal convictions, I had two supervisors. I discussed all family and marriage counseling cases with the psychologist-marriage counselor and all individual and group therapy cases with the nurse psychotherapist supervisor. This supervisory process continued for approximately two years on a regular basis. (I now receive supervision occasionally, when a problem client or situation arises.) Dual supervision was an excellent learning experience for me. The variety enhanced and broadened my own therapeutic style and resolved my philosophical conflict.

Business Details

After we discussed the marriage counseling route and supervision, we examined some basic mechanical and environmental details, such as scheduling, the office layout, and the difference between private and group clients.

Layout. In the office there are a group room, two bathrooms, a waiting room, a receptionist/record-keeping area, and four private offices. The site for the office had been chosen before I joined the practice. Several features of the location contributed to our success. The building is adjacent to the local highway and near several popular stores. Clients under the stress of a first visit are likely to become lost so it is important for the office to be easily reached. Parking poses no problems to potentially distraught clients.

One drawback is poor ventilation in some offices. Another was our inability to regulate the temperature in the office because at first the thermostat was in another part of the building. This became a constant source of irritation between the landlord and us. However, this problem helped to bring us together during the early phases of our interdisciplinary mistrust.

Private and Group Clients. An important decision was our establishing two categories of clients. Private clients are those who request a specific therapist during their first visit. All arrangements, including fees, are strictly between the client and therapist. However, the therapist pays a small flat fee for each hour of private client therapy. Because there are two full-time therapists, a monthly fee ceiling was established for them.

Group clients contact the group for an appointment without naming a specific therapist. These clients are assigned to therapists on a rotating basis. The distribution of group clients was a potential problem, and we did not discover a universally acceptable approach until after we tried numerous methods, such as screening and then assigning according to client problem and therapist approach. This method sounds good theoretically; however, many clients who when screened were in crisis and amenable to therapy did not return to see the assigned therapist.

A modified rotating method was the final solution. Clients were assigned to therapists on a rotating basis; however, if the client or couple were in great distress, they were assigned to the therapist who was available within the next 24 hours (often one of the full-time therapists). It was not uncommon for us to be able to schedule appointments for clients in distress after a few hours of their initial contact with the group. It was this ability to schedule clients soon after their first call plus our availability for evening and weekend appointments that promoted the growth of a positive reputation in the community.

Professional Differences

Soon after I joined the group, one of the questions that was, and continues to be, asked of me is "How does your practice as a nurse psychotherapist differ from the practice of a psychologist or psychiatrist?" There are many aspects of my role that do not differ greatly from those of other mental-health professionals. However, within an interdisciplinary setting there were a number of areas in which my role did differ significantly. As the practice developed and we experimented with the range and scope of our services, both the similarities and differences became more apparent.

One of the earliest and most visible differences was my ability and willingness to make home visits. (We have since discovered that I am the only mental-health private practitioner in the community who makes home visits.) Nurses have the experience and skills to conduct home visits; professional nurses are taught to enter the client's home and provide holistic care. Also to my advantage were my doctoral preparation in nursing and years of experience in community mental health in inner-city New York.

The first home visit was the result of a request by a local vocational rehabilitation agency that we assess a group of quadriplegics. They specified that the assessments needed to be conducted in their respective homes.

My first visit brought out some biases among male colleagues—biases I had not noticed before. On the day I was to make the visit I noticed their concern about checking schedules to determine who was available to accompany me. I was amused by their solicitude because I was used to making unaccompanied visits to the Bowery and other crime-filled and drug-ridden areas; this visit seemed exceptionally tame to me. The discussion that ensued illumined sexual biases and stereotypes held but unrecognized.

The first visit reinforced the unique characteristics of the nursing role. The young man, age 16, was a quadriplegic whose injury was the result of a diving accident. The agency requested a total evaluation of the boy, including social, environmental, family, emotional, and cultural attitudes. While hospitalized, the boy had been administered a standard intelligence test (WAIS) on which he scored low/average. This score ran counter to previous reports, and the concern was that his vocational direction would be restricted as a result of the lowered score.

My visit was very revealing. There were no books, art work, or any other evidence of cultural or intellectual stimuli. The television blared and the family bickered during the interview. The boy was not only unable to study but was made to babysit for his sister's 3-year-old daughter. The decline in the boy's intelligence score could easily be attributed to the environmental situation. Upon further investigation and discussion with my colleagues we discovered that before the accident the boy had compensated for the cultural and intellectual deprivation in the home by immersing himself in school and extracurricular activities, such as the language club and church groups. After the accident he was literally a prisoner and no longer able to compensate. I recommended several measures to restore some of the stimuli he had previously sought. In addition, I suggested family counseling to improve the home situation and increase the family's understanding of the boy's needs and problems.

At first glance, home-visit fees seem quite high; however, the fee often includes a considerable amount of travel time and expenses. For instance, I charged $110 for one visit. The client, recovering from neurosurgery and unable to travel, had become extremely depressed. The surgeon referred his client to me. She lived over thirty miles from the office; my travel time was prorated at 75% of a therapy hour. She was charged the equivalent of two therapy sessions for the therapy. At that time, the general fee was $40 for a fifty-minute hour. Due to the relative expense and the insurance companies' unwillingness to reimburse for travel time plus expenses, home visits are reserved for emergencies.

Another obvious difference between myself and the other therapists was my ability to provide career counseling about nursing. A book I wrote on nursing as a career received local newspaper coverage. Both the book and the article emphasized that improper counseling and inappropriate schooling are costly mistakes. As a result, I received many requests from prospective nurses for interviews to discuss educational avenues in nursing and their implications for future career advancement. It was not uncommon for physicians, dentists, or even other nurses to request an appointment for a relative and then ask to be present during the discussion. It seems that preparation for a career in nursing, given the present state of educational upheaval and transition, requires careful guidance and counseling. We had not anticipated this community need; however, we met it gladly.

The fact that I was a nurse—and not a psychologist—also seemed to influence how the clients were distributed among us. I began to notice that I handled the lion's share of certain kinds of cases—cases in which physical illness was a component of mental problems and cases of sexual dysfunction. I accepted the former role differentiation as an unconscious awareness by my colleagues that nurses, because of training, aptitude, and traditional role, might handle the infirm mental patient more successfully than others. I rejected the latter role differentiation—I think I received a large share of patients with sexual dysfunction because of a myth that nurses have more sexual knowledge than others. I pointed out to my colleagues their unconscious bias; after our confrontation, marital clients are shared with more diversity.

Discovering and discussing biases and stereotypes is an enjoyable aspect of interdisciplinary practice. Another is exploring our respective modes of treatment and their psychological underpinnings.

Therapeutic Modalities

We discovered during early discussions that no two of us had similar views about motivation and behavior. The group consists of a humanist, a behaviorist, an analytic psychotherapist, an eclectic (now known as multimodal), and a Bowen-based family therapist.[3,4] Because of this variety, it was approximately a year before we understood and accepted each other's therapeutic approach. During the first year we were often critical of each other's style. It was difficult for the humanist to understand the "rigid, cold" time and payment rules of the analytic psychotherapist. Likewise, it was difficult for the analytic psychotherapist to understand the brief therapy and token-reward system of the behaviorist or the lack of time restrictions of the Bowen-based family therapist.[5] Over time, however, our different modalities became our strength. Our ability at first only to understand, then admire and respect, the competence of our colleagues allows clients to receive a wide range of services.

Referrals

Because we value the differences in therapeutic approaches, we developed the ability to make in-house referrals. It is not unusual for a therapist to refer a client to another therapist because of the nature of the client and his presenting symptom suggested the referral or because of a recognized need for a change in therapist or therapeutic approach.

Most textbooks advocate selecting a therapist who best meets the individual needs of the client. This approach is logical; however, it is sometimes difficult to be objective when one has an emotional and professional investment in a practice. The tendency would be to keep the client in the practice rather than to refer them elsewhere. We combat this tendency with frequent case discussions and group critiques.

Referrals are particularly common between the therapists who did individual counseling and those who were primarily marriage counselors. As individual clients worked through their conflicts, it often became apparent that there were serious problems in their marriage. Growth in therapy by one spouse often produces an imbalance in the relationship. The couple is warned of this potential. The critical period seems to occur about two years after individual therapy begins. After two years of treating one half of a couple, the therapist may find it difficult to remain unbiased when counseling both people. The group practice provides easy referral to a known competent marriage counselor.

After the initial phases of the practice had been resolved and stereotypes and role distinctions had been clarified, referrals to me from the group and from the community were related to my general reputation, the treatment modality I employed, and my sex and nursing background—characteristics that contributed to the strength of our interdisciplinary practice. Referrals from outside the practice came from individuals and other professionals. Specifically, new clients came to me from lawyers, friends, a psychiatrist, psychologists, other nurses, a local chapter of the National Organization for Women (NOW), a women's center, newspaper publicity about the practice, a workshop I did on crisis intervention, and general word of mouth.

One of the most important advantages of interdisciplinary practice is the opportunity for collegial discussions and exchange of ideas across disciplines. A related advantage is group support. In our situation, each therapist has an office of his own; but there are common rooms in which to share cases, problems, community issues or the common frustrations of private practice.

Community Education

As the group continued to develop and the practice grew, we expanded our services. One therapist developed an interest in parent-effectiveness training (PET) and obtained certification as a PET teacher. We offered the course to clients of the group. The response was so overwhelmingly positive that we also opened the course to the community at large. The course was very helpful to many clients not only because of the course content and related child-rearing techniques but also because of the group interaction. Many clients were surprised to discover other parents had similar problems with their children.

Psychological Testing

The convenience of psychological testing was another advantage of the group practice for both the client and therapist. As the number of clients grew, it became almost routine to administer an MMPI as a general psychological screening device. The test provided a general profile on each client, alerted the therapist to potential problem areas and strengths that might not be obvious during the early phases of therapy, and reduced the potential for hospitalization. In some instances, further and more extensive testing was required, particularly in the projective testing area. In these situations, the interdisciplinary model was ideal. Three of the members had preparation and skill in administering and interpreting a psychological battery. Clients remarked on the relative comfort of the testing experience. They were able to arrange for tests and be tested in a setting that was already familiar. Also, the therapist who requested the battery could get a verbal report almost immediately. The feeling of collegial responsibility was strong; the written report was seldom submitted more than a few days after the testing. Promptness was a distinct advantage for both the client and the therapist.

Professional Education

During the inception of the interdisciplinary group, we had envisioned a multiservice, multipurpose practice. Early in the formation of our identity we designed (and advertised through the local newspapers) a number of workshops and seminars for professionals in health-related fields. These offerings included "Concepts of Family Therapy" and "Strategies of Marriage Counseling." The fees charged were minimal. After two unsuccessful attempts, we decided that a market did not exist, and we discontinued the idea. However, at a later date we did initiate a monthly continuing-education meeting for ourselves. This meeting addressed a variety of theoretical and practical issues of counseling. At the meetings we discussed such topics as "Crisis Addiction and Perceptual Reactance," "Family Therapy—The Bowen Model," and "Sensory Deprivation as a Treatment Modality." The seminars had a very positive effect on the group; they were a vehicle for intellectual interchange and a forum for exploring our differences.

Conferences

Occasionally we held case conferences to stimulate intellectual exchange and communication. Through conferences, we shared approaches to client behavior, problematic situations, and complex cases.

A shared problem was the demanding, manipulative client. Frequently a manipulative client would disrupt the work of several therapists. We learned to assist others to set limits and to avoid engaging in power struggles with such patients. We planned a seminar to explore methods for approaching the manipulative client in general and the dynamics of such behavior; each therapist contributed his or her understanding of the dynamics of such behavior.

Conferences were helpful in certain complicated cases. When, for instance, a complex network of relationships emerged among our respective clients, we held a conference. During the conference, we found ourselves making pleas for our respective clients and their individual causes. Our final realization was the subtlety of countertransference and the ease with which we had become enmeshed in the rivalries and struggles of our clients. Conferences in which we shared incidents of countertransference and reactions to a variety of client behavior were exceptionally valuable.

Cotherapy and Peer Support

As we understood and accepted each other's therapeutic approaches, we began to see the advantage of conjoint therapy or cotherapy. This approach was especially useful in marriage and family therapy situations. For instance, I counseled a married couple who had difficulty maintaining order in their family of six children. I decided family therapy was the best approach, so I requested the help of another therapist, a family therapist for many years. The outcome was very satisfying. During the postsessions we shared our ideas, attempted to validate the chaos we observed, and designed a treatment plan that we used for the following sessions. It was an excellent learning experience and provided the family with some new insight into their behavior and strategies for living together under the same roof.

On another occasion, my services as cotherapist were requested. During that collaboration, I was able to show my colleague that his empathy with his client's children clouded his perception of the children's contribution to the problem. In exchange, I learned another way to approach family therapy, using Bowen's theory and techniques.[3,4,5]

We usually did not charge colleagues for our services during conjoint sessions. However, when the cotherapy lasted longer than one or two sessions we shared the fee. This arrangement seemed to prevent a feeling of exploitation and was equitable for all of us over time.

Another advantage of group practice is that peer support and assistance are available in unpredictable situations. In emergency situations—for instance a threatened suicide—the help of another therapist can be invaluable: to help calm the patient, to remain with the patient while arrangements for further care are made, and so on.

The issue of hospitalization and medication was addressed early in the practice. Three of us established relationships with different psychiatrists. When clients required hospitalization or medication they were referred to one of three psychiatrists. The screening procedure with the MMPI helped to keep the number of clients requiring hospitalization and/or medication to a very small minority.

Low Fees

Another asset of the group practice is its ability to absorb a number of low-paying clients. Often the beginning practitioner cannot afford to see clients who are living on a limited income or clients whose income suddenly decreases as a result of separation, a return to full-time education, or unemployment. During the early phases of private practice, the basic capital outlay for rent, security, and phone service and installation, in addition to the slow build-up of clients, usually prohibits accepting clients who are unable to pay full fees.

Unfortunately, our ability to lower or defer fees resulted in a major problem: we achieved the reputation of a low-cost or budget-therapy group. Until we recognized and accepted that we were not subsidized as a state or government agency, we continued to weaken our own financial base with low fees. We rarely turned anyone away. It took time for us to accept our limitations and learn to limit the number of reduced-fee clients we accepted. Experience taught us to require clients to remit most or all of the fee prior to collecting reimbursement from the insurance company. This policy resulted in a better financial base, earlier analysis of the classical client-therapist struggle over money with the therapist, and fewer terminations because of fees.

Research

Another asset of the interdisciplinary and group nature of the practice is the potential for designing and implementing research. We initially discussed a variety of projects, which unfortunately were never started due to lack of time. We did undertake the development of a tool to describe our client population: presenting symptom, referral source, treatment modality (group, individual, or family therapy), length of treatment, and final outcome (relocation or termination with validation of therapist). As we collected and analyzed our data one exceptionally interesting finding emerged: the extent of the referral network was greater and more complex than we imagined. One client was either directly or indirectly responsible for 17 of our cases. This data demonstrated that reputation in a suburban setting is critical.

Time Demands

A private group interdisciplinary practice demands commitment and time. Any practice is time consuming; clients take up only part of the time that must be spent. Additional difficulties result from combining a part-time practice with a full-time job. In our practice, only two of us were part-time, and three were full-time members. The fact that three of us were full-time therapists was a great asset. We were usually able to schedule appointments with little or no wait and were able to maintain office hours of 9:00 A.M. to 10:00 P.M., including Saturdays. Part-time members were required to make a commitment to the practice that included time over and above the time scheduled for clients—time for participation in case conferences and group seminars and time for informal discussion with colleagues. In my case, arrival well in advance of my first scheduled appointment answered this informal need. I found it important to let the secretary know where I could be reached in case of client emergencies. A general spirit of commitment and consideration was probably the characteristic that contributed most to the practice's growth and success.

At the end of four years, the group separated. One member moved away. At the same time, our office lease expired. The problems with the building and landlord had increased, so we did not renew the lease. Following great deliberation, the group divided. The psychologist and social worker went into practice together, and the marriage counselor and I rented office space in a nearby restored Victorian house. Although the larger interdisciplinary group is no longer intact, the mutual respect we developed for each other as members of different disciplines, representing the range of therapeutic theories, remains.

Author's Biography

Dr. Keville Frederickson obtained a baccalaureate degree in nursing at Columbia University School of Nursing and a master's of education in psychiatric/mental-health nursing and a doctorate in nursing education at Teachers College, Columbia University, New York City.

She has held positions of staff nurse, head nurse, and in-service instructor in various clinical areas at Columbia-Presbyterian Medical Center, New York City. Dr. Frederickson has also taught nursing in a variety of clinical settings and was an assistant professor of psychiatric/mental-health nursing at Columbia University School of Nursing undergraduate program.

She is currently an associate professor and director of continuing education in the Division of Nursing, New York University. In addition, she maintains a private practice as a nurse psychotherapist and is a licensed marriage counselor in New Jersey.

REFERENCES

1. State of New Jersey, Department of Law and Public Safety: Board of Marriage Counselor Examiners laws. *NJSA* **45**:8B-1, 1969.
2. Rouslin S, Clarke AR: Commentary on professional parity, based on an interview with Fay F Korn, RN, MS. *Perspectives Psychiatric Care* **16**:115–117, 1978.
3. Wolman B, (ed): *The Therapists Handbook: Treatment Methods of Mental Disorders*. New York, Van Nostrand Reinhold, 1976.
4. Guerin PJ (ed): *Family Therapy: Theory and Practice*. New York, Gardner Press, 1976.
5. Bowen M: The use of family theory in clinical practice. *Comp Psychiatry* **7**:345–374, 1966.

CHAPTER
9

Nursing on the Prairie

Elaine R. Neilson, RN

As a member of the 1950 graduating class from Sioux Valley Hospital School of Nursing in Sioux Falls, South Dakota, I had no idea that 28 years later I would be describing the experiences of a nurse practitioner in a midwestern community. The move from hospital nursing to private practice was gradual, brought about largely by experience and the needs of the community.

My practice today is general. I am the only nurse practitioner in South Dakota. I travel over 25,500 miles each year within a 20-mile radius, seeing 50–75 patients per week and making an average of 2500 home visits a year. I have traveled up to a distance of 150 miles in one day and have made as many as 18 calls in a 24-hour period. Much variety of experience was necessary to reach this satisfying professional goal.

After receiving my diploma in 1950, I remained at Sioux Valley Hospital as a general nurse specializing in orthopedic and medical nursing. I then worked as a general duty nurse in the emergency room of a large hospital in Phoenix and later served as the administrator of a small South Dakota hospital. During the three and a half years at that post I administered anesthesia, took histories, gave physical examinations, and did laboratory work. My administrative duties included purchasing, record keeping, determining personnel, and general management. My marriage to a midwestern farmer opened the door to my career as a private practitioner.

My home near Beresford is in a four-county area in the southeastern corner of South Dakota, midway between Sioux Falls, South Dakota and Sioux City, Iowa. Twenty-one years ago, two of these counties had public-health nurses, and two did not. One county had been without a public-health nurse for 27 years. My county's elderly doctor, who had served the community for 40 years, was ill. This family physician was a key person in the community, and his death left a big gap. Many of the new physicians in Sioux Falls were specialists, and people missed the special confidential relationship to the family doctor.

Aware of my profession, neighbors began calling for my help in caring for elderly patients, patients with terminal illnesses who preferred to spend their last days at home, and people without transportation who needed professional services. I found myself giving enemas, B-12 and insulin injections, and bed baths and taking blood-pressure checks. It became obvious that the area sorely needed direct nursing care, and my calls to various homes soon became a full day's activity. As a result, my husband and I worked out a schedule of charges paralleling those fees received by private-duty nurses. I included mileage in my charges.

My business is operated from my home. Ninety percent of my practice is in the home of the patient; the other ten percent are treated at my kitchen table. Of course, I do not admit patients with contagious diseases into my home. My services are available seven days a week, 52 weeks a year. My charges began at $2 a visit; now, twenty-one years later, the fee is $6 for a day visit and $12 for a night, Sunday, or holiday visit.

These reasonable fees enable anyone to afford nursing care. Some patients pay for their own nursing service; most private insurance companies pick up the tab on a fee-for-service basis. Medicare, welfare, and Blue Cross are the only agencies that do not pay. Medicare reimburses only nurses employed by a public-health agency. Medicare patients are referred to the public-health nurses if those patients so desire. Many welfare patients learn that with proper budgeting they can pay at least part of their bill, which is indeed sufficient.

I carry malpractice insurance—$100,000 to $500,000—with the St. Paul Fire and Marine Insurance Company and employ part-time legal assistance. I have had no legal or malpractice problems. I also employ a part-time bookkeeper to keep my records and files completely in order at all times.

I believe that the nurse practitioner can best perform his or her services in the patient's home. This gives the nurse the opportunity to see the family in action. Sometimes problems are due to living habits. I ask myself these questions, among others: Are they eating properly? Are their activities too sedate or too vigorous? Are medications out of reach of small children? Is the home reasonably clean? Is the home overrun with pets? Many times treatments and procedures can be changed to suit the family and home, making home care more acceptable to everyone involved. I also feel very strongly that the elderly can and should remain in their own homes for as long as possible rather than move to a nursing home prematurely. When it becomes obvious that care for the elderly is becoming a hardship for those involved, I do then advise nursing-home care.

My only advertising is word of mouth by my patients. I have never advertised my services. The work has had to sell itself. My services have developed into an occupation with a $25,000 yearly income. I believe there are several reasons for my success. The patients seem to feel secure when the evaluation is done in the home; they also appreciate that, when they need nursing care, they do not have to go out into the variety of weather that South Dakota can produce.

I feel certain frustrations, however. Despite the fact that I have received national recognition, sometimes I feel taken for granted by local organizations. I have never had support from the State Nurses' Association, the State League for Nursing, or the State Board of Nursing Examiners. It appears very difficult for people in our own profession to accept the independent nursing practice. They think the nurse practitioner is trying to "play doctor." I am not in competition with anyone and am very careful to stay within my own boundaries.

Developing a practice of this magnitude is the result of working *with* the physicians of the community. Physicians have advised me on the equipment I should have and the examinations and procedures I should follow. Local doctors have been my biggest boosters in most instances. This support has been very beneficial because there are always innovations in patient care to assess. I hold many telephone conferences with physicians and often visit doctors to receive advice or compare records. If, during a home visit, I believe medication is necessary, I can call a physician. I recommend to each patient a complete physical examination by a physician at least once a year, more often if necessary. When specimens are sent to the clinical laboratory, I make sure copies of the results are sent both to the physician and to me. The X-ray center also sends out duplicate results. I have a back-up gynecologist, psychiatrist, internist, orthopedic specialist, and surgeon to whom I refer patients. I consult with licensed physiotherapists frequently to update any innovations in this field. I can contact a private psychologist who will visit the home if necessary.

How did I achieve the confidence of physicians? This confidence has taken years to build. Perhaps one case will illustrate: I had an elderly patient who was slowly becoming very weak and frail; she had been in very robust health, an aggressive individual. She was referred to a leading internist and hospitalized for extensive testing. These tests were all negative and the patient was sent home. I was advised to send her to a psychiatrist; I did not agree with this advice. For the next two weeks at home she continued to deteriorate. I asked the internist to refer this patient to a neurosurgeon. The internist refused at first and then consented reluctantly. The neurosurgeon thought I was "out in left field" also until a brain scan and skull studies were done. There was a growth in the right occipital region. Surgery, which I attended, revealed a carcinoma. This elderly patient lived only six weeks after surgery; however, the family was satisfied that they had done everything possible to help their mother. Those two attending physicians have become very close friends of mine. There have been many cases similar in nature to this incident.

I have hospital privileges and make hospital rounds once a week to check on my patients. I have been asked many times by the family to make these visits as the patients have confidence in my judgment. Because patients know me, they will tell me things they will not tell a stranger. Of course, I do not make any decisions; I simply point out the pros and cons of each situation and answer the patient's or family's questions about care or treatment. I have to assess the situation very carefully to decide what the physician is attempting to achieve and what the policy of the hospital is. I have access to hospital records and consult with the nurses. I communicate to the nurses the problems I have witnessed during my home visits.

One incident that illustrates this communication concerns an elderly gentleman who was very grouchy during his stay in the hospital. He appeared very uncomfortable. I finally realized that I had never seen him not wearing his sweater, winter or summer. I suggested to the nurses that they let him wear his sweater. The patient settled down; and the sweater worked as a security blanket does for a small child. The patient was happy with his hospital care from then on.

I am welcome to accompany the doctors on their rounds to witness their observations firsthand. I recently was invited to a large hospital's Saturday morning seminar to give physicians information on a case of Pasteurella multicida. We had a case of this disease in our area. I diagnosed a second case of this disease because I noticed the same odor in both cases. This odor was apparent as soon as I opened the door to the home of my second patient. It is gratifying to me that doctors and hospitals within the area are open to this kind of information and will work to disseminate it.

Counseling is an important part of the practical-nursing profession. The alcoholic, the depressed, and the troubled frequently need help. It is amazing how many of these problems are found in a community. If necessary, I will transport a patient to a doctor or place where specialized help can be obtained. I have found attendance at Alcoholics Anonymous meetings very helpful in gaining understanding in the methods of dealing with those particular problems. I carry Alanon and AA literature. I have counseled alcoholics in jail and have spent long hours trying to get—and at times posting—bail for people with alcohol problems.

I counsel couples about family planning. The home seems to provide a relaxed atmosphere where goals can be discussed freely.

The police in our city and surrounding area make use of my services. I transport blood specimens for alcohol testing, supervise extraction of accident victims, and give emergency medication. When called in on cases of child abuse, child neglect, drug problems, or family problems, I frequently give my services without financial charge. At times, I have consulted the state attorney general's office to determine proper procedures in handling the insane and in handling cases of child abuse.

I now use an answering service. I wear a pager that is answered by the dispatchers of the local police department. A rental fee for the pager and answering service is paid to the city of Beresford. I have used these services for six years. Before then, I had a mobile phone in my car, but, because I was not always in my vehicle, I found it quite inefficient. I receive as few as one and as many as 15 calls on my pager each day. The majority of these are day calls, although patients call at night in emergencies. This reliable service fields calls from up to 35 miles away.

Records are kept on all patients; presently over 3000 names are on file. Hospitals send me copies of the discharge summaries of referred patients. Doctors, too, send notes updating the progress of these patients. These notes include such information as X-ray readings, lab reports, and recommended treatment.

I travel in a red Toyota equipped to meet almost any emergency. I carry a small tank of oxygen, a mask unit, dressings of all sizes, cotton balls, thermometers, a stethoscope, an otoscope, splints of various sizes, a sphygmomanometer, Fleets enemas, needles, and syringes, sterile rubber gloves, hand soap, towels, sutures, airways, injectible medications, irrigating solutions, catheters, and catheter trays. In addition, I carry, at times, equipment needed in the care of specific patients. I own much equipment and supplies the patient can rent: whirlpool baths, apparatus for Buck's extension, intermittent positive-pressure breathing machines, commodes, walkers, crutches, hospital beds, and other supplies not available in the community.

When the snow in South Dakota is so high that the car won't plow through, I have a snowmobile equipped with a sled. During the winter of 1968–1969, when over 130 inches of snowfall were recorded, my husband and I hauled food, fuel, and other supplies to patients and others who were isolated for weeks at a time by blocked roads. The sled can accommodate a stretcher, should the need arise.

My religious faith is strong, although my "church time" is spent frequently in my red Toyota en route to a patient's home. I have studied the various religious faiths of our community to acquaint myself with the beliefs and customs practiced in the homes I enter.

Providing physical therapy for patients has been extremely rewarding. A very gratifying experience was my treatment of 54-year-old Karen Jones. Karen had been confined to a wheelchair or bed; she was unable to move without her husband's assistance. She had been injured in an automobile accident in 1949. In 1971, I was asked to care for Karen by a doctor who had been the family physician for many years. Karen had developed a bladder infection. I was called for assistance because the temperature was 20° below zero with blowing snow and transportation to a hospital would be very difficult. I stayed on the case during the infection, and Karen learned to trust me. Gradually I began to ask Karen questions about the paralysis: why she was unable to sit up without a back rest, why she was unable to roll over in bed, why she was unable to get from the wheelchair to the bed without assistance. Her legs, although unused, had been well exercised and did not show the signs of atrophy of a true paralytic. With Karen's consent, I undertook treatment, starting with lessons in rolling over, which Karen learned quickly. With a rail installed for leverage and after further strengthening of the leg muscles, Mrs. Jones learned how to maneuver from the bed to the wheelchair and back again. Restoring and revitalizing muscles that had been idle for 22 years was a challenge for the both the nurse and the patient. Massages, whirlpool baths, and a strenuous routine of exercises were employed. I then designed braces for her and had them made by the National Limbs Company of Sioux Falls. Because Karen could not afford her first pair of shoes for those braces, I purchased them for her. Later that year, Karen walked through the door of her home for the first time in 22 years. I am certain that Karen's is not a miracle cure of paralysis, but a case of encouragement, determination, and a workable goal. Early in the case, I also discovered that Karen's vision was not keen and made arrangements for glasses to be fitted. Glasses were a must if equilibrium was to be restored, a help in housework, and a way for Karen to enjoy television, cards, and reading. Karen now does all of her housework and lives a satisfying life. I continue to see her twice a week. As with all my patients, care for Karen ends by mutual agreement when the service is no longer needed and can be resumed when the need arises.

Postoperative care can be supervised from the home for many patients. For example, certain surgeries, dressing changes, and injuries of various kinds require procedures that can be done at home. There is an advantage to not leaving home during recuperation. The youngest diabetic I have instructed at home was 9 years old; the oldest was 85.

This concept of home care is an old one that only recently is being positively reevaluated. To see a patient in his or her domestic surroundings reveals much about his or her needs and strengths. Physical, mental, or spiritual problems can be so interrelated and seem easier to discover in the home situation. A balanced, whole person is the private practitioner's objective.

To a great extent, I keep abreast of the latest nursing techniques through self-study. A standing invitation from physicians to attend their seminars is appreciated, and, when time permits, I take advantage of this opportunity. I participate in continuing education by monitoring television nursing courses. Personal conferences with physicians are very beneficial.

I am a voluntary ambulance attendant and also teach classes to the emergency medical technicians. I instruct the health-related classes for the Girl Scouts. I have recently begun lecturing to various college classes and nursing-school groups. Some nurses currently starting general practice have done internships under my guidance. One student, studying under this program, spent a week with me and then chose to amplify her studies by taking a four-year registered nursing program. This student commented that, to be successful, she needed more training than she was being given.

Each year during Christmas, I make a list for the service organizations. This list names people who could use food, gifts, or just a little bit of joy during a time that can be so depressing for those in need. During Christmas, too, I give my nursing services free of charge to the needy.

As the lone nurse practitioner in South Dakota, I plan to continue for a few more years in the work I love so much. I hope to teach another nurse to take over this service eventually. There is nothing more rewarding than to see the results of one's labor at the tranquil end of a long day on the prairie.

Author's Biography

Elaine Nielson received her diploma from the Sioux Valley Hospital School of Nursing in Sioux Falls, South Dakota. She did general nursing in orthopedics and medicine and later worked in the emergency room of a large hospital in Phoenix. She served as administrator of a small South Dakota hospital for three and one-half years, during which time she administered anesthesia, did physical examins and laboratory work as well as fulfilling general management duties. She returned to South Dakota, where her private practice gradually evolved.

Suggested Readings

Magazine Articles

Agree BC: Beginning an independent nursing practice. *Am J Nurs* **74**:636–642, 1974.

Alford DM, Jensen JM: Primary health care: Reflections on private practice. *Am J Nurs* **76**:1966–1968, 1976.

Anders RL: Program consultation by a clinical specialist. *J Nurs Admin* **8**:34–38, 1978.

Archer SE, Flewhmann RP: Doing our own thing: Community health nurses in independent practice. *J Nurs Admin* **8**:44–51, 1978.

Ashley JA: This I believe about power in nursing. *Nurs Outlook* **21**:637–641, 1973.

Bakdash DP: Becoming an assertive nurse. *Am J Nurs* **78**:1710–1712, 1978.

Beletz EE: Is nursing's public image up to date? *Nurs Outlook* **22**:432–435, 1974.

Brown KC: The nurse practitioner in a private group practice. *Nurs Outlook* **22**:108–113, 1974.

Brunner NA, Singer LE: A joint practice council in action. *J Nurs Admin* **9**:16–20, 1979.

Bullough B, Bullough VL: Sex discrimination in health care. *Nurs Outlook* **23**:40–45, 1975.

Christman L: Accountability and autonomy are more than rhetoric. *Nurse Educator,* July-Aug 1978, pp 3–6.

Coletta SS: Values clarification in nursing: Why?" *Am J Nurs* **78**:2057–2063, 1978.

Edwards JA, Curtis J, Ortman L, et al: The Cambridge Council concept or two nurse practitioners make good. *Am J Nurs* **72**:460–465, 1972.

Fagin CM: Professional nursing—The problems of women in microcosm. *J NY State Nurses Assoc,* Spring 1971.

Fiske H, Zehring K: How to start your own business. *Ms,* Apr 1976, pp 55–70.

Gibson KW: If you've ever thought about being a nurse practitioner. *RN* **40**:38–40, 1977.

Goldstein S: The psychiatric clinical specialist in the general hospital. *J Nurs Admin,* March 1979, pp 34–37.

Greenridge J, Zimmern A, Kohnke M: Community nurse practitioners—A partnership. *Nurs Outlook* **21**:228–231, 1973.

Hunnings V: If you've ever thought about being a nurse practitioner. *RN,* **40**:35–38, 1977.

Jacox A, Prescott P: Determining a study's relevance for clinical practice. *Am J Nurs,* **78**:1882–1889, 1978.

Jones PS: An adaptation model for nursing practice. *Am J Nurs,* **78**:1900–1906, 1978.

Kelly D: One town's one-nurse service. *Am J Nurs* **73**:1536–1538, 1973.

Kolisch BJ, Kolisch PO: A discourse on the politics of nursing. *J Nurs Admin,* Mar-Apr 1976, pp 29–34.

Levine E: What do we know about nurse practitioners? *Am J Nurs* **77**:1799–1804, 1977.

Manksch I, Rogers ME: Nursing is coming of age . . . through the practitioner movement. *Am J Nurs.* **75**:1834–1843, 1975.

Muckolls KB: Who decides what the nurse can do? *Nurs Outlook* **22**:626–631, 1974.

Murray LB: A case for independent group nursing practice. *Nurs Outlook* **20**:60–63, 1972.

Norris CM: A few notes on consultation in nursing. *Nurs Outlook,* **25**:756–761, 1977.

Oberst MT: Fostering practitioner involvement in research. *J NY State Nurses Assoc* vol 9, no 1, 1978.

O'Connell K, Duffy M: Research in nursing practice: Its nature and direction. *Image* **8**:6–12, 1976.

Rafferty R, Carner J: Nursing Consultants, Inc—A corporation. *Nurs Outlook,* **21**:232–235, 1973.

Roy SC, Obloy SM: The practitioner movement toward a science of nursing. *Am J Nurs,* **78**:1698–1702, 1978.

Simms E: Preparation for independent practice. *Nurs Outlook* **25**:114–118, 1977.

Steel J: Joint practice with an MD: It's paying off for us and our patients. *RN* **40**:64–83, 1977.

Ujhely GB: The nurse as psychotherapist—What are the issues? *Perspectives Psychiatric Care* **11**:155–160, 1973.

Walker EO: Primex—The family nurse practitioner program. *Nurs Outlook* **20**:28–31, 1972.

Welch CA: Health care distribution and third-party payments for nurse's services. *Am J Nurs* **75**:1844–1847, 1975.

Whitson BJ, Hartley LM, Wolford HG: Complemental nursing. *Am J Nurs* **77**:984–988, 1977.

Wong DL: Private practice—At a price. *Nurs Outlook,* **25**:258–259, 1977.

Committee Reports

Extending the Scope of Nursing Practice: A Report of the Secretary's Committee to Study Extended Roles for Nurses. U.S. Government Printing Office, 1972.

National Joint Practice Commission (ANA and AMA): Excerpt from one of 24 case reports of primary care joint practices. *Am J Nurs* **77**:1466, 1977.

For Further Reading

Chin R, Bennis WG, Benne KD (eds): *The Planning of Change: Readings in the Applied Behavioral Sciences.* ed 2. New York, Holt, Rinehart & Winston, 1961.

Ehrenreich B, Ehrenreich J: *The American Health Empire: Power, Profits and Politics.* New York, Random House, 1971.

Jacos AK, Norris CM: *Organizing for Independent Nursing Practice.* New York, Appleton-Century-Crofts, 1977.

Kinlein ML: *Independent Nursing Practice with Clients.* Philadelphia, JB Lippincott, 1977.

Kneisl C, Wilson HS: *Current Perspectives in Psychiatric Nursing: Issues and Trends.* Saint Louis, CV Mosby, 1976.

Manksch HO: The organizational context of nursing practice. Davis D (ed): *The Nursing Profession: Five Sociological Essays.* New York, John Wiley & Sons, 1966.

Appendix I

PROSPECTUS
August 14, 1975

We propose a model of providing primary psychiatric services for adults with severe emotional problems, utilizing a collaborative approach between professional disciplines. The approach is an extension and expansion of the traditional office practice of psychiatry and conventional role definitions of traditional health professions.

Some attempts to expand conventional roles have resulted in role-blurring to the point of confusion and lack of responsible medical supervision for patient treatment. Traditional office practice of psychiatry has fostered a traditional one-on-one relationship between psychiatrist and patient within the confines of an office. Limiting the practice of psychiatry to the office has delineated the extent to which the psychiatrist's expertise can be applied. The types of presenting problems of patients are precluded by the setting and traditional time limits. The practice of nursing in psychiatric mental health is moving from its customary institutional and agency setting to the private health-service setting. Most of these practices are limited to those clientele who are not severely disturbed, because the necessity of medical intervention in the form of differential diagnosis and chemotherapy exists when treating those with major emotional problems. As a consequence, the private practices of professional nurses are emerging as duplicates of other mental-health professionals' private practice. Historically and realistically, the psychiatrist and nurse are the two professionals whose education and frame of reference prepare them to confront the complex issues of the serious emotional illness.

We propose collaboration between psychiatrists and professional nurses with specialized advanced training in psychiatry, toward the end of providing primary care for a selected, discrete patient population. Specifically, the target patient population are those with major psychophysiologic disturbances, most characterized by diagnoses of schizophrenia, manic-depressive illness, and psychotic depressions. This is a population subject to recurrent psychotic episodes, with frequent, expensive hospitalizations, and, in many community settings, not able to obtain quality, continuous care. We propose to provide maintenance and restorative therapy to forestall unnecessary deterioration from whatever base-line function a particular patient brings. Modalities of therapy will include individual, group, socialization, family, and chemotherapy.

A hallmark of the program will be comprehensive treatment for the select population, with integration of treatment modalities. That is, family, individual, group, and chemotherapy will be variably offered at the time of a patient's life and disease course that seems most appropriate, without need for referral to outside sources or change of basic treating staff.

A second hallmark will be emphasis upon long-term, continuous care, without the frequent changes of therapists, locations, etc., that all too frequently mar the treatment of patients in this target population. It is believed that such long-term care by the same basic staff will allow development of the necessary trusting relationships to provide treatment on a voluntary basis and allow consideration of less-restrictive alternatives, such as confinement and legal courses of action. Also, the long-term continuity will allow familiarity with past histories, which will help the therapist to predict impending psychotic episodes at an early stage and forestall both further deterioration and expensive inpatient hospitalization in many cases.

A third hallmark will be the mode of collaboration between professionals. Staff will maximally work together in order to enhance conjoint treatment planning and review and to foster familiarity of all patients by more than any one therapist. Specifically, both the psychiatrist and a nurse will participate in an initial patient interview and conference to establish a diagnosis, master problem list, and treatment plan. Medication adjustments will be done concurrently with a psychiatrist and nurse in attendance; family and group therapy will be provided by two persons. A monthly peer review will be made of all cases in individual therapy, with a personal conference between treating therapist and reviewer and a written notation in the patient's chart.

All therapists will use a uniform system of record keeping, utilizing a comprehensive data base, problem-oriented "SOAP" progress notes, and a quantified behavioral rating scale of symptomatology. This approach will extend the mutual expertise of psychiatrist and professional nurse and make maximum use of their critical skills.

A fourth hallmark will be a willingness to innovate and achieve maximum flexibility. Although the central base will be outpatient operation, treatment will also be extended to home visits, hospital consultations, and consultations with health agencies. The outpatient base is chosen with the assumption that the majority of patients, even with severe emotional problems, can be maintained without confinement and expense of hospitalization.

As stated, our plans are to have a small, highly trained, professional staff, with integration of procedures to obtain maximum flexibility. Such an organization is seen as necessary in order to remain relatively unhampered from the difficult requirements of coordination imposed upon public mental-health units that are obliged to provide comprehensive care by statute and that are responsible to many different bodies, sometimes with conflicting demands. To the current mental-health system, with its demonstrated advantages and limitations, we offer an alternative to both the traditional public agency and the traditional individual practice.

Appendix II

OWEN E. CLARK, MD

MARGARET GILMOUR, RN

MARGARET THOMAS, RN

announce the opening of the

WEST SEATTLE PSYCHIATRIC SERVICES

2620 S.W. Holden Street
Seattle, Wash 98126
937-0727

*Offering general psychiatric services
including psychotherapy, supervision of
psychoactive medication, mental health
consultation and home treatment*

OPEN HOUSE
October 1, 1975
3–6 PM

Index

Access to health care, 33-34
Accounting equation, 22
Accounting systems, 21-25, 150-151
   assets, 22-23
   balance sheet, 23-25
   credit 23-24
   debit, 23-24
   journal, 24-25
   liabilities, 22-23
Accounting T-form, 23-24
Accounting theory, 14
Advertising, 14-15, 123, 151-152
Alanon, 185
Alcoholics Anonymous, 186
American Bar Association Forum Committee on Health Law, 61
American Nurses Association, 103
ANA code, 32-48
ANA Commission on Nursing Research, 42, 69, 88
Analytic psychotherapist, 173-174
Articles of incorporation, 153
Assessment, 101, 126-129, 171-172
Attorney-client relationship, 61-62

Backup coverage, 130-131, 137
Behavioral data, 8-9, 15
Behaviorist, 173-174
Berlin, I., 44
Biases, 171-172
Bioethics, 30, 91-93
Biographical data, 104-105

Blue Cross, 159, 183
Blue Shield, 159
Board of governors, 107
Bookkeeping. *See* Accounting systems
Bowen-based family therapist, 173-174
Break-even point, 19-20
Business affairs, 169-170. *See also* Fees
   appointments, 170
   choice of therapist, 170, 174
   fees, 170, 172, 177-178
   office location, 169, 174
   parking, 169
Business legal advice, 61-62
Business profile, 18-19

California law, 52
California Medical Association, 59, 68
California Nurses Association, 59, 68
California professional corporation law, 59
California Senate Bill 666, 68
Campbell, C., 83
Capital stock, 21
Career counseling for nurses, 172-173
Case discussions, 174, 176, 179
Certification, 37-38, 124-125, 168
Certified registered nurse, 160
Chamber of Commerce, 10
Charter of incorporation, 21
Client advocate, 35, 46
Client plaintiff, 54
Clinical supervision, 158
Collaborative practice, 139-143, 157

197

Collections, 25–26, 133
Community Leukemia Fund, 99
Competence in nursing, 39
    continuing education, 39
    educational preparation, 39, 42–44
    individual study, 39
Comprehensive Health Planning Council, 10
Comprehensive service, 142
Conceptual model. *See* Theoretical framework
Concurrent diagnoses, 85
Confidentiality, 34–35, 92
Conflict of interest, 93
Consultation. *See* Referrals
Continuing education, 81, 132, 164, 176, 188–189
Contracts, 20
Corporations, 20–21, 56–61, 151–154
Counseling, 185–186
Crisis addiction, 176

Data collection, 9–12
Defining nursing practice, 5–6, 53–54
Defining private practice, 98–99, 118–119, 143–144
Demographic data, 6–8
Department of Business Licensing and Regulations, 26
Diabetic Association, 122
Diagnostic evaluations, 143
Dickoff, J. and James, P., 76–76
Director of nursing, 102–103, 105–109

Eclectic, 173–174
Economic theory, 13–14
Educational preparation, 124–125, 172–173. *See also* Preparation for practice
Ellis, R., 77
Employees, 135
Employment conditions, 44–46
Equipment, 144–146, 186–187
Ethical review board, 92–93
Etiology, 84, 86
Evaluation, 171–172
Expenses, 19

Family therapy, 148–149, 176
Fees, 13–14, 100, 124, 133, 149–150, 162–163, 182–183
Feinstein, A. R., 81–82
Fiduciary relationship, 62
Financing:
    grants, 16
    lending institutions, 17–18
    loans, 16–17
    Small Business Administration, 16–17

Fletcher, J., 35
Foundations, 153–154
Fringe benefits, 57

Gebbie, K. M., and Lavin, M. A., 84
Grants, 156–157
Group practice, 121, 166–180

Health-care system, 34, 48, 59, 89
Health Systems Agency, 10
Heidi, W. S., 44, 47
Holistic health practices, 54–55, 75, 171
Home visits, 156, 171–172, 182–183
Hospitalization, 175, 178
Hospital privileges, 185, *See also* Staff privileges
Humanist, 173–174

Income taxes, 21
Independent practice, 120–121
Independent practitioner, 98–99, 116–117
Individual Retirement Account, 58
Informed consent, 65, 91–92
Internal Revenue Service, 57, 151, 153–154
Intuitive behaviors, 73–75
Investigator bias, 93–94

Job description, 108–111

Keogh Plan, 58
Koltz, C. J., 9, 11
Krieger, D., 55
Kubler-Ross, E., 78

La Bell, D., 12
Letters of reference, 104–105
Liability insurance. *See* Malpractice
License:
    business, 26
    professional, 37, 167–168
Licensed marriage counselor, 167–169

MacLagan, W. G., 36
Macrotheories, 75
Malpractice, 63–67
    insurance, 54, 57, 183
    liability, 20–22, 54–55, 57, 64
    negligence, 63
    tort liability, 56, 64
Manic depressive, 142–143
Manipulative client, 176
Marketing mix, 13–16
    distribution, 13
    price, 13–14
    product, 13
    promotion, 14–16
Market survey, 3–13, 18

Market survey (continued)
  behavioral data, 8–9
  conducting survey, 9–12
  demographic data, 6–8
  objectives, 4–5
  services offered, 5–6
  survey analysis, 12–13
Maslow, A. H., 83
McBee Keysort®, 87
McLachlan, E., 74
Medicaid, 150
Medicare, 150, 183
Medications, 129–130, 143, 160, 178
Microtheories, 75
Missed appointments, 133–134
MMPI, 175, 178. See also Rating scales
Moorhead, Jean, 59
Multidisciplinary practice, 98, 166–180

National Health Lawyers Association, 61
National Organization for Women, 44, 174
New Jersey Board of Nursing, 168
New Jersey law, 168
New Jersey Nurse Practice Act, 168
Newman, M. A., 76
Nuse-defendent, 54
Nurse-physician relationship, 36–37, 40
Nurse Practice Act, 37–38, 147
Nurses Coalition for Action in Politics, 44
Nursing accountability, 37–38
Nursing interventions, 46–47
Nursing judgments, 37–38
Nursing process:
  assessment, 75, 82
  diagnosis, 75, 82–87
  goals, 76, 85
  intervention, 85
  symptoms, 85

Parent effectiveness training, 175
Partnership, 20–21, 58, 61–63
Pasteurella multocida, 185
Philosophy, 166–167
Physical illness, 173
Physical therapy, 187–188
Physician acceptance, 161–162
Political support, 67–69
Practice model, 77–79
Preparation for practice, 119–131
  credibility, 123–125
  educational preparation, 119–121, 182
  selection of patients, 129–131
  specific skills, 125–129
  visibility, 121–123, 132
Presumption of negligence, 54
Price setting, 13–14. See also Marketing mix
Primary provider, 32

Private practitioner, 98–99, 166–167, 181
Production unit, 19–20
Professional incompetence, 35–36
Professional relationships, 108–109, 132–133, 146–148, 184
Profit, 60. See also Corporations
Prospectus, 193–195
Psychological testing, 175, 178
Psychotherapy, 166
  adult, 143, 156, 166–167
  child, 167
  cotherapy, 177
  family counseling, 167, 177
  group, 167
  marriage counseling, 167, 177
Psychotic depression, 142–143
Public Health Nursing Department, 10–11

Quality of care, 33

Rating scales, 127–129
  Beck Depression Inventory, 127
  Denver Developmental Screening Tool, 128
  Hamilton Depression Scale, 127
  MMPI, 127, 175–177
Reach for recovery, 122
Records, 135–136, 184, 186
Referral network, 179
Referrals, 40, 67, 99–100, 111, 122–123, 130–133, 146–148, 174–175, 184
Reimbursement, 33–34, 124. See also Fees
Research climate, 81–82
Research in practice, 41–42, 73, 79, 82, 88–91, 93–94, 118–119, 178
Research tool, 178
Respondeat superior, 64–65
Resumé, 104–105
Role differentiation, 158–159, 170–173

Salaries, 150–151
Sample size, 93
Schizophrenia, 142–143
Scientific basis, 73
Self-help groups, 134
Sensory deprivation, 176
Sex therapy, 128
Sexual dysfunction, 173
Share holder, 56–60
Small Business Administration, 16–17. See also Financing
Sole proprietorship, 58, 61–63, 120–121, 163–164
Staff privileges, 99–112
  advantages of privileges, 111–112
  applying for privileges, 101–102
  approaching the institution, 102–105

Staff priveleges *(continued)*
  granting of privileges, 105–111
Standard intelligence test, 171
Standard Metropolitan Statistical Area study, 11
Standards of nursing care, 42–44, 66–67
State labor department census, 10
Statute of limitations, 63
Stereotypes, 171–172
Stevens, B., 75
Stocks. *See* shareholders
St. Paul Fire and Marine Insurance Company, 183
Student learning experience, 155
Subchapter S, 58–59
Supervision, 136–137, 164, 168–169
Symptoms, 85

Target population, 143
Taxes, 57–58, 61, 63
Termination, 129–131
Theoretical framework, 75
  Berne's transactional analysis, 118
  eclectic, 118
  Levine's wholistic nursing, 118
  medical model, 142–143, 160–161
  Roy's adaptation model, 118
Theories, 75–79
  factor isolating, 76
  factor relating, 76

Theories *(continued)*
  situation producing, 76
  situation relating, 76
Theory construction, 76
Therapeutic relationship, 148–149
Time demands, 179
Tort liability, 56, 64. *See also* Malpractice
Transportation, 7, 10
Treatment program, 148–149

Undercapitalization, 56
Unethical practice, 35–36
University of Washington School of Nursing, 155
U.S. Department of Commerce, 10

Values clarification, 31
Variable expenses, 19
Variable income, 134–135
Variables, 76, 86
Visiting Nurse Services, 10

Washington State Joint Practice Commission, 147
Welfare, 183
West Seattle Psychiatric Service, 139, 144, 153

Yarling, R., 35